Fundamentals of Industrial Problem Solving

Fundamentals of Industrial Problem Solving

A Practitioner's Guide

Zdravko I. Stefanov
Dow Chemical
Lake Jackson, TX, USA

Eldad Herceg
Pearland, TX, USA

Carla Schmidt
Angleton, TX, USA

David M. Jacobson
Union, KY, USA

Dana Livingston
Sanford, MI, USA

J.P. Chauvel
Lake Jackson, TX, USA

Sunil Kumar Chaudhary
Missouri City, TX, USA

Christopher Paul Christenson
Seguin, TX, USA

WILEY

This edition first published 2022

Registered Office
John Wiley & Sons, Inc., 111 River Street, Hoboken, NJ 07030, USA

Editorial Office
111 River Street, Hoboken, NJ 07030, USA

For details of our global editorial offices, customer services, and more information about Wiley products visit us at www.wiley.com.

Wiley also publishes its books in a variety of electronic formats and by print-on-demand. Some content that appears in standard print versions of this book may not be available in other formats.

Library of Congress Cataloging-in-Publication Data

Names: Stefanov, Zdravko Ivanov, author.
Title: Fundamentals of industrial problem solving : a practitioner's guide / Zdravko Ivanov Stefanov [and seven others].
Description: Hoboken, NJ : Wiley, 2022. | Includes bibliographical references and index.
Identifiers: LCCN 2022017570 (print) | LCCN 2022017571 (ebook) | ISBN 9781119543183 (cloth) | ISBN 9781119543213 (adobe pdf) | ISBN 9781119543190 (epub)
Subjects: LCSH: Problem solving. | Chemical industry.
Classification: LCC HD30.29 .S75 2022 (print) | LCC HD30.29 (ebook) | DDC 658.4/03–dc23/eng/20220610
LC record available at https://lccn.loc.gov/2022017570
LC ebook record available at https://lccn.loc.gov/2022017571

Cover design by Wiley

Set in 9.5/12.5pt STIXTwoText by Straive, Pondicherry, India

Contents

List of Figures *xvii*
List of Tables *xxi*

Section I Overview of First Principles Industrial Problem Solving *1*

1 **Introduction** *3*
1.1 A Familiar Story from Real Life, Or Why Do We Need FPS? *3*
1.2 When Does FPS Usually Get Involved? *4*
Reference *10*

2 **First Principles Problem-Solving – Why First Principles?** *11*
2.1 What Is a Zone 4 Problem? *17*
2.2 How Does First Principle Focus Help with Zone 4 Characteristic Problems? *19*
2.2.1 Multifactor Problems Are Helped by First Principle Focus *19*
2.2.2 Nonlinear Helped by First Principle Focus *23*
2.2.3 Managerial Interference Problems Are Helped by First Principle Focus *25*
2.2.4 Multifunctional Team Problems Are Helped by First Principle Focus *26*
2.2.5 Costly Symptoms Problems Are Helped by First Principle Focus *26*
2.2.6 Life-of-the-Plant Problems Are Helped by First Principle Focus *27*
2.2.7 Counterintuitive in Nature Problems Are Helped by First Principle Focus *28*
2.2.8 New Science Problems Are Helped by First Principle Focus *29*
2.2.9 Perceived Personal Loss Problems Are Helped by First Principle Focus *29*
2.3 Final Thoughts on First Principles Problem-Solving – Why First Principles? *30*

3 Opportunity Identification/Definition – Initial Team Formation *31*
3.1 Chronic Issue or Crisis? *31*
3.1.1 Crisis *31*
3.1.2 Chronic Issue *32*
3.2 History *32*
3.3 Charter/DQO Writing *33*
3.3.1 Problem-Solving Aspect of the DQO *33*
3.3.2 Team Dynamics Aspect of the DQO *34*
3.4 Team Structure *35*
3.5 Examples *35*
Reference *40*

4 Hypothesis Testing *41*
4.1 Application of Hypothesis Testing in Problem-Solving *41*
4.1.1 A Good Hypothesis *42*
4.1.2 Qualified Hypotheses *44*
4.1.3 Facts vs. Hypotheses *46*
4.1.4 Hypothesis Tracking: The Onion *47*
4.1.5 M.E.L.D. Hypothesis Ranking *49*
4.1.6 Bringing Tools and Methodology Together *49*
Reference *50*

5 Scenario Development *51*
Reference *52*

Section II Project Management *53*

6 Value of Project *55*
6.1 Crisis Control *57*
6.2 Time Control *57*
6.3 Money Control *57*

7 Team Management and Roles *59*
7.1 Crisis Project Team Building *61*
7.1.1 Team Building in a Crisis Project *61*
7.1.2 Build Understanding of the Consequences of the Crisis *64*
Go to the Source *65*
Make Sure It Is a Real Crisis *65*
Summarize the Crisis into Just a Sentence or Two *67*

7.1.3 Using the Three Best Questions (The Three Main Pillars of the Charter/DQO Process) *67*
Key Points from the Example on Technique of Using the Three Best Questions *67*
7.1.4 Using the WIIFM Question (What's in It for Me) *68*
Key Points from the Example on Technique of Using the WIIIFM Question *68*
7.1.5 Leverage Situational Interaction Opportunities to Build Connections *69*
Key Points from the Example on Technique of Leveraging Situational Interaction Opportunities to Build Connections *70*
7.1.6 Utilize Low Tolerance of Second Guessing *70*
Key Points from the Example on Using a Low Tolerance of Second Guessing *71*
7.1.7 Take Action at Every Opportunity and Engage the Team *72*
Key Points from the Example on Taking Action at Every Opportunity and Engage the Team *73*
7.1.8 Practice Regular Progress Reminders *73*
Key Points from the Example on Practicing Regular Progress Reminders *74*
7.1.9 Build Respect for Team Members and Lead by Example *75*
Key Points from the Example on Building Respect for Team Members and Leading by Example *76*
Crisis Project Team Building: Summary *76*
7.2 Life of the Plant Project Type Team Building *76*
7.2.1 Crisis Project Techniques Also Work for Life of the Plant Project Work *77*
7.2.2 Critical Nature of a Face-to-Face Kickoff Meeting *77*
7.2.3 Generalized/Typical Team Building Approach *78*
7.2.4 Asymmetric Workload Balance for Outsiders Versus Insiders *79*
7.2.5 Opportunity Pairing of Outsiders with Insiders *80*
7.2.6 Carefully Revisit Past *80*
7.2.7 Something New *81*
Life of the Plant Project Type Team Building: Summary *81*
7.3 Roles and Responsibilities for an FPS Project *82*
What Kind of Work do we Expect to be doing on this Project? *84*
7.3.1 Roles and Responsibilities for an FPS Project: Summary *86*
7.4 The Challenge in Maintaining Parallel Processing Within the Team *86*
7.4.1 The Challenge in Maintaining Parallel Processing Within the Team: Summary *88*
7.5 Strategies in Team Management for Very Large Projects *88*

7.5.1 When to Form a Steering Team 88
7.5.2 Strategies in Team Management for Very Large Projects: Summary 90
7.5.3 Team Management and Roles: Summary 90

8 Stakeholder Identification, Analysis, and Communication Plan 91
8.1 Stakeholder Identification 93
8.1.1 The First Step, Stakeholder Identification-make a List 93
8.1.2 Typical Stakeholder List Information 94
8.1.3 Technique for Generating a Stakeholder List Within a Project Meeting 94
8.1.4 Where to Store Stakeholder Lists 94
8.1.5 Tips on Making Stakeholder Lists Effective 95
Using Route Lists 95
Shared Systems Approach 95
8.1.6 Typical Stakeholders in an FPS Project 96
8.1.7 Special Case of Key Stakeholders-Steering Team 98
8.1.8 Changing Nature of Stakeholders 99
8.1.9 Main Root Causes of Having an Incomplete List of Stakeholders 99
Initial Stakeholder Analysis Incomplete 99
Scope Change Due to Problem Location Understanding 101
Phase Changes on Project such as Moving to Solution Implementation 102
Company Churn as People Change Jobs 102
Problem Severity Shifts Due to Outside Forces 103
8.1.10 Last Thoughts on Stakeholder Identification 104
8.2 Stakeholder Analysis – Deepen Your Understanding of the Stakeholder List 104
8.2.1 Initial (Current) Bias Toward Project 104
8.2.2 Needs vs. Initial Bias 106
8.2.3 Needs 107
8.2.4 Typical Stakeholder's Needs 107
Rate-determining Step – What Is Slowing Down Progress? 108
Timeline – When Is This Going to Be Done? 108
Problem $ Cost $ Benefit Analysis – What Is This Problem Costing and What Is It Worth to Us? 111
Impact to Employees – Will Our Employees Be Affected by the Problem? 115
8.2.5 Last Thoughts on Needs and Bias Gathering 116
8.3 Acting on Your Stakeholder Analysis – The Communication Plan 117
8.3.1 5Ws and an H. Who, What, Why, Where, When, and How? 118
8.3.2 WHY – Purpose and Desired Result 118

8.3.3 WHAT *121*
8.3.4 WHAT 1: In Progress Project Work *122*
8.3.5 WHAT 2: Team Working Meetings *123*
8.3.6 WHAT 3: Team Working Meeting Minutes *123*
8.3.7 WHAT 4: Standardized Stakeholder Project Updates *126*
8.3.8 WHAT 5: Formal Stakeholder Project Milestone Review *127*
8.3.9 Communication Tools to Use *129*
8.3.10 FPS Storyboard *129*
8.3.11 Final Thoughts on Storyboards *133*
8.3.12 Standardized E-mail Project Updates *133*
8.3.13 Elevator Speech *134*
8.3.14 WHERE *135*
8.3.15 Face to Face *135*
8.3.16 Mixed *135*
8.3.17 Remote Web or Telecom-based Systems *136*
8.3.18 Other WHERE Decisions *137*
8.3.19 WHEN *137*
8.3.20 HOW *138*
8.3.21 Pulling It All Together in a Communication Plan *139*
8.3.22 Final Thoughts on Communication Planning *140*
8.4 Managing Difficult Stakeholders *140*
8.4.1 Final Thoughts on Managing Difficult Stakeholders *145*

9 Generating, Selecting, and Implementing Solutions to the Problem *147*
9.1 Introduction *147*
9.2 The Joy of a Solution that Just Works *148*
9.3 What Happens During the Solution Phase? *150*
9.4 How Do We Generate Solutions? *150*
9.4.1 Bridging Solution *151*
9.4.2 Dead Easy/Obvious Solutions *153*
9.4.3 Standardized Solutions *154*
9.4.4 Benchmarking Solutions *155*
9.4.5 Innovative Solutions *156*
Channeled Brainstorming *157*
Functional Analysis *158*
How Do You Mitigate the Risks Created During Solution Generation? *159*
9.4.6 When Do We Select Solutions? *161*
9.5 How Do We Select Solutions? *164*
9.5.1 Solution Selection: What Is Success? *165*

9.5.2 How Do You Determine a Solution's Quality? *165*
Theoretical-Paper Exercises (Computer Modeling) *165*
Empirical-Piloting *167*
9.5.3 How Do You Determine a Solution's Cost and Timing? *170*
9.5.4 How Do You Do Final Solution Selection? *171*
Decision Timing *171*
Keeping Options Open and Closing Them *172*
9.6 Solution Implementation *173*
9.6.1 Solution Implementation in Crisis Project Type: What Is Success? What Is Necessary for Success? *174*
Cost Strategy Is Chosen. Now What? *177*
9.6.2 Solution Implementation in Chronic Problem Project Type: What Is Success? What Is Necessary for Success? *178*
9.6.3 Solution Implementation Planning *179*
Success Criteria Definition (What Is Success?) *180*
Overall Plan Generation, Monitoring, and Adjustment *180*
Two Errors of Planning *180*
Natural Planning *181*
Lean/Just in Time Planning *181*
Theory of Constraints Planning *181*
Stress Testing Planning *182*
Change Management Plan *182*
Solution-phase Stakeholder Communication Plan *183*
Control Plan Generation and Validation *183*
9.6.4 Solution Validation *184*
Solution Adjustment vs. Going Back to Your List of Solutions *184*
9.6.5 Solution Phase Summary *184*

10 Sustaining the Gains and Control Plans *185*
10.1 Why We Fail to Sustain the Gains *185*
10.1.1 The Perception Problem *186*
10.1.2 The Reality Problem *190*
High-risk Problem Types *191*
Low Frequency *191*
Counter Intuitive *193*
Reward Conflicting *196*
High-risk Solutions *197*
High-energy Solutions *197*
Soft Changes *198*
Lone Wolves *199*
Poor Solution Implementation *200*

Poor Change Management and Buy-in *201*
Roles and Responsibility Breakdown *202*
High-Skill Training Breakdown *203*
Outside Forces Conspiring Against Sustainability *204*
Rotten Apples *205*
Business Cycle Failures *206*
Reorganization Disease *206*
Summary of the *Reality* Problem Components of a Control Plan *207*
10.2 Building Robust Control Plans for Solutions to Fundamental Problems *208*
Typical Control Plans *209*
FPS Control Plan Emphasis *211*
10.3 Document, Document, Document *211*
How to Motivate Teams to Document *211*
5Ws and an H *213*
Summary *215*

11 Project Wrap-up *217*

Section III Examples *221*

12 The Big World of Problems *223*
12.1 Existing Products and Processes *223*
12.1.1 Common Problems *223*
Plant Product Is Out of Specification *223*
Color *224*
Odor *224*
Trace Components – Evils of Recycles, Accumulation, and Trace Reactions *225*
12.2 New Products and Processes *226*
12.2.1 Products Are Often Developed Over Time Without Fundamental Understanding *226*
12.2.2 Product Doesn't React Properly *227*
12.2.3 Shelf Life *227*
12.2.4 Polymers *228*
12.2.5 Process and Plant Capacity *229*
12.2.6 Catalyst *229*
12.2.7 Reactive Chemicals *230*
12.2.8 Materials Failure *231*
12.2.9 Waste Water Treatment *231*

12.3 Analytical Challenges to Problem Solving 233
12.3.1 Key Measurements Are Missing 233
12.4 Rare Problems 234
12.5 Improving the State of Existing Processes That Are Running as Expected 235
12.5.1 The Reference "Good State" 235
12.5.2 New Product/Process Development (All Plant Modifications Are a Subset of This Class) Where the Fundamentals Not Well Understood 235

13 Corrosion Problems 237
13.1 The Case of the Corroded Drum 237
13.2 The Case of the Corroded Reactor Tubes 238
13.3 The Case of the Corroded Rotating Equipment Seals 239
13.4 The Case of the Corrosion Catalyzed Fouling 239
13.5 Summary 240

14 Mechanical Issues 241
14.1 Rotating Equipment Examples 242
14.2 Materials of Construction 246
14.3 Heat Exchangers 247
14.4 Summary 248

15 Color 249

Section IV Problem-Solving Tools 259

16 Emphasis Is on Speed, Low Cost, Productivity, and Right Answer 261

17 Literature 265
17.1 Why Literature Search 265
17.1.1 Problem Understanding 265
17.1.2 Literature Search Is Economical 265
17.1.3 Lost Knowledge 266
17.2 How to Do Literature Search Properly 266
17.2.1 Sources 266
Peer-reviewed Internal Sources 266
Peer-reviewed External Sources 267
Internal Never Published/Paper-only Sources 267
Plant Libraries 267

Patents 267
The Unpublished Works 268
17.2.2 The Search Process 268
17.2.3 Pitfalls 269
Massive Errors Propagated Throughout Literature 269
Source Age 270

18 Scientific Method *271*

19 Process Data Analysis *275*
19.1 Preprocessing 275
19.1.1 Sampling 275
Process Dynamics Considerations 276
Historian Compression Considerations 276
Transitions, Shutdowns, Missing Data 277
Data with Large Differences in Sampling Frequency 278
Linearization and Other Transforms 278
Lags and Boxing/Grouping 278
Bad Sensors 279
19.2 Univariate Methods for Data Analysis and Exploration 279
19.2.1 Trending 279
19.2.2 Univariate (Shewhart) Control Charts 280
19.2.3 Statistical Distribution Analysis 280
19.2.4 Some Advanced Statistics Aspects 281
19.3 Multivariate Methods for Data Analysis and Exploration 281
19.3.1 Principal Component Analysis 281
19.3.2 Partial Least Squares 284
19.3.3 Some Batch Process Considerations 285
Lack of Steady State 285
Batch Data Is 3D 285
The Batch Durations Are Usually Not the Same Among the Different Batches 285
19.4 Example of Plant Data Analysis 286
19.4.1 Multivariate Analysis of the Plant Data 286
19.4.2 Summary of Data Analysis Results 289
References 290

20 First Principles Modeling *291*
20.1 Selecting the Right Theoretical Modeling Approach 292
20.2 Fundamental Models 294

20.3 Examples of Theoretical Modeling Application *297*
References *302*

21 Analytical *303*
21.1 Definition of Analytical Chemistry and Data *303*
21.2 The Relationship Between Hypotheses, DQOs, and Analytical Data *305*
21.3 Preparing an Analytical or Experimentation DQO *308*
21.3.1 What Are the Best Means of Collecting the Sample? *311*
21.3.2 What Are the Best Ways to Prepare the Samples? *312*
21.3.3 What Are the Key Considerations for the Analysis? *312*
21.4 How Analytical Data Are Used to Solve Problems *313*
21.5 Conclusion *315*

22 Engineering Disciplines Examples *317*
22.1 Typical Engineering Disciplines in FPS *317*
22.1.1 Reaction Engineering *317*
22.1.2 Separations *317*
22.1.3 Fluid Mechanics and Mixing (FM&M) *318*
22.1.4 Solids Processing *318*
22.1.5 Reactive Chemicals *318*
22.1.6 Heat Transfer *319*
22.1.7 Process Control *319*
22.1.8 Process Analytical *319*
22.2 Examples *319*
22.2.1 Unwanted Material Messes Up a Separation Train *319*
22.2.2 Isomer Tower Suffers from Poor Performance *320*
22.2.3 Continuous Crystallizer Is Having a Hard Time *321*
22.2.4 Plant Makes Smelly Solvent *322*
22.2.5 Batch Solvent Stripper Optimization *322*
22.3 Summary *323*

23 Six Sigma, LEAN, and Continuous Improvement *325*
23.1 LSS Comparison with FPS *325*
23.2 Criticisms of LSS in Industry at Large, Beyond FPS *330*
23.3 Recommendations for the Problem-Solving Community in a Company *332*

24 Theory of Constraints *335*
24.1 TOC – Recommended Reading List *345*

25 **Visual Presentation** *347*
25.1 The Napoleon Chart *348*
25.2 The Hypothesis Matrix *349*
25.3 The Issue Map *350*
25.4 Photos and Videos *351*
References *353*

Appendix: Scientific Method Practice Levels *355*
Index *361*

List of Figures

Figure 1.1 The problem solving continuum concept. *5*
Figure 1.2 Phases of an FPS project. *6*
Figure 1.3 Present state – Black Swan events trigger transient development of problem-solving capacity. *8*
Figure 1.4 Desired state: Black Swan events trigger persistent development of problem-solving capacity. *8*
Figure 1.5 Optimum long-term staffing model concept. *9*
Figure 2.1 Empirical vs. theoretical split. *13*
Figure 2.2 Spectrum of plant problem solving. *14*
Figure 2.3 Two-factor interaction. *19*
Figure 2.4 Three-factor interaction. *20*
Figure 2.5 Tank levels over 35 hours. *21*
Figure 2.6 Correlation between corn syrup tank level and oil tank level. *22*
Figure 2.7 Non-linear relationship. *24*
Figure 2.8 Non-linear relationship with operating window. *24*
Figure 4.1 Example template for a hypothesis onion. *47*
Figure 4.2 Hypothesis onion with refuted hypotheses. *48*
Figure 6.1 Value/cost dynamics. *56*
Figure 8.1 Stakeholder identification chart. *95*
Figure 8.2 Storyboard picture. *130*
Figure 9.1 PDCA cycle used in piloting a solution. *170*
Figure 10.1 Xmr plot of production data. *188*
Figure 10.2 FPS control plans vs. typical. *209*
Figure 10.3 Coffee process example SIPOC. *209*
Figure 15.1 Simplified diagram of an ethylene glycol process. *250*
Figure 15.2 Observed vs. predicted color. *252*
Figure 15.3 Pure unsaturated aldol condensation product. *254*
Figure 15.4 Nitro group added to double bond. *255*

Figure 15.5 Low- and high-color plant samples – UV spectra. *256*
Figure 15.6 Fitted peaks from calculated spectra and difference spectrum from real samples. *256*
Figure 18.1 The steps of the scientific method. *272*
Figure 19.1 Principal component analysis. *282*
Figure 19.2 Batch unfolding methods. *286*
Figure 19.3 Observed vs. predicted color. *287*
Figure 19.4 Observed vs. predicted color without UV readings. *288*
Figure 19.5 Observed vs. predicted color – validation set. *289*
Figure 19.6 Main contributors to the major color event. *289*
Figure 19.7 Main contributors to another particular color event. *290*
Figure 20.1 Levels of fundamental modeling (multiple length-scales are present). *295*
Figure 20.2 Chemical mechanism for tar formation. *298*
Figure 20.3 Before and after pictures of the fouled equipment. *299*
Figure 20.4 Adsorption of ammonia in the SAR32 mordenite. *299*
Figure 20.5 Estimated Gibbs energy for NH_3 adsorption from theoretical calculations and from equilibrium constant estimated from miniplant. *300*
Figure 20.6 Effect of water/primary reactant ratio on different NH_3-level feeds. *301*
Figure 20.7 Adiabatic reactor scheme. *302*
Figure 21.1 Ideal process for analytical data collection to test hypotheses. *305*
Figure 23.1 Phases of the FPS process. *326*
Figure 23.2 Compare typical Six Sigma work with FPS. *330*
Figure 23.3 Comparison of problem-solving approaches. *330*
Figure 23.4 Problem-solving selection process. *333*
Figure 24.1 Theory of Constraints overview. *336*
Figure 24.2 ToC place in the FPS framework. *336*
Figure 24.3 Simple linear process. *336*
Figure 24.4 Process capacity baseline condition. *338*
Figure 24.5 Results after Round 1 – Step A reliability improvement to 90%. *338*
Figure 24.6 Results after Round 2 – Step C reliability improvement to 90%. *339*
Figure 24.7 Results after Round 3 – Step B sprint increase. *339*
Figure 24.8 TOC applied to problem-solving speed and efficiency. *342*
Figure 24.9 List of steps to follow in case of tangible and intangible rate-limiting step. *343*
Figure 25.1 Map of Napoleon's Russian campaign. *348*

Figure 25.2 Nominal flowsheet. *349*
Figure 25.3 "Napoleon" chart flowsheet. *349*
Figure 25.4 The hypothesis matrix is also a graphical tool. *350*
Figure 25.5 Issue map example. *352*

List of Tables

Table 8.1 Example communication plan. *141*
Table 9.1 Questions driving premature decisions. *162*
Table 10.1 Actual production rate data. *188*
Table 15.1 Hypothesis list. *251*
Table 15.2 VIP metric values for the most important variables. *252*
Table 19.1 VIP metric values for the most important variables. *288*
Table 21.1 Examples of implied hypotheses. *306*
Table 23.1 Comparison of DMAIC key deliverables with FPS. *327*
Table 23.2 Analyze – Realization steps compared with FPS. *328*
Table 23.3 Specific DQO example. *329*
Table 24.1 Summary of Improvements during three rounds of improvement work. *340*

Section I

Overview of First Principles Industrial Problem Solving

1 Introduction

The authors have found complex problem-solving projects to be very stimulating, challenging, and rewarding work. We feel companies should plan to sustainably have complex problem-solving capabilities, and we have found that becoming an advanced problem solver is an excellent technical career. The successful resolution of complex problems is necessary for the optimum performance of the company and creates significant value. In this book, we will teach our collective experience on best practice in approaching these problems and resolving urgent complex problems permanently.

There are two target audiences for this book. First are the people who will learn and apply the methods and will work on advanced problems to solve them. The entire book is targeted predominantly at them – the problem solvers. Another, very important second audience, are the senior managers who need to accept the long-term need for advanced problem-solving capacity and therefore work to sustain it within their company. They must be sponsors.

1.1 A Familiar Story from Real Life, Or Why Do We Need FPS?[1]

"It was a dark and stormy night" . . . that was how my operations team felt. The plant was down again and we were bleeding money. This problem has been a reoccurring event for decades, but not at a predictable time. In the last year, K-403 had failed over and over, or fouled or plugged. Struggling through the many restarts was costing our plant over half the expected annual capacity of the plant

1 Inspired by Eliyahu M. Goldratt's book, *The Goal.*

Fundamentals of Industrial Problem Solving: A Practitioner's Guide, First Edition.
Zdravko I. Stefanov, Eldad Herceg, Carla Schmidt, David M. Jacobson, Dana Livingston, J.P. Chauvel, Sunil Kumar Chaudhary, and Christopher Paul Christenson.

each of the past two years. As the Technical Leader, it was my responsibility to fix the problem. We had lots of outside help from Corporate Engineering and Technology Experts. The Business Vice President has taken to calling my cell phone at home in the evenings for a status update. She wants to send in a consulting team from a famous business consulting firm, known for business turn-arounds. At the same time, the Continuous Improvement group has restarted a nested team of Six Sigma/LEAN Master Black Belt Leaders to try and fix this . . . again . . . the last fix apparently did not work. My people are overloaded with conflicting demands from their normal responsibilities and the added work to support the many nested and transient visitors and provide the boots on the ground for their work. It just felt hopeless.

Then after one of those late-night phone calls you remember when years ago a colleague told you of a similar experience he went through.

> ***Wisdom Tidbit – Common sense, experience, and hard work are all required to solve many complex problems.***

In the chemical and process industries, some problems occur or periodically reoccur in plants or in large projects that are very severe and valuable. These problems resist solution by more commonly used approaches and can grow to become central to the business' viability. We have collected and systematically organized our best practice that we have developed when faced with these complex urgent problems. By integrating science and engineering with our project management methods, seemingly unsolvable complex problems can be permanently resolved quickly and affordably. The fundamental problem solving (FPS) process is compatible with other well-documented and established problem-solving approaches including Six Sigma, LEAN, Theory of Constraints, and sound engineering and chemistry. Usually one or more of these approaches has been previously used on many of the problems that are subsequently worked on using an FPS approach.

1.2 When Does FPS Usually Get Involved?

The common experience in the occurrence and resolution of all problems is that the resource that ultimately solves a particular problem follows an exponential curve as depicted in Figure 1.1. With a competent staff of sufficient size, training, and experience, many problems are fixed or at least improved by the first person who recognizes the need for an improvement. Thankfully, this is true and must be true for a firm to be viable and continue to exist. However, a fraction of all issues are complex enough that they need to be escalated beyond the first person who originally recognized the problem. A good portion of these issues are in turn

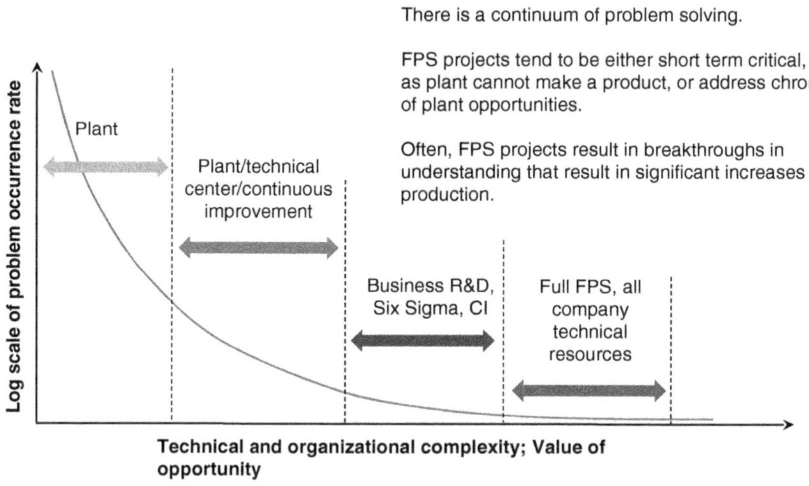

Figure 1.1 The problem solving continuum concept. (*See insert for colour representation of the figure.*)

resolved by small informal teams who are collectively able to work on the problem. Again, there are a fraction of problems that prove difficult and must be escalated to a formal team. That team is normally chartered or formed by outside stakeholders, and the problems are often more complex, valuable, and urgent problems. At the end of the continuum are the severe or chronically reoccurring problems that have often been worked on and declared solved by earlier teams but have in fact not been permanently or acceptably solved to meet the long-term needs of the organization. Or the problems are high value, severe, and urgent. People working to more effectively resolve the problems in the far-right portion of the continuum are the primary subjects for this book. However, many of the methods introduced can be helpful for more effectively resolving issues all along the continuum.

The X-axis in the continuum for Figure 1.1 can be and usually is increasing simultaneously in many factors, going left to right. Normally, the complexity, the value lost per time, records of previously being worked on, urgency all increase moving to the right on the X-axis. The root causes are often found to have multiple factors and intermittent factors.

Examples of problems, which call for the use of the FPS approach, can include leakage, chronic problems or step change performance issues. Leakage here describes a gradual loss of knowledge where the performance has gradually declined of a period of time. Often this is associated with people turn over and the loss of experience and knowledge. The impact can be worsened if the discipline to document and retain necessary knowledge is not adequate. Chronic problems can include fouling, corrosion, poor quality issues where a work-around has been implemented. The work-around is preferable to the problem that was addressed since it gives a *step in the direction of goodness*. However, the root cause still exists. This is treating the symptom rather than curing the disease. Examples can include installed spare pumps, reactors, or columns, so unit A is in service while the fouled unit B is repaired. The whole process operates by alternating units A and B, while the failure rate of individual units continues without understanding or improvement.

The methods taught in this book were developed and are targeted at problems in the chemical industries, but the techniques and the overall approach can be used widely for manufacturing problems, outside work, diagnosis and treatment in medicine, and even political decisions. Overall key steps in the process are depicted in Figure 1.2. What and when to do the key activities are summarized in Figure 1.2, and how to do the work most efficiently and quickly are taught throughout the book using the tools and methods of the FPS process.

Wisdom Tidbit – Define success, don't describe the problem.

One distinguishing aspect of this process appears in the first step of Figure 1.2:

"Define Success"

Which is much preferred vs.:

"Describe the Problem or Define the Defect."

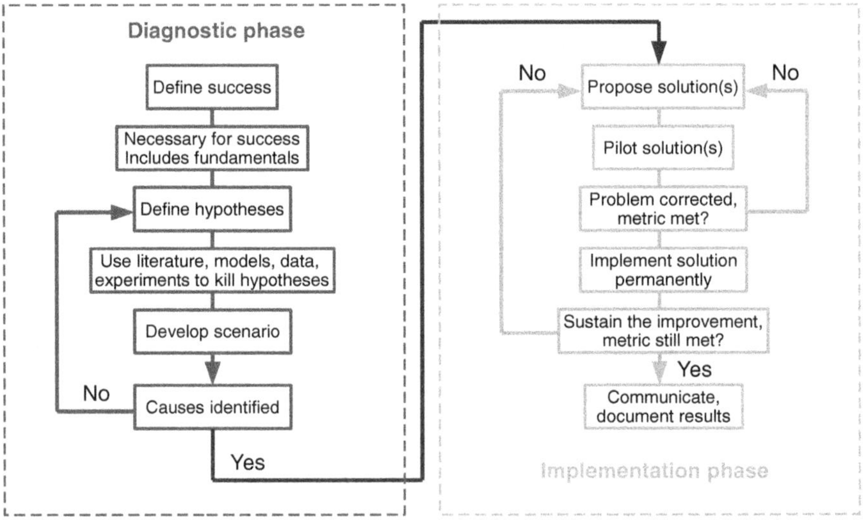

Figure 1.2 Phases of an FPS project. (*See insert for colour representation of the figure.*)

Often, if people focus on a problem description, they limit the approaches considered and simultaneously they don't limit the deliverables to what is high value. Whereas a clear agreement on "what is success" helps focus the team on only delivering "gotta have" work and not working to deliver "nice to have" results. By using "what is success" in the chartering work, extra lower valued work can be minimized.

For example, compare the two descriptions below.

> ***Describe the Problem*** – *The plant is running at 50% capacity and 180% of budget. K-403 is failing frequently.*
>
> Vs.
>
> ***Describe Success*** – *Within six months, the root causes for K-403 failures are known, and both short-term three-month remedies and long-term remedies are being implemented. Within one year, the plant is running at 100% capacity and on budget. There are no EH&S events during this work.*

Demographics and employee population changes within the chemical industry have resulted in a strong trend to less employees in most traditional roles and equally importantly with less work experience of the people in these roles when compared to earlier decades. With a minimally staffed and inexperienced technical staff, a company can maintain operations when there are no issues. However, the technical staff often lacks the ability to recognize and effectively respond to unplanned events as well as would be needed for the optimum long-term value of the firm. In some cases, inexperienced people create problems through their actions or inactions.

In industry, we have seen the buildup and subsequent decline of organizational capabilities and capacity to work on complex problems and quickly resolve them. The occasional occurrence of extremely urgent events, so-called Black Swan (Taleb 2007) events, lead to the transient development of more capable advanced problem-solving capacity within an organization. Since these very severe events are thankfully rare, when the company has not seen a Black Swan event recently, the advanced problem-solving capabilities decline through people change over and intentional reductions if the long-term value of complex problem-solving capability is not well understood and supported.

To illustrate this concept, Figure 1.3 shows a multiple-year trend of technical problem-solving capabilities and the occurrence of rare yet severe events. Also plotted on the second Y-axis is the cumulative value loss due to the severe events. As depicted, in response to the first event, a significant capability develops and the rate of loss declines as the problem is solved. Then if a short-term staffing cost focus is used, the capabilities will decline. If a second severe event occurs, the entire cycle of building capabilities and accepting losses will be repeated.

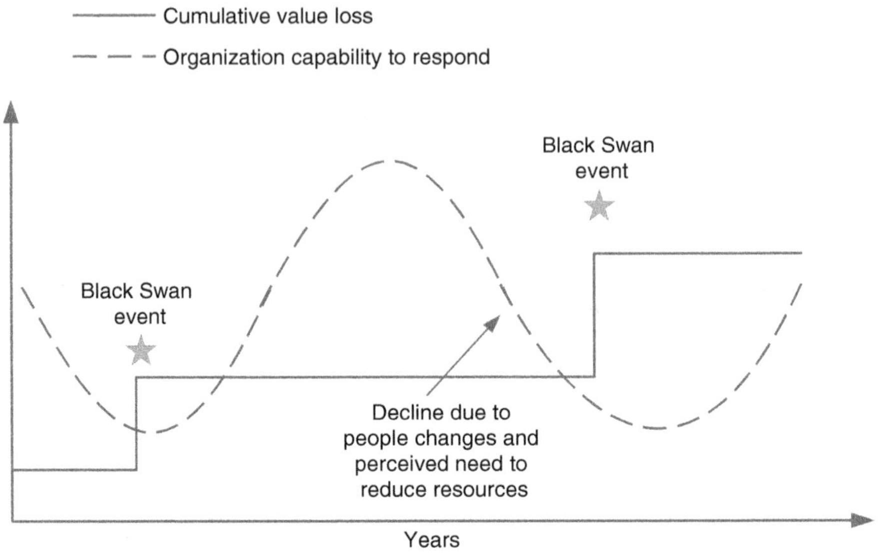

Figure 1.3 Present state – Black Swan events trigger transient development of problem-solving capacity.

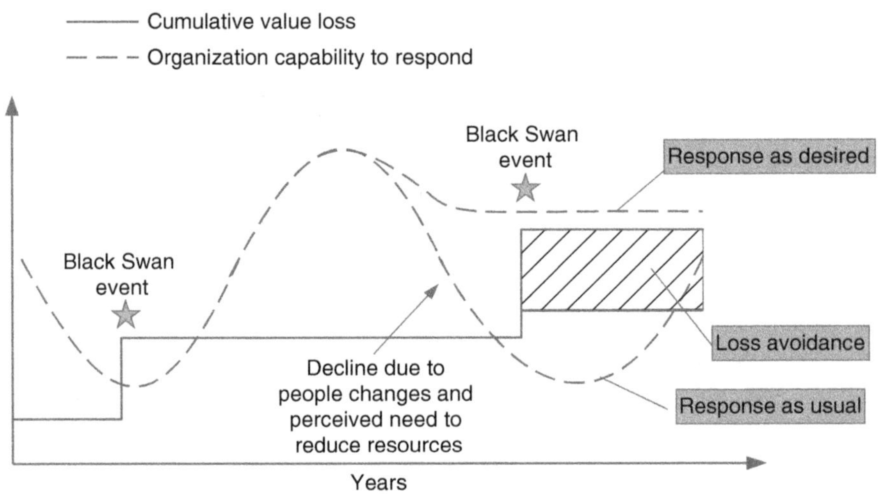

Figure 1.4 Desired state: Black Swan events trigger persistent development of problem-solving capacity.

This cycle of building and losing capabilities and suffering avoidable losses can be undesirable to the company. The alternative is illustrated in Figure 1.4. If we view the period after the first Black Swan event as an opportunity to strategically act to preserve the problem-solving capacity within the organization and develop new people, an orderly succession can be provided. The payback is that the losses due to the second Black Swan event are much less.

The alternative approach to consider in strategic technical staffing management is introduced in Figure 1.5. In any organization practicing and managing technology, there is a variable return on the work that the people can deliver. Thinking of the boundary conditions, when you have no technical capacity and add some capacity, the incremental value delivered is very large. In fact, that continues to be true as you add staff who brings added capacity for a substantial range of added people. Considering the opposite boundary condition, we can also expect that at very large staffing, the incremental value returned by the next person added is small or even negative. It is possible to be over staffed. In general, this is an example of the microeconomics optimization of return on resource costs vs. value contributed. The difficulty is that a large portion of the value is created during infrequent events while the cost to sustain the staff is steady.

Also, typical rates of people turnover and attrition can lead to departures of the more experienced people. Good long-term succession planning in the technical staff includes investing time to develop inexperienced people to have the abilities required by the severe events. The developing technical people will benefit from working on complex projects that are less urgent, valuable, or complex. A challenge to the whole organization is that there is significant uncertainty in the true

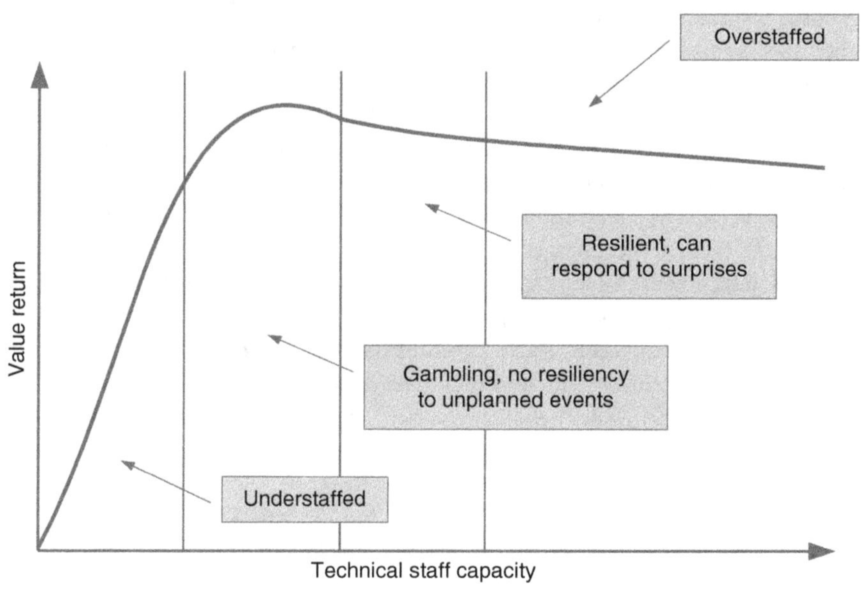

Figure 1.5 Optimum long-term staffing model concept.

location of the value line. As a consequence, there is a frequent tendency to reduce people into the region labeled as the non-resilient gambling region. Two problems with that approach are 1) the organization is not resilient and 2) the more subtle problem can be that the people in place experience persistent overwork conditions. As a result of being overloaded, they tend to not develop valuable technical skills, and the increased stress can negatively correlate with job satisfaction and employee retention.

For a number of the actual problems we have worked on requiring outside help, what has happened is that the known and demonstrated effective process knowledge has been lost or misunderstood by newer process owners after a people change or a series of people changes. While this seems surprising, it is not an unheard-of scenario. There are cases where the present problem has been solved before by predecessors of the current process owners. If the correct way to operate has been sufficiently internally documented, the problem can be quickly resolved by understanding how the operation has deviated from the recommended operating conditions and returning the process to those conditions. In other cases, where the solutions were not documented internally, outside literature can have direct application and lead to a resolution.

> ***Wisdom Tidbit – On a new plant startup that is not working, look for where the project manager had to remove 10–20% of the total capital in order to reduce costs.***

Another common scenario is that the process owners often partnered with the companies' subject matter experts (SMEs) have been working on the problem continuously or periodically for months or even years.

In certain cases, people hesitate to take new help into an existing improvement team. Human nature for many people is that the people who have been working on the problem can view outside help in these cases as unwelcome. Taking outside help can be viewed as a reflection on the persons own abilities. While this is rarely accurate, the feeling is common. The benefits of having outside people to bring fresh eyes and increased focus can be effective. The more structured approaches we advocate are key to speeding up the work. Working through the resistance to help from the process owners is an important issue to recognize and plan for when bring in an outside team to join and established team.

Reference

Taleb, N.N. (2007). *The Black Swan: The Impact of the Highly Improbable*. New York: Random House.

2

First Principles Problem-Solving – Why First Principles?

Is it possible to solve a problem without understanding the science behind the problem? In our experience, the answer is a definite YES. Really understanding a problem to the point of a strong understanding of the science behind why the problem is occurring takes time and often considerable expertise. Industry is often short on time and expertise, so shallow understanding is surprisingly fine at being successful and is often the preferred method.

To illustrate the principle, here is a simple example. An engineer was asked by his boss to try and get some additional capacity out of a reactor system. The engineer knows that increasing temperature generally will increase reaction rates, so asks the operators to increase the reactor temperature a couple of degrees. Reaction rate does go up and the capacity needed was achieved. The engineer never pulled out a calculator, did not do the kinetics calculation, did zero research, and used very little understanding to solve the problem; but the problem got solved nonetheless.

The same shallow understanding can happen even when more complex analysis is done. For example, an engineer was charting product quality on a statistical process control (SPC) chart. The engineer noticed that of the three suppliers of raw materials for the product, supplier A-Company consistently produced product that was out of control on the control chart and produced poorer quality material. The engineer decided to use the other two suppliers and stop buying from A-Company. The product quality improved due to the decision. The engineer did not understand the science behind why A-Company raw material was worse. They did not need to and did not choose to dig that deep. Still the problem was solved.

Big Data analysis also can work in this manner of shallow understanding of the theoretical portion of a problem. For example, a team wanted to decrease the

Fundamentals of Industrial Problem Solving: A Practitioner's Guide, First Edition.
Zdravko I. Stefanov, Eldad Herceg, Carla Schmidt, David M. Jacobson, Dana Livingston, J.P. Chauvel, Sunil Kumar Chaudhary, and Christopher Paul Christenson.

energy needed to produce a product because energy costs were driving most of the product costs. The team took everything that was measured on the process such as flow rates, pressures, temperatures, and so on for the past several years and fed them into a software program that did statistical analysis to find correlations. The analysis showed the lowest energy usage could be achieved by operating a certain pressure–temperature combination on the feeds to the system. The team had some theories on why that could be, but just shifted the process to the better operating regime and called it a win. Big data analysis provided the answer, but the science was not fully understood. The problem was solved.

In all these examples, digging deep may have cost time and money. It would likely have slowed the improvement down. The examples were shallow on theory and deep on experience and observation. This method is the empirical approach to problem-solving. The empirical approach to problem-solving does not rely on theory. Instead it uses the practical observations and experience to solve problems. Contrast that to a theoretical approach.

First principles, the underlying laws of physics, chemistry, and other sciences are used to solve problems in the theoretical approach. To illustrate this form of problem-solving, here is an example.

A satellite had a close call with a small meteorite. A meteorite nicked a solar cell, resulting in little physical damage. Unfortunately, the impact caused a rotational shift and a slight deviation from the intended orbit. A team of engineers and physicist tasked with correcting the problem, set about doing the theoretical calculations of what thrust was needed on which vector to correct the problem. The team completed the calculations using a combination of computer simulations and independent hand calculations. The burn was executed putting the satellite back to where it belonged.

The theoretical approach can be used on earth as well. For example, a tanker transporting a chemical was building pressure. The relief valve designed to relieve the pressure was plugged in the line to the relief. Concern over the pressure build led to a frantic call. If the tanker over pressurized, it would effectively be a bomb. The chemists and engineers tasked with the problem got to work. The pressure build was known to occur due to a chemical reaction that produced a gas. The chemists and engineers calculated the total gas production and consequential pressure build. The total theoretical pressure build was well within the maximum allowable working pressure of the tanker. The worry about the tanker being a bomb was alleviated and the problem of the plugged relief could be addressed without concern of explosion danger to the folks tasked with fixing the issue.

The theoretical approach is most frequently done in the design phases of equipment and process design. Simulations, calculations, first principle work is core to new designs. New design does have some amount of empiricism as well. Experience does matter and theory at the design phases is tempered with some empirical analysis. For example, in designing a chemical reactor, an empirical approach may be used to make sure the reactor can be maintained. As designs are completed for processes or products, then more empirical methods are used to optimize and perfect running the processes or using the products. Because orders of magnitude more time is spent running processes or using products vs. the designing of them, there is more opportunity to use empirical approaches to problems.

To illustrate this concept, Figure 2.1 at a high level describes the problem-solving techniques of empirical approach and theoretical approach to problem-solving. If conceptually we could measure how much of each technique was used in solving any given problem, we could say that a problem was solved using X% empirical approach and 100%–X% theoretical approach. On the left-hand side of the chart, the beginning of the lines shows problems solved using 10% empirical approach and 90% theoretical approach. On the far right-hand side of the chart, the problem is being solved using 90% empirical approach and 10% theoretical.

In our experience, the empirical approach to problem-solving is by far the most common approach that is used in industry. To see why and how fundamental problem-solving fits in, we will reiterate Figure 1.1 here, as Figure 2.2.

In Figure 2.2 on the left-hand side in Zone 1 is where most of the plant problems are being solved. In this region, the folks running the plant on the front line operations will solve the problem. Operators, technicians, and mechanics that run and maintain plant operations spend the most time day to day with the plant and therefore solve plant problems frequently. Engineers that are also directly working in the plant also spend significant time-solving plant problems. Because the

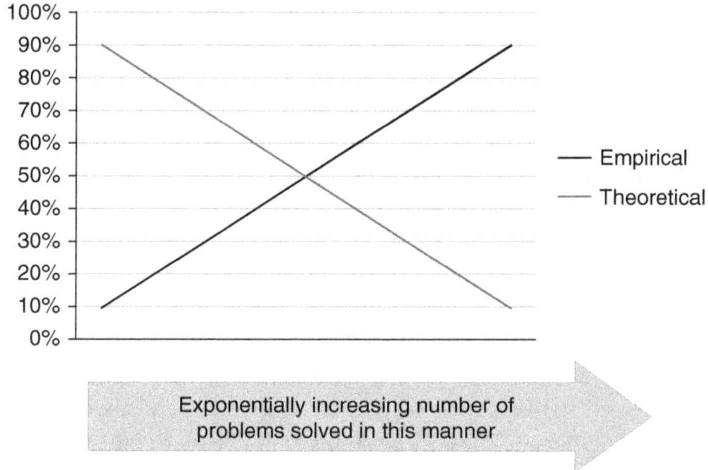

Figure 2.1 Empirical vs. theoretical split.

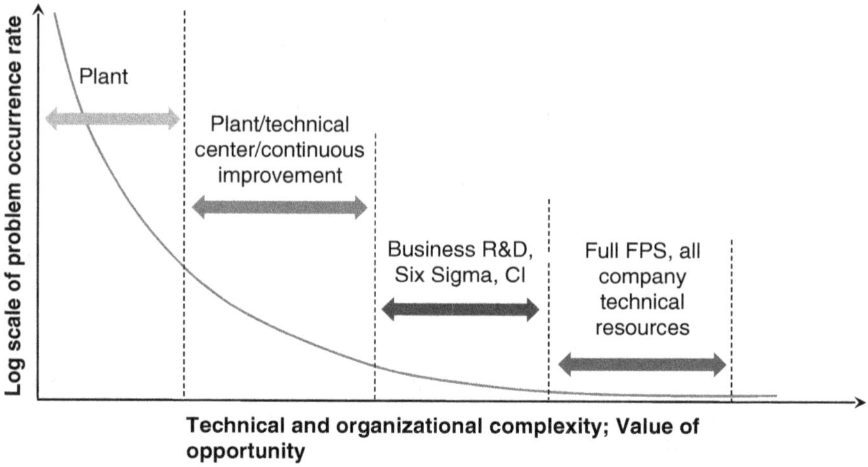

Figure 2.2 Spectrum of plant problem solving.

problem-solving is on plants that may have been working for many years or many decades, the original design intent have been modified and suit current operations. One does not need a PhD in reaction kinetics to run a chemical reactor. One needs to know how, standard operating procedures, technical background, and a heavy dose of experience. The problem-solving follows the same needs. Those on the front lines running equipment do need theory, but more importantly need focus on experience is true for most systems. Here are a couple of examples of front-line problem-solving done at the plant by plant people.

A distillation column was not doing the proper separation and was struggling to produce in specification product at rates it typically ran. The lead operator called the engineer assigned to the plant to discuss a solvent washout. The operator told the engineer that this column sometimes gets high–molecular weight polymer build up on the reflux tray. The engineer agreed and they together wrote a temporary procedure to do the clean out. The column was cleaned out and restarted. After the cleaning, the distillation performed as normal.

A mechanic noticed that a compressor was running higher than normal vibration levels. After checking the maintenance history, the mechanic noticed that during the last compressor service, the bearing replacement was skipped due to missing parts. After confirming that the replacement bearings were available, the mechanic recommended a bearing service within the next month or so, as the existing bearings appeared to be at the end of their service life and fixing them now could avoid further damage to the entire compressor. The plant management supported doing this work, the bearings were replaced, and the compressor was restarted and returned to normal vibration levels.

The Zone 2 from Figure 2.2 is where the problem is more difficult and is beyond the experience, skills, and often beyond the time available for the front-line workers. They are very busy people and as problems get more complex or difficult, sometimes their time availability starts to become a limitation to completing a problem resolution. These problems will get escalated to others. Sometimes it will be to senior engineers, senior technicians, senior mechanics, or senior operators. Different companies are structured in different ways, but usually there is a backstop of experience that are flexible resources able to help troubleshoot across many different plants or operations. This is also where Lean Six Sigma teams can fit into problem-solving. Usually they lead Greenbelt Lean Six Sigma projects (smaller scope), but may also have the occasional Black Belt project. Dedicated teams can provide the focus and rigorous methodology to systematically fix problems. Often the teams are needed because of the organizational complexity of the problem. Organization complexity occurs when multiple functions or teams of people have a problem to solve where the problem falls between the cracks of responsiveness. For example, the production team and the quality assurance team can have a problem that is a combination of work in both areas. Teams help those functions figure out the problem by working across functions. Root causes of the problems can be validated and the problems can be solved. Fewer problems are solved in this zone when compared to the front-line problems that the plant solves itself. In our experience, it is between 10 and 100 times less frequent. Here are a couple of examples of the front-line plant folks working with either senior technical folks or with a continuous improvement Lean Six Sigma team to solve a problem.

A production engineer and operator were discussing a problem. Ever since demand was down and rates were low, one of the heat exchangers kept fouling. Because demand was low, the fouling was not stopping production; but it was a pain to periodically clean. After initial troubleshooting efforts, they decided to ask a couple of senior engineers. One of the engineers with design experience in heat exchangers and experience on that process and other similar processes suggested it could be low flow rates allowing deposits that could then cook to foul the heat exchanger. Increasing recycle rates through the heat exchanger would maintain the velocity, but could still be run at the lower overall rates. The production engineer and operators trialed the solution and found it worked, and problem solved. The standard operating procedures were updated to include higher recycle rates during low production periods.

A Lean Six Sigma Green Belt team was formed to address an unacceptable odor in the product. The team formed had a quality assurance technician, a production engineer, and an operator. After completing an Ishikawa fishbone diagram, the team focused on trying to understand what exactly was the odor-causing species. The team did some head space analysis of the odor-containing product.

The analysis pointed to O_2 contamination causing the odor. The process was run under vacuum, so the team looked for vacuum leaks. After extensive searching and some pressure testing of various process systems, the team found the leak and fixed it. The process restarted odor free. The team established a control plan and closed a successful project.

The Zone 3 from Figure 2.2 is where the value, organizational complexity, and or technical complexity of the problem make for a complex project requiring a high level of expertise to solve or at least to coordinate. These projects are often Black Belt Lean Six Sigma projects or similar methodology. They are typically characterized by full-time project leaders, often will involve some specialized expertise, and often a corporate-level resource such as Research and Development. Sometimes outside consulting is brought in for these more complex problems. These projects will typically have more of a theoretical approach when compared to Zone 1 or 2, but still by and large be mostly empirical approach. The organizational complexity of these problems is indicated by the number of separate functions/work groups assigned to tackling the problem. Usually there are at least two, but often three to six different functions involved. The value impact is typically an order of magnitude (10×) or more than Zone 2 type projects. For example, if Zone 2 projects were typically $100,000/year benefit, Zone 3 would be more in the range of $1 million/year. When Zone 3 projects are tackled without a formal team, there is almost always a very senior technical expert who is leading the work and calling most of the shots as an informal team. These projects are the heart and soul of most continuous improvement efforts. They make money and are effective. Here are some examples of Zone 3 projects.

Multiple trains of the same core process of a chemical company are run in multiple geographies. Because of this, a corporate consultant position was in place for the technology. A problem was occurring in one of the units, the main reactor was not performing as efficiently in one location when compared with the same reactor in other locations. The conversion was low. The problem escalated beyond the local team up to the corporate consultant. There was no formal team formed, but key technologist around the world was brought together to try and solve the problem. After gathering input from the plants and the technologists, the corporate consultant pulled together detailed bench-marking of every reactor performances from around the globe, analyzed the data, and looked for the possible causes. Many theories were proposed, but the consultant with their experience realized that catalyst quality was critical to the successful reaction. The data that the consultant analyzed supported this theory, so follow-up meetings were aimed at exploring this possibility. The catalyst was supplied by the local geographies and was a natural substance mined at local mines. The team decided to air freight catalyst from a higher performing reactor site and try that to try and either kill the theory or validate it. The trial returned the reactor to normal performance. Testing of the poor performing catalyst showed trace impurities that are possible poisons.

The team requested more mine samples and ultimately had the catalyst supplier provide catalyst from another location in the mine known to have fewer impurities. The problem was solved with the top technical talent of the company leveraging a well-known and respected scientist in the corporate consulting role.

A customer was having serious performance problems with an adhesive. The adhesive was not bonding properly and therefore causing part failures. A Lean Six Sigma Black Belt project team was formed. A full-time certified Lean Six Sigma Black Belt was brought in to lead the effort. The team followed the Lean Six Sigma methodology and used DMAIC (Define, Measure, Analyze, Improve, and Control). During the measure phase, the team discovered that the adhesive's physical properties were drifting over time. This indicated an unwanted chemical reaction occurring in the adhesive as it ages. The failing material was analyzed and screened for probable catalysts. Retained good material was also tested for baseline comparison. Small particles of residual catalyst originating from one of the feedstocks for the material were found in the bad material. The team identified this as the root cause of the drifting material. They improved the filtration on the feeds to remove most of the residual catalyst. Accelerated aging testing showed that improved filtration did fix the drifting problem. The team institutionalized the solution and put controls to validate the filtration was working. The problem was solved by a cross-functional team highly focused on the problem.

This chapter will focus on what it takes to attack Zone 4 problems. This chapter concentrates on the importance of the first principles in problem-solving. Many problems are solved many times, again and again, which means they are not really solved. The main theme is that fundamentally understood problems tend to stay solved and the gained understanding help solve future problems much faster. Understanding the governing physics-based laws and variables enable permanent solution. Many Zone 4 problems involve multiple variables, and the variables are often interactive, intermittent, and nonlinear. This means that there is no "single root cause." Using the first principles does not mean it will take longer and cost more. Approaches to move faster are discussed, such as using the literature, models, order of magnitude estimates, identification of key variables, involving subject matter experts that have near world-class skill in required topic areas, both internal and external. There is also a discussion on other problem-solving methods used in industry and why they often prove LESS effective than the FPS approach.

2.1 What Is a Zone 4 Problem?

The empirical approach is more intuitive for most folks, so it tends to dominate problem-solving. Even as babies, experience is used in solving problems. We try things, we fail, we try again, and then eventually the right course of action presents itself, and we solve whatever problem is in front of us. It starts as things like getting peas to stay on our spoon where we notice horizontal curved side down is the only configuration that works, but continues through life to some of the world's toughest problems. The theoretical approach is dry and for most folks, pretty unappealing. Sitting in front of a computer reading through a bunch of published science, calculating for pages and pages, or running an experiment in a lab are not the stuff of fun for most people. As illustrated in Figures 2.1 and 2.2, Zone 1–3 problems represent are most of the problems people encounter in their work, so mostly an empirical approach is incredibly successful and often becomes a model that forms a rut in our improvement road that forces our wheels of problem-solving to follow the same path. Unfortunately, there are problems that when they are tackled with mostly the empirical approach become illusive, perhaps unsolvable. What is the nature of these tough problems? They have the following key characteristics:

1) **Multifactor** – Often Zone 4 problems are multifactor interaction problems. Many variables are working together to cause the problem.
2) **Nonlinear** – Sometimes the factors and effects are nonlinear to the extreme. This will make solving the problem complex and difficult to solve. Empirical methods will often struggle in these types of problems.
3) **Managerial interference** – Often Zone 4 problems have extraordinary burn rates (money being lost/period of time). The high burn rate ramps up the political and managerial interference and can take a simple problem to solve and make it extraordinarily hard to solve.
4) **Multi-functional** – Often Zone 4 problems cross many different workgroups and workflows. This leads to confusion and difficulty in understanding enough of the problem to fully realize what is happening.
5) **Costly symptoms** – Often Zone 4 problems manifest themselves with many negative symptoms, some are **VERY** costly. When the burn rate is high, then the cost of making mistakes and being wrong is also high.
6) **Life-of-the-Plant** – Often Zone 4 problems have been around so long that they are as accepted as the fact that the sun will rise. Before there were light bulbs, it was accepted that to have light at night, a person needed a fire in some form. Life-of-the-plant problems have been unsolvable long enough to be the way things will always be.
7) **Counterintuitive in nature** – Some Zone 4 problems will need solutions that require exactly the opposite of conventional wisdom. The problems do not get solved by a Zone 1–3 approach because the people have a very strong belief opposed to the required change. Solving a problem like this requires doing the opposite of what in the past has been the effective. Flipping the spoon over to curve side up to hold the peas stops getting to be tried because everyone knows

it will not work. But, if the peas are covered in sticky material, curve side up may actually hold the most peas. The solution is counterintuitive in nature and therefore a blind spot that plagues solving the problem.

8) **New science** – Rarely, but sometimes Zone 4 problems are new science. Using old science and experience can never solve a new science problem. This will pose a problem because developing the science becomes part of solving the problem.
9) **Perceived personal loss** – The prior person or team who worked to solve the problem and concluded the problem cannot be solved can often feel humiliated if others come in and do solve the problem.

2.2 How Does First Principle Focus Help with Zone 4 Characteristic Problems?

2.2.1 Multifactor Problems Are Helped by First Principle Focus

In some ways, all problems have multiple factors involved. This is not what we mean by multifactor in nature. Most problems have one or two factors that dominate to cause the problem. The Pareto principle applies; 20% of the factors cause 80% of the problem. These are Zone 1–3 problems. Zone 4 problems require that multiple factors work together to cause the problem. Alone, the factors do not influence the problem. In the language of Design of Experiments, these are interaction effects causing the problem. Two-factor interactions are about as complex as most empirical approaches can handle. Three, four, five, or more factor interactions will often be unsolvable with the empirical approach. A quick explanation of an interaction effect is hopefully helpful in understanding the complex nature of multifactor interaction.

In Figure 2.3, there is a two-factor interaction. The factors working together are a raw material supplier and the weather, more specifically if it is raining or not. The product being produced is great with both suppliers when it is not raining. But supplier B makes terrible product when it is raining. Interaction effects are

Figure 2.3 Two-factor interaction.

		Supplier	
		A	B
Raining	No	*Great*	*Great*
	Yes	*Great*	*Bad*
		Quality	

relatively easy to find in a controlled lab experiment, but often become more difficult in full-scale production situations. This is because in a design of experiment, the data being collected is designed to cover the entire design space and to try and do that in an even manner. In other words, maybe half the data is with rain, half without; half the data with supplier A and half the data with supplier B. The production world does not work this way. Instead it is often very uneven. Perhaps the plant is in an area that only rains 70 days a year, so relatively only a small set of data there 70/365 – 19% of the data set. Perhaps Supplier A is the preferred supplier, so is run most of the time (70%). Also, the product in question is one of many products made with similar feedstocks; it may only be in production 25% of the time. That makes the bad quality event due to this combination of causes a rare configuration; 19% × 70% × 25% = 3% or around 12 days a year. Think about how difficult this makes the two-factor interaction problem to understand a production unit. Most of the time with many different products, Supplier B material makes great stuff; why would anyone look there to solve a problem? The rainy days when other products were being run or when they were using Supplier A do not cause problems. Why would anyone look to the rain to be part of the problem? Two-factor interactions in combination with the naturally occurring frequency of the data make these difficult to solve. This gets worse when the number of factors in the interaction increase. For example, what if it takes three factors to cause the problem?

Figure 2.4 is a three-factor interaction. It takes a very specific set of events to cause this problem. Supplier B has to be the feedstock, it needs to be raining, and it needs to on a day with a high winds from the south. Three-factor interaction problems make two-factor interaction problems seem like a walk in the park. Hopefully, it is easy to see how this problem would be unsolvable by most methods. Multifactor interaction problems are very tough to solve just using experience.

		High-wind direction			
		South		All other directions	
		Supplier			
		A	B	A	B
Raining	Yes	*Great*	*Great*	*Great*	*Great*
	No	*Great*	*Bad*	*Great*	*Great*
		Quality			

Figure 2.4 Three-factor interaction.

Many folks try and attack the issue of multifactor problems with purely empirical use of big data analysis. In big data analysis techniques, interaction effects are hunted by putting dozens, hundreds, or even thousands of factors into various software packages designed to do the math. The software will spit out statistics that tell the user which factors might be important and will even indicate interactions that might be important. These big data hunts can yield fruit to solve the problem; unfortunately, they also produce a lot of garbage to sort through to find the fruit. To understand the garbage side of the problem, think of a system with only four factors. How many interaction effects are there? Four factors would have 6 two-factor interactions, 4 three-factor interactions, and 1 four-factor interactions. Eleven interactions with just four factors involved! Imagine the big data analysis hunting for interaction effects with hundreds or thousands of factors! There are so many potential interactions the system becomes unwieldy with many false positives. Teams end up chasing interaction effects that are just random chance signals. Correlation does not equal causation. There will be false positives in big data analysis, lots of them. To understand how false positives occur, just think of something simple like two tank levels. Tank levels go up and they go down. If you plot any two tanks in the world, they for some part of the data set will very likely have a correlation. If an oil tank in Saudi Arabia is slowly building inventory and a corn syrup tank in Iowa, USA, is slowly losing inventory plotting them over the same time interval can lead to interesting conclusions (Figures 2.5 and 2.6).

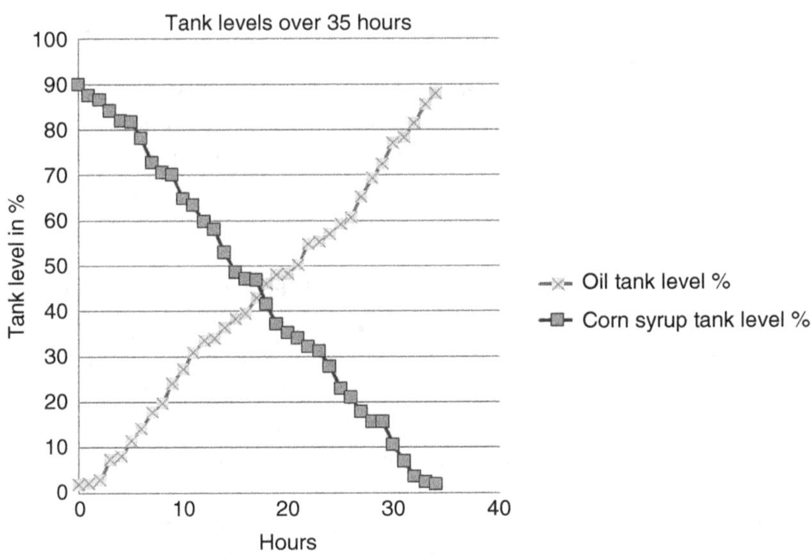

Figure 2.5 Tank levels over 35 hours.

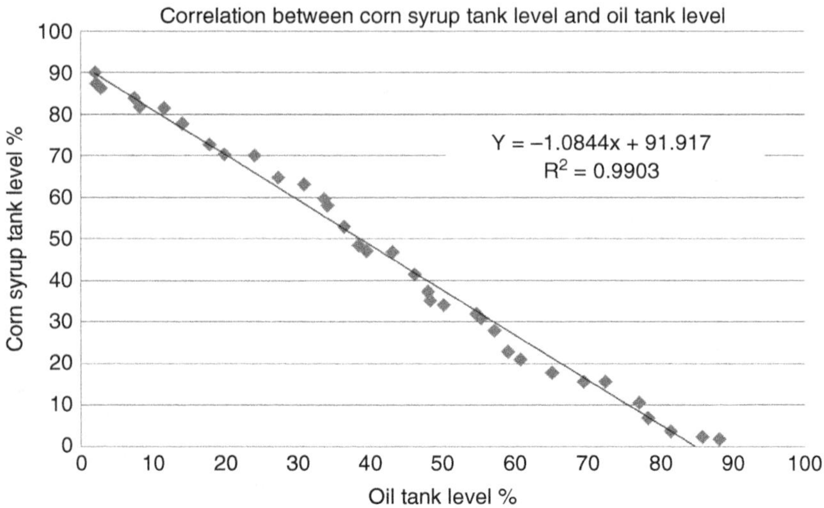

Figure 2.6 Correlation between corn syrup tank level and oil tank level.

Is There a Correlation?

Plotting these tank levels against each other in an X–Y plot (Figure 2.6) will show a wonderful correlation! Obviously, as the oil tank level in Saudi Arabia increases, the corn syrup tank in Iowa HAS to go down. The R squared is fantastic and there is plenty of data to support this conclusion! And . . . it is all complete non-sense. Obviously, there is no causation relationship between these completely unrelated tanks. Exactly, the same happens when doing big data analysis with hundreds or thousands of variables. There frequently will be enormous numbers of false positive hits on not only main effects (single X–Y factors) but also on all of the multifactor interactions. Empirical methods break down and fail not because you are looking for a needle in a haystack, but because you are looking for a specific needle in a needle stack.

Fundamental problem-solving is successful in multifactor problems because it starts with first principles, the underlying laws of physics, chemistry, and other sciences. To see the difference in this theoretical approach vs. the empirical approach, let's look at our three-factor interaction from Figure 2.4 where rain, a strong wind from the south, and using Supplier B causes bad quality product. In this example, the right combination only happens 12 days a year. The problem at first blush appears random.

Starting with first principles means NOT doing any big data analysis initially. First hypotheses need to be formed based on solid science. In this case, expertise in chemistry is needed. What molecule or molecules are causing the product to be bad? What chemistry can form those molecules? These questions become the hypotheses to be used in the hypothesis onion. The hypothesis onion will be covered to a detailed level in other chapters but suffice it to say there will be a number of theories based in chemistry on ways the molecules of concern are formed. Those theories will need to be tested to see which one is actually happening and causing the problem. In the case of Figure 2.4, the chemical equation to cause the bad molecule in the product that ultimately caused it to produce bad quality product was produced by a trace contaminant from supplier B and water. There was no water in the production unit, so the hunt for water was on. Ultimately, a horizontal vent facing south was found to be the problem. Data analysis was then used to confirm the interaction effect of rain, strong south wind, and Supplier B.

To think of the difference between

A) Analyze, find, and then theorize (fishing)
B) Theorize, analyze, and then find (hunting)

it is useful to picture them as fishing with a net and hunting. In fishing with a net, you cast the net in an area you think there might be fish. What comes up in the net can be a lot of what you don't want (by catch) and what you want (fish). Data analysis prior to having theories being formed is tossing a big net out and seeing what is in the net. It then becomes sorting out to see if what you want is in the net. In hunting, the prey is tracked, and if it is there it can be caught. Starting with the first principle theory, you know exactly what you are looking for when you do the data analysis. Both methods can and do work. For hard problems, in our experience, hunting (theoretical) on the average works much better, especially in systems that are multifactor in nature. There is just too much by catch and that slows down solving the problem.

2.2.2 Nonlinear Helped by First Principle Focus

Sometimes the factors and effects are nonlinear to the extreme. This will make solving the problem complex and difficult to solve. Empirical methods will often struggle in these types of problems. To understand this, it is helpful to understand why most processes are effectively linear. In Figure 2.7 is an X–Y plot of an X variable that as it changes causes changes in the Y variable with a nonlinear relationship.

Nonlinear relationships are often the case in industry. Curves outnumber lines, until one very important piece of information is added. In industry, operating windows have been optimized and are often very tight. In Figure 2.8, we have the same chart, but with the operating window shown. The operating window is where the process is run 24 hours a day 365 days a year. Almost all of the data ever seen is in this window.

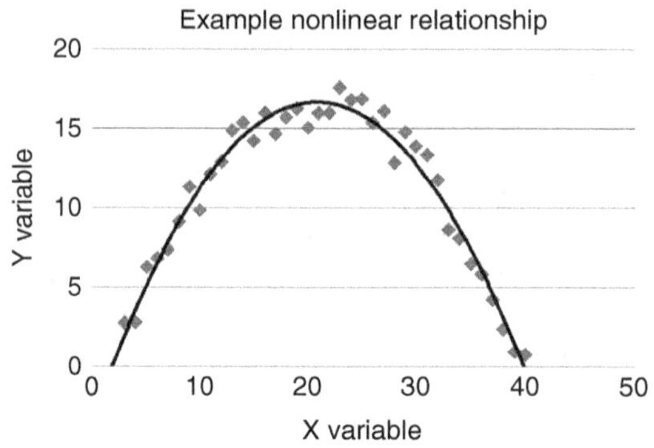

Figure 2.7 Non-linear relationship.

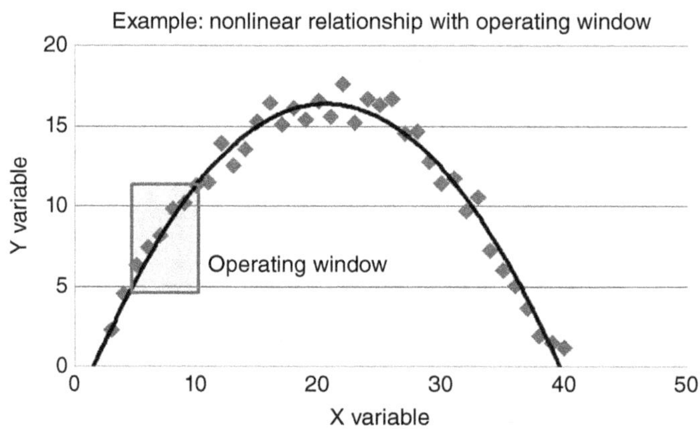

Figure 2.8 Non-linear relationship with operating window.

In Figure 2.8, the example of a nonlinear relationship with the operating window shown, it is clear that where the plant runs is effectively a line. This is a common example seen in industry where a relationship that is nonlinear over a wide range is practically linear over the narrow range that is normally used in a particular operation. With empirical approach, data are generally only collected at or very near where the plant is currently operated. Going beyond that can add risk. For example: If in Figure 2.8, a lower Y value indicated better product quality, while X was important for other reasons. If X was production rate, then more is better. With an empirical approach, there would be a great deal of reluctance to increase production because of the expectation that product quality will keep getting worse linearly. The theoretical approach can do better. A first principles model built on governing laws of physics, chemistry, or engineering could predict the curve. Lab studies or well-planned plant scale trials could validate the prediction.

With the empirical approach, problem solvers can be locked into only knowing what they have seen or experienced. Moving to the theoretical approach that is the basis for FPS moves beyond what has been experienced and into the realm of what could be experienced.

2.2.3 Managerial Interference Problems Are Helped by First Principle Focus

It may seem counterintuitive that first principle thinking helps solve problems when the pressure is on from upper managers. The managers are working hard at trying to be helpful, but when it is not a managerial problem, the help can be counterproductive. One way to think of this is to do a simple experiment. Have someone saying random numbers while you try and do a calculation. You will find it hard to focus and are highly likely to make errors. The trouble with high burn rate projects that are costing a lot of money is that there will be lots of this kind of background noise. Focus is the best way to combat this background managerial excitement noise. Here are some of the questions the project leader and teams may get as a part of this noise, making it hard to concentrate on solving the problem.

Example Managerial-Focused Questions:

- When will this be done?
- Do you know how much this is costing us?
- What are we going to do?
- What am I going to tell our customers?

All of these types of questions are actually not directed or even helpful at solving the problem. They are important for different reasons, but not important problem-solving questions. Part of the project leader's job will be to manage these questions while keeping the team **FOCUSED** on problem questions. Using first principles pushes science into the discussion. It provides the focus on the right questions to ask to solve a problem. The right questions are science-based eses aimed at solving the problem itself not the symptoms and fall out the problem is causing. Contrast problem-type questions to the

Example Problem-Focused Questions:

- Is Fe(III) acting as a catalyst in this system resulting in our run away reaction?
- Is water adsorbing on our catalyst surface resulting in blocking the reaction sites?
- Are the bearings being corroded by residual HCl in the feeds resulting in our high compressor vibration?

First principle questions help teams focus efforts finding answers that will have a chance at solving the problem. The high burn rate problems will have managerial interference and benefit greatly with first principle focus and therefore are great candidates for FPS techniques.

2.2.4 Multifunctional Team Problems Are Helped by First Principle Focus

Often Zone 4 problems cross many different workgroups and workflows. This leads to confusion and difficulty in understanding enough of the problem to fully realize what is happening. Focus on hypotheses helps defuse the political in-fighting that happens between groups. Once a hypothesis is agreed upon, the team can focus on and rally behind. To understand this, it is helpful to contrast the focus improvement.

In-Fighting Issue:
Quality assurance is dropping the ball on testing again!

Hypothesis Focus:
One of our main hypotheses is that the product is good, but we have a measurement system issue. A gauge R & R study will be used to test this hypothesis. The desired result is that if it is the problem, we can fix it; if it is not the problem, we can cross it off the list and stop putting any energy into it.

Often with serious problems much of the effort is wasted on finger pointing. Groups and individuals seem to gravitate to trying to blame someone. The proverbial scapegoat seems to be a part of the human condition. Problems cannot be solved with blame. Blame will not find root causes and cannot find solutions. The shift from personal or team attacks to a science focus is very powerful. Focus on the problem at hand is the only way to solve it. One of the most powerful parts of the FPS methodology is how effective a well-written hypothesis goes to driving in-fighting out of the system.

2.2.5 Costly Symptoms Problems Are Helped by First Principle Focus

Often Zone 4 problems manifest themselves with many negative symptoms and some are VERY costly. When the burn rate is high, then the cost of making mistakes and being wrong is also high. Empirical and experiential methods can be supplemented with first principle methods to increase the certainty of a proposed change.

To understand the concept of why sometime a high degree of certainty is needed and therefore FPS methods need to be used, let us look at an earlier example that was solved using empirical methods.

A distillation column was not doing the proper separation and was struggling to produce in specification product at rates it typically ran. The lead operator called the engineer assigned to the plant to discuss a solvent washout. The operator told the engineer that this column sometimes gets high–molecular weight polymer build up on the reflux tray. The engineer agreed and they together wrote a temporary procedure to do the clean out. The column was cleaned out and restarted. It then performed as normal.

This is a classic example of the simple and effective use of the empirical method. What if shutting down the process was going to cost the company $10 million? Do you think the engineer's boss or their boss's boss would support this approach or maybe want something more? Most folks would want more work and proof of what is going on before investing money on a proposed remedy. First management would want the best information possible to support the conclusions. For example, a first principles model validating that the behavior seen in the plant could be caused by 25% of the holes in distribution tray being plugged. Maybe they would need a gamma ray scan of the column that validates the hypothesis of plugging to convince them. Additional hypotheses should be checked also to help reduce the risk. It is high risk to only have one cause of a problem being tested. With the high price tag of a shutdown, follow-up work to really address the root cause of the polymer forming should also be on the table.

FPS improves certainty by validating empirical problem-solving and move closer to the true root causes, which is exactly what is needed when the money is flying down the drain.

2.2.6 Life-of-the-Plant Problems Are Helped by First Principle Focus

Life-of-the-plant problems are problems that have not been solved through repeated efforts and have hung around and are essentially part of normal operations. Often there will have been multiple efforts to improve the problem and often because the true root cause of the problem has not been found, the work centers around treating symptoms. Here are a few life-of-the-plant style problems that are examples of what is being discussed:

- Compressors are put on short rebuild cycles because people know they will fail frequently and it is better to do that as planned.
- Heat exchangers are very frequently cleaned because the fouling root causes are not understood.
- Expensive filtration steps are added to processes to deal with unknown solids.
- Reactors underperform for unknown reasons.
- Products will have a high defect rate causing rework or scrapping.

There are many more examples of life-of-the-plant style problems. The reason the theoretical science-based approach really performs superbly on these problems is that by definition, the empirical approach has been used repeatedly on these without a truly satisfactory solution. The Zone 3 approach has failed. Doing the Zone 3 approach again is highly unlikely to solve the problem. In this scenario, it is recommended to switch to an FPS approach that is mostly theoretical in nature, going to first principles. The empirical approach is a pragmatic way of fixing first principle problems. You do not need to deeply understand something to fix it; only sometimes you actually do need to deeply understand something to solve it. Life-of-the-plant problems are one of the best places to apply FPS techniques.

2.2.7 Counterintuitive in Nature Problems Are Helped by First Principle Focus

This is a very strange class of problems out there that may actually not be terribly difficult to solve, but solving them takes doing the **OPPOSITE** of what has worked in the past or to do something that does not make intuitive sense or sense from most engineering or science first principles. The counterintuitive nature of these problems stem from complicating effects that push secondary factors into a condition of being the more dominant effect vs. the primary typical factors driving systems. Here is an example to clarify what these problems look like.

A distillation column was struggling to separate and make high purity products, and this problem was ultimately limiting the capacity. The column was run under a vacuum and was prone to flooding. Flooding is a condition where the velocity of the gas in the column sweeps the liquid reflux running down the column and makes it run up the column; this causes high pressure drop in the column and makes it impossible to run. Unfortunately, the typical way to improve purity in distillation is to increase the reflux ratio (ratio between boil up rate and take off rate). Increasing that ratio drives any column toward flooding. This column was already near flooding, so it often pushed it over the edge. Attempting to fix the problem caused more problems. Several "accepted wisdom" changes to improve column operations had been tried. The teams tried increasing the pressure to buy more room on flood point. They tried changing temperature profiles. In short, they tried many other standard ways of solving the problem. None of them worked.

The more fundamental answer was hidden in a secondary physical property characteristic that was documented in the technology report for the chemicals involved. In general, ability to separate to cuts with different boiling points gets easier as pressure is decreased. The specific chemicals in this system had these characteristics in the extreme; lowering the pressure made it more than an order of magnitude easier to separate. The answer to improve the column was counterintuitive to the way nearly every distillation column in the world behaves. To improve purity, reflux ratio had to be cut WAY back to avoid flooding and the pressure had to be lowered to a deep vacuum. These changes not only improved purity, they also increased capacity. Empirical analysis techniques did not solve this problem; it took a deep dive into first principles and hunting down lesser known drivers for the system.

Some problems have a very unexpected, surprising, and counterintuitive nature that will blind the teams from solving the problem. First principle focus changes that game. The FPS theoretical approach solves these problems with a significantly higher success rate vs. traditional more empirical approaches.

2.2.8 New Science Problems Are Helped by First Principle Focus

Rarely, but sometimes Zone 4 problems are new science. Using old science and experience can never solve a new science problem. This will pose a problem because developing the science becomes part of solving the problem. Nobody wants to develop new science in the middle of solving a problem. It is significantly easier to research existing science or use experience. New science can take years and problems often need to be solved in hours or days. To understand how new science can become part of problem-solving, here is an example.

A team was investigating bulging drums. The chemical being stored should have been stable, but for some unknown reason, it was decomposing and releasing a gas that was pressurizing the drum causing it to bulge. After investigating many avenues of chemical reactions that could have caused the problem, the team ran out of ideas. A hallway conversation with a microbiologist about the problem prompted them to do something new and actually strange. The microbiologist looked for microbes that could survive and thrive in the environment of a chemical drum and ultimately found them. At the time, none of the known microbes would "eat" the chemical in question; it was thought impervious to microbe attack. These microbes were new to science. The work went to the lab to figure out how these bugs grew and what it would take to kill them and prevent them from having the chemicals in the drum for lunch! The work had to go to developing new science to get solved.

Hopefully, it is obvious why theoretical methods are the basis for solving problems that are only solved by developing new science.

> ***Wisdom Tidbit – New science developed during problem-solving can be valuable intellectual property. That contaminant that is catalyzing your product causing a problem could become a best-selling commercial catalyst for your company.***

2.2.9 Perceived Personal Loss Problems Are Helped by First Principle Focus

The folks who have been working at solving the problem will often be worried about outsiders coming into the situation to solve the problem. People most often are very proud of the work that they do and not solving a problem can feel like failure. First principles focus dives deep into the science, numbers, and research.

Focus in these areas helps teams quickly realize how difficult the problem is to understand (this is the shift from empirical to theoretical). It quickly becomes difficult to feel like a failure when the complexity and difficulty of the problem is understood.

2.3 Final Thoughts on First Principles Problem-Solving – Why First Principles?

Most problem-solving does not require a first principle focus. The reason behind this has to do with how very effective empirical problem-solving can be for many problems. Unfortunately, the empirical solving techniques begin to unravel and be less effective for some of the most important problems. That leaves many problems unsolved, and these can become tolerated as unsolvable. FPS can make a big difference when the problem has significant organizational complexity, and/or technical complexity, and/or high costs.

The techniques in this book will work for Zone 1–3 problems that are at their heart easier to solve. They are sound techniques, but can feel like overkill on simpler, less complex problems. The answer to using them on simpler problems is to adjust them accordingly. Instead developing and testing 20 hypotheses, maybe just pick the two or three most likely. The authors intent is not to influence the reader away from empirical techniques they work. Instead our intent is to help the reader understand the limitations of the empirical approach and recognize when those problems are unlikely to be solved using that approach.

3

Opportunity Identification/Definition – Initial Team Formation

The phase of opportunity identification and initial team formation includes four important aspects: determination whether the project is a crisis or a chronic issue, the history of the problem, the creation of the project charter or the Data Quality Objective, and the determination of the team structure.

3.1 Chronic Issue or Crisis?

Usually plant problems fall within one of these two categories. As it will be discussed below, whether the problem is a crisis or a chronic issue determines to a great extent how the project will be run.

3.1.1 Crisis

When the problem involves force majeure, fire, explosion, it is usually a crisis. Same can hold true if the plant is not able to produce enough product to fulfill contract obligations. Such problems require rapid resolution, and this impacts many things about how the project is led.

Crises usually justify themselves and therefore it is usually not a problem to get resources, manpower, money and whatever else the team needs to solve the problem. The problem solver needs to be prepared to a rapid pace of the project, with multiple meetings and activities, possibly daily, during weekends, etc. Crises are very taxing in the short term, but provide great opportunity to get visibility and solve significant high value problems.

Fundamentals of Industrial Problem Solving: A Practitioner's Guide, First Edition.
Zdravko I. Stefanov, Eldad Herceg, Carla Schmidt, David M. Jacobson, Dana Livingston, J.P. Chauvel, Sunil Kumar Chaudhary, and Christopher Paul Christenson.

3.1.2 Chronic Issue

When the problem involves things that have become part of the environment, so to speak, it is usually classified as a chronic issue. Such are regular catalyst changes, reactor cleanups, etc. These events usually happen at a regular time interval and have been embedded in the plant life. Such problems require resolution at slower pace, but nevertheless need to be solved if the company wants to maintain competitive advantage.

The chronic issues can be much harder to justify with the stakeholders, as they themselves often need to be convinced that there is a problem at all. After all, "the plant has been running like this for years and there is nothing that can be done" is a usual argument. The problem solver needs to be prepared for a slow-pace project that can take long time to solve. Everyone will have much more urgent day-to-day activities and the perseverance is the key here. It maybe counterintuitive, but chronic issues are much harder to solve compared to crises, as the problem solver needs to maintain pressure over long periods of time, keep everyone on track, continue to provide resources, keep stakeholders happy with stage gate reviews, showing quick wins and progress and so on.

3.2 History

In both cases, crisis and chronic issue, there is usually a past history that can be much more pronounced in the case of the chronic issue. There may be a team already in place that has tried to tackle the problem multiple times. It may be the case that the existing team is very reluctant to get outside help or does not realize they need help. This situation is to be approached with extreme caution. Many times the problem solver is expected to both solve the problem and not alienate everyone on the current team. Unfortunately, there is no one-fits-all solution here, but there are several courses of action, depending on the current situation. The problem solver can be the formal leader of the team in the case of no resistance. If there is resistance from the existing team, the problem solver can rescind formal leadership to someone from the local organization, with whom the team members feel more comfortable. As long as the problem solver negotiates ahead of time that the condition is to follow the proper fundamental problem-solving process (i.e. use science) and the stakeholders reinforce this, it is not of great importance whether they are a formal project lead or not. If the problem solver can enter the project so the existing people welcome him rather than feel insulted or threatened, everyone is set up for success.

In any case, it is imperative for success that the history of the problem is examined and addressed appropriately.

3.3 Charter/DQO Writing

So why do we call our project charter/contract a DQO? The abbreviation DQO means Data Quality Objective. According to (USEPA 2006), "The U.S. Environmental Protection Agency (EPA) has developed the Data Quality Objectives (DQO) Process as the Agency's recommended planning process when environmental data are used to select between two alternatives or derive an estimate of contamination. The DQO Process is used to develop performance and acceptance criteria (or data quality objectives) that clarify study objectives, define the appropriate type of data, and specify tolerable levels of potential decision errors that will be used as the basis for establishing the quality and quantity of data needed to support decisions." Our project chartering process has many parallels with the EPA DQO process, and the abbreviation got embedded in the FPS terminology.

The importance of the Charter/DQO lies in two aspects – problem-solving and team dynamics.

3.3.1 Problem-Solving Aspect of the DQO

From problem-solving perspective, the issue of having a proper definition of success is absolutely critical. In Chapter 1, we already introduced briefly the important distinction between defining a problem and defining success. To elaborate, this distinction is very similar to two major approaches to improvement – to learn from mistakes and to learn from success. Historically, both approaches have been utilized; however, it appears to be more natural to learn from mistakes. For example, one famous example is a story about Abraham Wald and his investigation of the damages of allied bombers during the Second World War. Wald was tasked to analyze the damages on the airplanes and decide how to improve armor. Initially people looked where the planes were getting most bullet holes and thought those areas are to be reinforced. Wald however realized that the planes that have holes are *those that have returned*. They did not get holes in vital areas, in other words. So the areas without holes are *precisely the areas to be reinforced*. This view has been developed over the years to hit its current apogee in the resilience engineering (RE) discipline. One of the main pillars of RE is the idea that we need to "move from ensuring that 'as few things as possible go wrong' to ensuring that 'as many things as possible go right'." The principle of Define Success (in contrast to Define the Defect) will be reinforced again and again throughout the book. Defining Success is in our experience and opinion much wider concept that allows the problem solver to escape the box of the problem and look around. Many times the solution of the perceived problem is not even required to achieve success.

Wisdom Tidbit – Defining success is equivalent to thinking outside the box, in contrast to defining the problem, which is similar to thinking inside a box.

Similarly of great importance is to define what is necessary for success. This part of the charter/DQO process allows the team to really concentrate on what they need to obtain success in a structured fashion. Developing the necessary for success criteria is the main prerequisite to writing a set of good hypotheses and properly apply the scientific method later on in the process.

Finally, defining what is rate limiting is of equal importance. It boils down to try and figure out what potentially can prevent the project from getting done and the problem getting solved for good.

Wisdom Tidbit – The three main pillars of the charter/DQO process in FPS are:

- ***What is success?***
- ***What is necessary for success?***
- ***What is rate limiting?***

3.3.2 Team Dynamics Aspect of the DQO

The charter is the contract that is made between the team and the stakeholders. Getting this contract written and approved by all parties is of utmost importance, as it sets up the expectations, the boundaries, the available resources, and the deliverables in advance. This way there is a much smaller chance of misunderstandings and conflicts later on in the project. The problem solver should do everything in their power to make sure the charter/DQO is written and approved before any further work starts. Fortunately, it is usually well understood by everyone why this is important so there is not much resistance doing without much delay. It also gives a chance to the problem solver to understand the objectives and the objections the team members and stakeholder may have. Also it helps to disarm any future contradictions, as the objecting party had a chance to speak during the charter/DQO phase.

The summary here is to get the charter/DQO written and agreed upon as soon as possible. Be very stringent about this and do not allow deviation. It is better to even decline involvement in the project if there is no clear charter/DQO, as the potential negatives of proceeding without one can have career destroying consequences.

3.4 Team Structure

There can be two teams, the technical team and the steering team. The technical team is directly involved with the technical aspects of the problem solving – the literature search, the hypotheses killing, the science, etc. The steering team helps to address rate-limiting step (RLS) requests. The steering team helps to adjust resource investments (value creep) and plan status communication timing and format (not content).

The technical team has three main components – sponsor, stakeholders, and core team members. The sponsor is a person who has financial accountability to get the problem solved – often a senior manager. The stakeholders include plant leadership, commercial function, site leader, and the sponsor. The core team can have several components:

- Facilitator/scribe
- Technical lead
- Plant representative
- Several subject matter experts (SME)

It is quite common that some of these roles will overlap. Sometimes it is important to prevent overlap, such as in urgent projects. For these, for example the facilitator and the technical lead should not be the same person, so that each can concentrate on separate duties and progress faster.

Some of the team members will have to own the hypothesis matrix and manage it. Subteams can be formed for each hypothesis and ad-hoc members can be drawn in when needed. The team might change composition as the project evolves – depends on the hypotheses being killed. As this happens, a key team activity is to periodically look for at the RLS and escalate requests to the steering team. The RLS will regularly shift.

The steering team can overlap with the technical team, or not. The key deliverable of steering team is to steer. It must avoid problem-solving, but help to address RLS requests.

3.5 Examples

Two examples of project charters/DQOs that include the team compositions are given below.

The first example is from a project dealing with ethylene glycol color problem.

Project Charter/DQO
Business: Ethylene Oxide/Ethylene Glycol

<table>
<tr><td>Title:</td><td>Eliminate High Yellowness Index in Ethylene Glycol in Asian Plant</td></tr>
<tr><td>Project Process:</td><td>Fundamental Problem Solving</td></tr>
<tr><td colspan="2">Strategic Alignment</td></tr>
<tr><td colspan="2">This project is fully aligned with ethylene oxide/ethylene glycol's business strategy to sell to higher value polyester markets in South East Asia and China.</td></tr>
<tr><td colspan="2">Opportunity Statement</td></tr>
<tr><td colspan="2">Since start-up six years ago, ethylene glycol produced in the new Asian plant has had intermittent high yellowness index. The yellow color is observed in the two shore storage tanks near the harbor, not in the process itself. The intermittent nature of the color disrupts supply chain for our company and for our customers. High yellowness index ethylene glycol is not suitable for the polyester market and must be sold into the less profitable antifreeze market. The annual average margin loss exceeds $XX million.</td></tr>
<tr><td colspan="2">Burn Rate – is the present status creating $ losses, e.g. plant down
Should project be schedule driven, budget driven, or results driven?</td></tr>
<tr><td colspan="2">The project is schedule driven.
Recently, the storage tanks have contained mainly high yellowness index ethylene glycol. This has significantly reduced margins and caused damage to customer relationships.</td></tr>
<tr><td colspan="2">Fundamental Problem-Solving (FPS) Definition of Success</td></tr>
<tr><td colspan="2">Success is to understand why the ethylene glycol exceeds the yellowness index specification and then to use that knowledge to implement permanent solutions so that the product never exceeds this specification.</td></tr>
<tr><td colspan="2">Necessary for Success (FPS)</td></tr>
<tr><td colspan="2">1) There exists a core team with necessary skills and focus to solve the problem. The core team is chartered and supported by the sponsor and stake holders (listed below).
2) The absolute cause of the yellowness is identified.
3) The reaction pathway to produce it is understood both theoretically and experimentally. Use of external published literature is believed to be essential.
4) The impact of plant and shore tank operations on the formation (or lack of formation) of high yellowness index product is understood.
5) The core team develops a set of fundamentally reasonable hypotheses that explain the yellowness, including chemistry and conditions that facilitate formation. The team then prioritizes and kills the hypotheses.</td></tr>
</table>

6) The core team has the needed expertise in chemistry, engineering, the operations of the plant/shore tanks, project management, . . .
7) The sponsors/stakeholders are:
   - Sponsor: Global Improvement Leader, EO/EG
   - Sponsor: Commercial Director, Oxygenates
   - Stake Holder: Site Leader
   - Stake Holder: Production Leader, EO/EG Operations
   - Stake Holder: Engineering Technology Center Leader, EO/EG
   - Stake Holder: R&D Leader, EO/EG
8) The team is:
   - Fundamental Problem-Solving Project Manager
   - Plant Operations Run Plant Engineer
   - Process Data Mining Subject Matter Expert
   - Theoretical Chemist (Density Functional Theory)
   - Physical Organic Chemist/Analytical Chemist

Rate Limiting (FPS)

The RLS is formation of the team, collecting main hypotheses and organizing them to begin killing them.

The structure of the yellowness and the mechanism of formation are unknown.

Scope and Boundaries

Identification of the cause of the yellowness (molecules), the mechanism of their formation, how this is influenced by the operation of the process, and storage tanks are all in scope. Design of a proposal to remediate is also in scope, including potential plant trials.

Deliverables

The key deliverable is a path forward to produce, store, and ship low yellowness index ethylene glycol for the polyester market.

Timeline (suggested)

Target to complete project and implement process changes is four months.

Leverage

- Internal literature
- External literature
- Theoretical chemistry
- Analytical chemistry
- Process data mining
- Synthetic organic chemistry – to make the same yellow color compounds in the lab.

The second example is from a project dealing with fouling of tar flash evaporator.

Project Charter/DQO
Business: Agricultural Chemicals

<table>
<tr><td>Title:</td><td>Eliminate Fouling on Tar Flash Evaporator</td></tr>
<tr><td>Project Process:</td><td>Fundamental Problem-Solving</td></tr>
<tr><td colspan="2">Strategic Alignment</td></tr>
<tr><td colspan="2">This project is fully aligned with the Agricultural Chemicals business strategy to increase sell margins and asset utilization of our manufacturing assets.</td></tr>
<tr><td colspan="2">Opportunity Statement</td></tr>
<tr><td colspan="2">This plant, which was constructed more than 30 years ago, has always had intermittent fouling of the top of the tar flash evaporator. The median time between events is about 25 days. The longest run without fouling was 81 days. There have been intervals of frequent fouling – about every 8–10 days – for months at a time, which is very disruptive to the production schedule and stressful for the people in operations. The opportunity is to fundamentally understand what the tar is, how it is formed, why it deposits on top of the tar flash evaporator. Then, it is to use this knowledge to eliminate the fouling permanently.</td></tr>
<tr><td colspan="2">Burn Rate – is the present status creating $ losses, e.g. plant down
Should project be schedule driven, budget driven, or results driven?</td></tr>
<tr><td colspan="2">The project is schedule driven.
The last four months have been very difficult for operations with frequent fouling events. Production is lagging the production plan significantly. The problem needs to be understood and solved as quickly as possible or customers will be placed on allocation.</td></tr>
<tr><td colspan="2">Fundamental Problem-Solving (FPS) Definition of Success</td></tr>
<tr><td colspan="2">Success is to be able to run the tar flash at least 90 days between cleanings.
Long-term success is to never have to clean the tar flash again.</td></tr>
<tr><td colspan="2">Necessary for Success (FPS)</td></tr>
<tr><td colspan="2">1) There exists a core team with necessary skills and focus to solve the problem. The core team is chartered and supported by the sponsor and stake holders (listed below).
2) The absolute cause of the tar(s) is identified.
3) The reaction pathway to produce them is understood both theoretically and experimentally. Use of external published literature is believed to be essential.</td></tr>
</table>

4) The core team develops a set of fundamentally reasonable hypotheses that explain the tar, including chemistry and conditions that facilitate formation. The team then prioritizes and kills the hypotheses.
5) The core team has the needed expertise in chemistry, engineering, the operations of the plant, project management, . . .
6) The sponsors/stakeholders are:
   - Sponsor: President, Agricultural Chemicals Division
   - Sponsor: Commercial Director, Agricultural Chemicals
   - Stake Holder: Site Leader
   - Stake Holder: Production Leader
   - Stake Holder: Engineering Technology Center Leader
   - Stake Holder: R&D Leader
7) The team is:
   - Fundamental Problem-Solving Project Manager
   - Fundamental Problem-Solving Lead Technology Fellow
   - Plant Operations Run Plant Engineer
   - Agricultural Chemicals R&D Engineer
   - Reactive Chemicals Subject Matter Expert
   - Theoretical Chemist (Density Functional Theory)
   - Physical Organic Chemist/Analytical Chemist

Rate Limiting (FPS)

The RLS is formation of the team, collecting main hypotheses and organizing them to begin killing them.

The structure of the tar and the mechanism of formation are unknown.

Scope and Boundaries

Identification of the cause of the tar(s) (molecules), the mechanism of their formation, how this is influenced by the operation of the process are all in scope. Design of a proposal to remediate is also in scope, including potential plant trials.

Deliverables

The key deliverable is a path forward run the plant at full rates, safely, without ever having to clean the tar flash again. It is important to have understanding of the fundamentals of the problem including documentation of the findings, as this problem has been "solved" about six times in the last 30 years, only to reappear again.

Timeline (suggested)

Target to complete project and implement process changes is four weeks.

Leverage
• Any necessary resources – budget is for all practical purposes unlimited • Internal literature • External literature • Theoretical chemistry • Analytical chemistry – in particular high-resolution LC-MS (liquid chromatography – mass spectrometry) • Process Data Mining – in particular principal component analysis and related techniques • Synthetic Organic Chemistry/Reactive Chemicals Group – to make the same tar molecules in the lab

Reference

United States Environmental Protection Agency (2006). *Guidance on Systematic Planning Using the Data Quality Objectives Process*, EPA QA/G-4, EPA/240/B-06/001. Washington, DC: Office of Environmental Information.

4

Hypothesis Testing

Rigorous hypothesis testing is the core of the fundamental problem-solving methodology. This chapter provides the context for how this scientific tool combined with strong inference can be used to solve problems with much greater efficiency. It discusses what is a good hypothesis, as well as how to acquire facts to test the hypotheses. Several tools are also presented to highlight how to develop and manage hypotheses.

4.1 Application of Hypothesis Testing in Problem-Solving

A person's first introduction to hypothesis testing is generally during their early education science classes when the scientific method is taught as a key learning requirement. However, the formal application of hypothesis testing is rarely discussed at the university level or beyond in the work environment. Perhaps it is the use of simple problems when teaching hypothesis testing that has led many to not use it in their own scientific work. This tool may be perfect for guiding an eight-year-old budding scientist to investigate whether watering bean sprouts with canned soda, milk, or tea causes them to grow taller than watering with plain water, but can it handle complex problems such as how to prevent medications from degrading during storage on the store shelves?

The scientific method and hypothesis testing has been used to tackle problems of enormous complexity. For example, several structures were proposed for DNA by Watson and Crick. Rigorous prediction and testing were used to refute many of these structures and narrow the focus down toward the double-helix structure that is known today. Further, when the movement of Uranus was found to deviate from that which would be predicted by Sir Isaac Newton's theory of universal gravitation, a hypothesis was offered and later confirmed that there was another planet,

Fundamentals of Industrial Problem Solving: A Practitioner's Guide, First Edition.
Zdravko I. Stefanov, Eldad Herceg, Carla Schmidt, David M. Jacobson, Dana Livingston, J.P. Chauvel, Sunil Kumar Chaudhary, and Christopher Paul Christenson.

Neptune, affecting its orbit. Given the two scales of these examples (molecular to astronomical), it should be easy to conclude that hypothesis testing is a valuable tool for all levels of problem-solving. In addition, it can be an invaluable tool when delving into new areas that are unknown to the researcher/investigator. Trying to find an answer in the unknown is like looking for your car keys in a completely dark room. A system is required if there is any hope of reaching your goal. The scientific method and hypothesis testing provide a system for searching through this darkness.

However, the application of hypothesis testing in fundamental problem-solving requires a certain rigor that is not generally taught in early education. In 1964, John R. Platt published a paper titled "Strong Inference" in Science (Platt 1964). Platt proposed that the most efficient way to progress in a scientific endeavor is to propose several alternate hypotheses to explain a phenomenon. Next, determine and conduct the best experiment that can be used to exclude or invalidate one or more of the hypotheses. Then draw conclusions, eliminate the hypotheses that were excluded, and repeat the process by refining current hypotheses or developing new ones. Platt likened this process to climbing a tree and reaching a fork. The experimental test guides the climber on which branch to follow next. When a hypothesis has been tested many times and cannot be excluded, then it is deemed the most likely explanation for the phenomenon. This method of hypothesis testing is based on writings by Francis Bacon who is credited with saying that we must "proceed at first by negatives, and at last to end in affirmatives after exclusion has been exhausted."

Strong inference hypothesis testing is efficient because it allows the problem solver to take a broad field of hypotheses and devise tests to exclude large sections of them, which ultimately leads to a keener focus and better understanding of the most likely hypotheses. It also has the benefit of showing progress and accomplishment in an investigation even if the conclusion has not been reached. It gives the feeling of satisfaction similar to completing and crossing off a task on a "to do" list.

4.1.1 A Good Hypothesis

The key characteristics of a good hypothesis are:

1) Rooted in fundamental scientific principles or concepts
2) Clearly stated (Note: Complex hypotheses are more difficult to test, but so are those that are too vague.)
3) Has a clear cause and effect relationship
4) A test can be devised that would invalidate the hypothesis

The above characteristics are typically not achieved in the first few drafts of a hypothesis. Newly offered hypotheses often require much manipulation and refinement to get them to the point of being a good hypothesis. Therefore, one of the first requirements in this methodology is that hypotheses must be written down. Often a team working on a problem will have lots of ideas about what is causing the problem. The process of documenting these theories on paper, a dry-erase board, or on a computer screen causes the author to refine the statements more clearly. The author is forced to add more detail for completeness so if others find the document, they can understand the context. The act of writing hypotheses can also lead to the generation of alternate hypotheses. For example, a common failure mode in trayed distillation columns is the collapse of a tray onto the one below it. One person on the investigative team proposes that the tray collapsed because it is more than 20 years old and the bolts have corroded. Another person proposes that the collapse was caused by plugging of the weep holes, which allowed more mass to accumulate on the tray than the design specification. Yet another person suggests that the tower was unknowingly overfilled with liquid during startup and that initial heating generated enough vapor to lift liquid in the column. The force of this rising liquid raised the tray from its supports. Writing down the hypotheses, especially in a team environment, gives everyone a chance to refine and clarify each hypothesis. Testing can only begin once the hypothesis is clearly understood.

Further exploring the example of the tray collapse in the distillation column, the lead investigator asks the team members to write down their hypotheses on the flip chart in the conference room. The first person writes down that "Corrosion caused the tray collapse". If we compare this hypothesis to the list of key characteristics above, you will note that there are some that have not been satisfied. As this hypothesis was initially written, it is too vague. There are many aspects that need to be clarified:

1) Corrosion of which pieces?
   a) The bolts on the tray?
   b) The tray itself?
   c) The support ring that the tray sits upon?
2) When did the corrosion occur?
   a) During normal operation?
   b) During maintenance activity when the column was open to the environment?

As the team begins to discuss these questions, they generate more theories that lead to alternate ideas that could be turned into good hypotheses.

1) Is it possible that the wrong bolts were used during the recent maintenance activity?
2) Is it possible that the maintenance team forgot to install the bolts on this tray?
3) Is it possible that the bolts were installed but no nuts were installed?

Setting these alternate hypotheses aside for the time being, the team returns to the corrosion hypothesis and refine it to the following statement: "Corrosion of the bolts during normal operation over the past 20 years led to the collapse of the tray." At this point, they have achieved a hypothesis that is clearly stated with a cause-and-effect relationship, but it lacks information on what caused the corrosion. Now the team must decide how to proceed with this hypothesis. Should they try to figure out the fundamental science behind the corrosion or open the distillation column and look at the bolts to see if they are corroded? The fundamental science might be easily found in literature or by consulting with an expert metallurgist. It would require knowing the metallurgy of the bolts and the properties of the process fluids and gases that are normally present on this tray in the column (pH, composition, temperature, etc.). On the other hand, if the column has already been cleared of process liquids and is ready for entry by maintenance personnel, then simply going in and looking at the bolts may be the fastest way to kill this hypothesis. The point in this example is that the hypothesis should be written well enough that the investigative team can understand it clearly and devise a test that could exclude it. The team should only invest enough time in the hypothesis to get it to this point and not strive for perfection. If the team can devise a test to kill the hypothesis quickly, then further refinement may not be necessary unless the testing fails to invalidate the hypothesis.

4.1.2 Qualified Hypotheses

One of the tools that can be used to refine and develop good hypotheses is the Qualified Hypothesis Template. The tool is simply a document that is used to capture the relevant information about a specific hypothesis or family of related hypotheses. It can be documented in any software program of the authors choosing (spreadsheet, information map, slide presentation program, discussion board, etc.). The key is to gather specific information about the hypothesis into a cohesive document that the whole team can access. There are four sections to the template:

1) Statement of the hypothesis as clearly and succinctly as possible
2) An answer to the question "What has to be true for this hypothesis to be the root cause of this problem?"
3) Any background information on why the hypothesis is deemed plausible by the team. For example,
   a) Relevant fundamental science or literature.
   b) Previous occurrences of the problem and established root causes.
   c) Observations that were already made during the event.
4) A definition of what fact would have to be true for this hypothesis to be invalidated and a clear explanation of how that fact should be obtained.

A clear and succinct hypothesis statement is needed to help distinguish it from alternate hypotheses that have slight variations (Section 1). For example, a liquid-level measurement is needed to properly operate the reboiler on a distillation column. In the situation above for the tray collapse, one person suggested that the column was overfilled with liquid. This could arise from a faulty liquid-level measurement. One hypothesis is that the level transmitter was out of calibration, but another is that the level transmitter valves were not open to the process. Distinguishing between these hypotheses is critical to determining the proper testing methods and ultimately the best corrective actions to prevent reoccurrence in the future.

The question "What has to be true for this to be the root cause of the problem?" (Section 2) gives the team a chance to consider whether the hypothesis truly fits all the observed occurrences of the problem under consideration and if all necessary conditions are met for this to be the true root cause. It is not uncommon for people to draw a correlation based on one instance of a problem and then try to apply it to all other circumstances. By asking the "What has to be true. . ." question, the team is challenged to look for situations where the hypothesis doesn't fit the observed behavior. If enough of these are found, then the team is forced to conclude that the hypothesis is not sufficient, in its current form, to explain the problem. The question also challenges the team to consider whether there is a combination of factors that are not yet included in the hypothesis that are required to sufficiently explain the problem. To continue with our example that overfilling the distillation column led to tray collapse, a failure of liquid-level sensing system would be required in combination with overfilling or else the high liquid-level alarm would have alerted the operators to the problem before heat was applied to the reboiler. Once these necessary conditions are defined for the hypothesis, then it becomes easier to answer the question in Section 4 on the Qualified Hypothesis Template. If any of the necessary conditions were not met for the hypothesis to be the root cause, then that hypothesis can be refuted.

Section 3 is included to gather background on the hypothesis. This information could be obtained from a variety of sources including literature, process design documents, modeling, previous investigations into the problem, maintenance records, etc. For example, this could be a previous gamma scan of the distillation column while in operation to use as comparison to current gamma scans to determine which of the trays have been damaged. Photos from previous internal inspections of the column could also be included in this portion of the template. This information is helpful when designing ways to test the hypothesis in Section 4.

Section 4 challenges the team to really delve into the details of the hypothesis and look for ways to refute it. This could be by gathering data and information already in existence, devising an experiment in the lab or while the process is running, determining what to investigate during a process shutdown, or using modeling. In any of these fact gathering exercises, the team should agree ahead of time on what the data must look like in order to refute the hypothesis. The main purpose of this is to force the team to define how to gather the data so that it is clear and to look for the simplest way to gather that information. In such situations for high-priority hypotheses, it is generally recommended that the team write a DQO specifically for gathering the data to test the hypothesis. There is more information on this in the Analytical Section of this book. For the column-overfilling scenario, the appropriate test could be to use the flow rate measurement (in and out) of the column and the vessel dimensions to estimate the liquid level in the column and compare to the level transmitter measurement. If agreement of the level transmitter reading and the calculation within the margin of error is found, then the hypothesis of overfilling can be refuted.

4.1.3 Facts vs. Hypotheses

For the purposes of this text, facts are any information that is used to test hypotheses. These can come from many sources, including:

- Literature
- Eye-witness accounts
- Videos or photos
- Process measurements
- Experiments, either on the actual process while running or in an alternate process model
- Computational models
- Consultation with experts

The one thing that cannot be used to test a hypothesis is an opinion. Any statement that begins with the words “I believe,” “I think,” “I was told,” etc. should be used with caution in testing hypotheses. These statements might very well be true, but they should be verified independently before they are accepted as a fact within the problem-solving effort.

Sometimes team members can get the concepts of hypotheses and facts confused. A hypothesis is a prediction or statement with a cause-and-effect relationship. A fact has no cause-and-effect relationship. It is simply information. For example, during an investigation to determine why a pressure relief device opened on a vessel, the engineer might discover in the process data that there was a temperature spike two minutes before the relief device opened. The engineer could communicate this finding to the team in one of two ways: as a fact: “At 10:53 a.m. on October 12, the temperature in the distillation column outlet line rose from 200 to 250 °C”; or as a hypothesis: “A rapid increase in gas outlet temperature resulted in a rapid increase in the gas pressure at the top of the column to the set point of the relief device which led to the device opening as designed." However, if the engineer had simply stated that there was a temperature spike right before the relief valve opened, then the statement is neither a fact because it is not specific enough in the details to be useful nor is it a hypothesis because it does not confer a cause-and-effect relationship.

4.1.4 Hypothesis Tracking: The Onion

As the team generates more and more hypotheses, it can become a very real challenge to organize and track these hypotheses, the facts that go with them, and the current status of each hypothesis. It is not uncommon to have more than 100 hypotheses generated for a particularly complex problem. Therefore, a system is needed to track the hypotheses. We call this tool the "Hypothesis Onion" because like an onion it can have many layers and if it is not managed properly it can make you cry. As with the Qualified Hypothesis Template, the software used to create the Hypothesis Onion can be chosen to fit the needs of the team. However, our common practice is to use a spreadsheet program to create the hypothesis onion. We have chosen this type of software because it is ubiquitous within the manufacturing world and everyone on the team no matter what role they fulfill (process operator, quality specialist, production leader, etc.) has access to the software and the knowledge to use it. Below is an example of how the onion is designed and used (Figure 4.1).

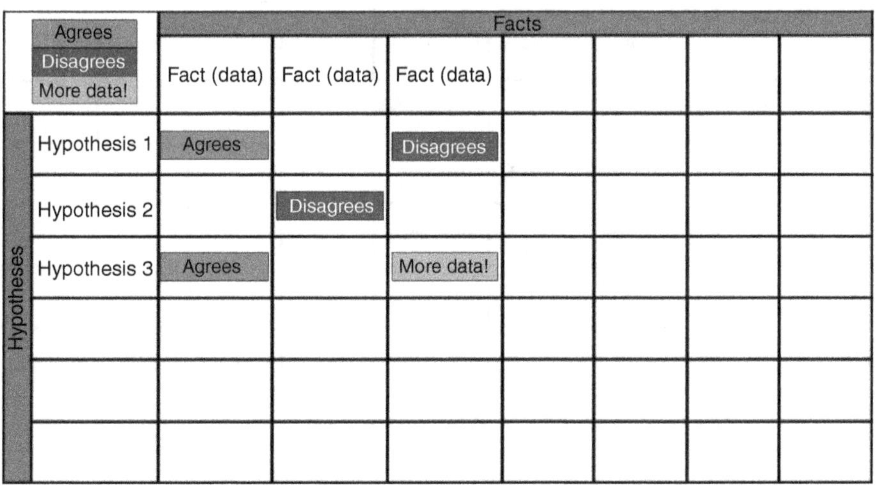

Figure 4.1 Example template for a hypothesis onion. (*See insert for colour representation of the figure.*)

Hypotheses are listed in the first column and facts are listed across the top row. The intersection of the hypothesis and fact can then be color-coded to indicate whether the fact has any bearing on the hypothesis. Typically, a red-yellow-green color code is used because it is simple and easily recognized by the team members. However, project teams can create more complex color schemes to fit their needs and care should be taken to accommodate team members that are color blind. Using the simplified color scheme, if the fact generally agrees with the hypothesis, then the intersection is colored green. If the fact refutes the hypothesis, then the intersection is colored red. If the fact is not sufficient to make a determination on whether the hypothesis is true or false and there is a need to get more data, then the intersection is colored yellow to indicate to the team that more work is needed on that section. Lastly, if the fact has no bearing on the hypothesis, it is usually left uncolored or coded with in such a way to indicate that it is not applicable.

One rule of using the hypothesis onion is that the team must agree that a fact is strong enough to refute or kill a hypothesis before it is turned red. Once this agreement is reached, the intersection is colored red and the hypothesis itself should be colored red, as in Figure 4.2, to indicate that it is no longer considered valid and not open for further debate. This helps streamline and focus team meetings because future team discussions should only focus on the hypotheses that have not been refuted.

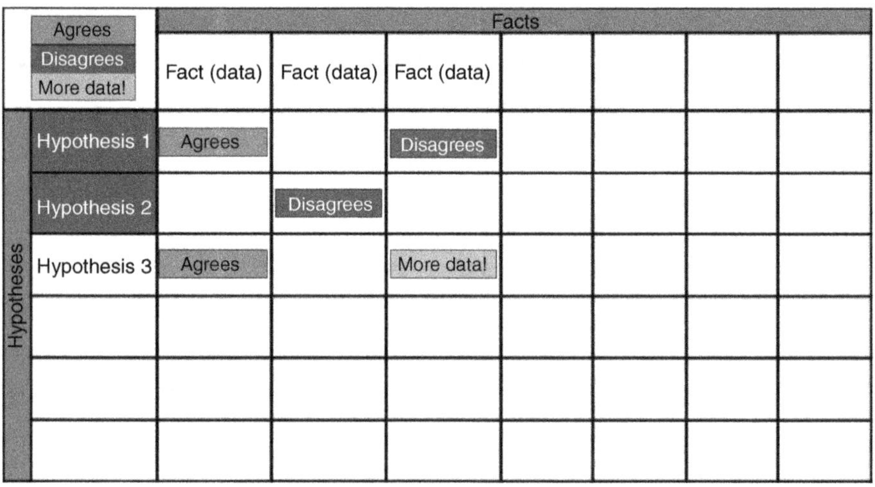

Figure 4.2 Hypothesis onion with refuted hypotheses. (*See insert for colour representation of the figure.*)

4.1.5 M.E.L.D. Hypothesis Ranking

Another technique that is used to manage hypotheses when there are too many to work on at one time is to rank them according to the following scheme:

M – Must work on, higher priority
E – Easy to refute or kill
L – Lower priority
D – Dead, killed, or refuted.

Taking the investigative team through this exercise of ranking the hypotheses encourages them to work on the ones that will most likely be the root cause of the problem, the M hypotheses. These are the hypotheses that require more work by the team and typically require a fully documented Qualified Hypothesis Template. The E hypotheses are the ones that can be tested with already existing data and very little effort so that they can then be marked as D if they are found to be refuted. Basically, the E hypotheses are ones that you can give team members to work on individually and that fit easily within their expertise area. If it is found that the E hypotheses cannot be refuted quickly, then they should be reevaluated against the M hypotheses and given the appropriate priority for further effort. The L hypotheses are the ones that the team feel are not likely to lead to a root cause but that they cannot be discarded immediately. This is the group of hypotheses that must be revisited if most, or all, of the M hypotheses are refuted.

4.1.6 Bringing Tools and Methodology Together

The previous sections reviewed what hypotheses and facts are and how to document, organize, manage, and test them. The question most often asked is which tool should the team start with in the beginning. Should they start with the hypothesis onion or the qualified hypothesis template? There really is no right or wrong answer to this question. It depends on the skill sets of the team members and how long they have been battling the problem. If the problem just manifested a few days before and the team is meeting for the first time, it is recommended to start with the hypothesis onion. This is a great tool for getting everyone's ideas and observations written down quickly, even if imperfectly. It can help ground the team and bring them to the same understanding of the problem. The hypothesis onion is also helpful if there has been an investigative team working on the problem for several months and now the team membership is being expanded to include "fresh eyes" from outside the original team. It can help everyone understand what hypotheses have been excluded and what data was used to exclude them. Once the hypotheses are shared among the team, then they can use the MELD ranking to determine where to begin working and develop the qualified hypothesis templates for the M hypotheses.

If, however, the problem has been a recurrent one for the past decade and there have been repeated attempts to solve it by many teams, then the recommendation is to start with qualified hypothesis template. In this situation, it is very likely that all the E hypotheses have been examined over the years and refuted, and perhaps even the L hypotheses. What is needed in this situation is to identify new hypotheses that are based in fundamental scientific understanding, usually with consultation of literature and experts in the field. Once new hypotheses have been developed and documented in the qualified hypothesis template, then the team can decide which ones warrant further investigation, the M hypotheses. This usually involves an assessment of the risk and resources required to test the hypothesis vs. the gains that could be achieved by solving the problem. Ultimately, if enough hypotheses are generated in this scenario, then it is best practice to put them in the hypothesis onion for easier tracking.

In conclusion, the methodology of creating the hypotheses and testing them to refute them, rather than prove them, is the most important learning from this chapter. The tools should be used to aid the team on their investigative journey and to limit confusion. Which tool the team starts with is much less important than the methodology of testing good hypotheses.

Reference

Platt, J.R. (1964). Strong inference: certain systematic methods of scientific thinking may produce much more rapid progress than others. *Science* 164(3642): 347–353. doi: https://doi.org/10.1126/science.146.3642.347.

5

Scenario Development

In the normal progression of the FPS process, the scenario development is based upon what hypothesis is left alive after the testing phase. Then the team develops the scenario, which is the logical sequence of events that led to the observed problems, or the lack of such events that prevents the plant to achieve success, as defined in the DQO. Many times this is a straightforward enough process, especially if all team members are well versed in the technology, come from the same organization, etc.

Sometimes, however, this process is not that straightforward. For example, many times in a project initially, there are some hypotheses that everyone just knows are true. Such hypotheses are often not well grounded in the laws of science. It is also common for team members to not be able to actually provide the data when challenged.

These hypotheses are called "sacred cows." Killing them is painful for the team – especially the extended team. These hypotheses also have the habit to resurface again and again, even after apparently being formally killed. Dealing with the sacred cows should be done in a careful manner. The approach includes two main aspects – the rigorous application of science and the full utilization of the soft people skills the problem solver has in their toolbox. The rigorous application of science is hard to object to. When the argument is framed as, for example, "This is a great hypothesis! Let's get the data, as it is needed to convince the stakeholders to move forward with financing, and to make sure everything is well documented." things tend to go much better. The team members supporting the hypothesis honestly will actually try to get the data, and in the failure of doing so will realize the hypothesis is false. Anyone else sticking to the sacred cow hypothesis "just because" will not be able to defend it. It is extremely unproductive to approach the sacred cows as "This is absolutely impossible and anyone

Fundamentals of Industrial Problem Solving: A Practitioner's Guide, First Edition.
Zdravko I. Stefanov, Eldad Herceg, Carla Schmidt, David M. Jacobson, Dana Livingston, J.P. Chauvel, Sunil Kumar Chaudhary, and Christopher Paul Christenson.

with half a brain could have realized that!" The sacred cows are favorites and such conflicting approach is the road to perdition for the team and the project. This is when the soft skills come in – to realize what the sacred cows are, who on the team is supporting them, and need to be approached with caution.

A thorough hypothesis onion with a well-written fact tab is critical – this reminds the team members that the team looked at that data/literature and concluded that the hypothesis is dead. If anyone tries to revive a dead sacred cow, they should be asked if anything has changed, please provide the new data, etc. (Conklin 2006).

After a while, the team will get used to killing hypotheses – it is fun. Working in a structured manner helps to show people where things are going. Clarity of the objective, the wins along the way (eliminated hypotheses), and a clear structure of the process in people's minds makes sure everyone stays engaged and on track.

And after a while, there will be some hypotheses that just do not want to die. This is actually a good thing, because these are the hypotheses that are possibly true and the scenario is grounded on them. Ideally, one wants to have only one hypothesis that is alive. If there are several hypotheses that are still alive, they should be united in one multivariate hypothesis. Many times the problem is multivariate and contains nonlinear variables, and this should be reflected in the way the final hypothesis is defined.

When most of the hypotheses are dead, and there are only few left alive and kicking, it is like the fog lifting – several team members will "see" what is going on – and a scenario will emerge.

If however, all hypotheses are dead, there are two options. Either, the team failed to generate the hypothesis that is behind the actual root cause, or some hypothesis is not really dead due to bad data, failed experiment, poor sample collection, wrong assumptions, etc. The existing dead hypotheses should be revisited and new hypotheses should be generated and the hypothesis testing process repeated.

Reference

Conklin, J. (2006). *Dialogue Mapping: Building Shared Understanding of Wicked Problems*. Hoboken, NJ: Wiley.

Section II

Project Management

6

Value of Project

Project management is critical component of the problem-solving process, and each project type from value perspective, described later in this chapter, needs a different style of project management in order to be successful.

The value of a project, which is one of the major metrics determining the severity of the problem, should be clearly articulated by a project sponsor(s) and/or stakeholder(s) in the charter document, and/or Data Quality Objective (DQO) document, which is more commonly used in the FPS process. Getting the project value agreed upon is one of the most important steps while writing the DQO. This helps to avoid a post-project completion value creep. May times the plant perceives a problem to be a huge emergency and potentially worth many millions of dollars. Part of this perception could be a value inflation due to strong desire of the plant to have the problem solved. When this happens, it is not uncommon for the plant to communicate to the problem solver, after the problem is already solved, that the problem was either not that serious and is actually worth an order of magnitude less money, or it is now immaterial that the problem was solved, as product margins dropped due to a cycle in the industry. Another reason to get the value agreed upon immediately is the ability to get resources from other functions, using the value statement as leverage. Finally, the value has to be agreed upon ***in writing***. Whether it is an email or project meeting notes that are agreed upon by everyone, again in writing, is sufficient. Just the fact that there is a very fast turnaround in cadre, particularly in leadership positions, means that people may move on and it can be hard to justify why all this work was done when new leadership shows up. Finally, pressing the process sponsor to agree to a high-value statement in writing is a litmus test of who is serious about the project and who is not convinced, but agrees just to avoid conflict. If a project sponsor is not serious about a project, they will never agree to a written value statement, and it saves everyone huge amount of time and money to not proceed with this particular project.

Fundamentals of Industrial Problem Solving: A Practitioner's Guide, First Edition.
Zdravko I. Stefanov, Eldad Herceg, Carla Schmidt, David M. Jacobson, Dana Livingston, J.P. Chauvel, Sunil Kumar Chaudhary, and Christopher Paul Christenson.

Wisdom Tidbit – Never start a project before you have agreed on the value statement with the project sponsor, in writing.

The project sponsor is the person who owns and provides funding, resources, and support for the project, whereas a stakeholder is an individual with an interest in the project and is affected by activity and outcome of the project. In many cases, sponsors and stakeholders are the same individuals or organizations/ businesses.

As can be seen from above, two main metrics, when deciding if a problem should be solved with FPS, are the technical/organizational complexity and the value of opportunity. If a problem comes and goes, if it has been "solved" many times in the past, if it is present for decades, etc., then most likely former metric is satisfied. The latter metric, the value of a project, should be clearly articulated by a project sponsor(s) and/or stakeholder(s) in the charter document, and/or Data Quality Objective (DQO) document, which is more commonly used in the FPS process. If the value is sufficiently high, an FPS project can be initiated with a high-enough sense of urgency that is needed to obtain resources and budget for the project to run smoothly and to ultimately be successful.

It is common that with the project progress, the remaining value declines, as the worst-case scenarios are eliminated (see Figure 6.1). This is known as a value creep. The value creep is not unusual. Visual representation of change in resources, estimated problem cost, and fundamental understanding of the problem, as the project progresses, is shown below in Figure 6.1.

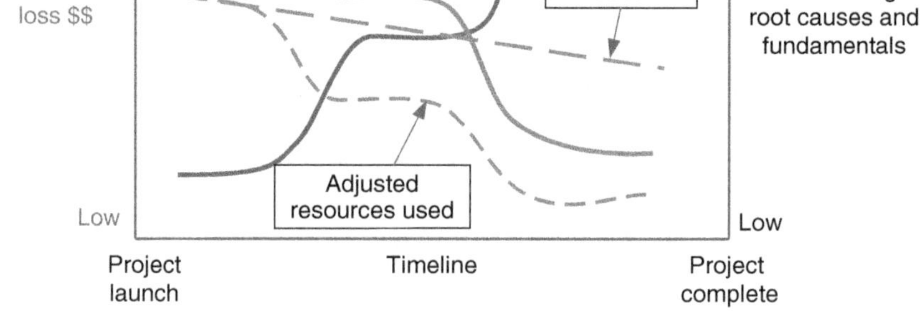

Figure 6.1 Value/cost dynamics.

With respect to value, there are three common types of FPS projects:

1) Crisis control – The problem has to be solved now! These projects are either of very high monetary value ($billions) or there is a critical safety component involved.
2) Time control – The problem has a deadline. These problems are causing a large financial loss ($millions) and time is of importance.
3) Money control – The problem is valuable, but needs to be affordably solved. Usually, these are chronic problems and solving these will increase productivity and profit.

6.1 Crisis Control

In the crisis control project, time is crucial. The problem is a major crisis, for example an extreme financial loss, damage to strategic customer relationships, and so on. The approach can be massively parallel, meaning the project team is working on all credible hypotheses in parallel. Resources are, in a practical sense, unlimited. At this level, the sense of urgency is high with the sponsors and driving a change into the system is not difficult. Major capital is probably off the table here because it takes time for major changes, such as equipment changes, to be implemented. However, if a capital solution can be implemented quickly, then the capital option is definitely considered. The key in this case is that the solution needs to be quick, but without bankrupting the company.

6.2 Time Control

Project management techniques used in a time-control problem are slightly different from a crisis type problem. These problems are causing a large financial loss and the time is important, but not as much as in a crisis mode problem. These are time-control projects, meaning, getting the problem solved dominates the expenses. The core team has significant resources, but it is working only on the highest priority hypotheses. The hypotheses are continuously being sorted in priority as some of them are being killed.

6.3 Money Control

Projects initiated to work on money control problems are usually dealing with chronic plant issues. In many instances, “selling” the project can be as much work as the technical work. If the problem is chronic, most likely there are individuals from the business/plant that believe that it cannot be solved. Solving the problem will increase the productivity and the profit. As such, these are not considered highly business critical. Hence, resources to work on these issues are limited and the project/work has to be done quickly but on a highly constrained budget and limited resources. Nevertheless, the pace of work is often slower for these compared to the crisis- and time-control projects. It is important to manage the resource spend as progress occurs. If the spend declines, it is often best to adjust the resources to match this.

Since the project management techniques, budget, resources, etc. needed to lead successful fundamental problem-solving projects are different for each FPS project type, it is important to understand from the sponsor which type of problem the team is working on. This should be captured in the Charter and/or DQO document, as well as clearly communicated in the Sponsor-Project Leader planning session.

7

Team Management and Roles

Team performance is a critical to success component baked into any FPS project. Unfortunately, the nature of FPS projects makes it extremely difficult to achieve a high-performing team.

One problem for FPS teamwork is the outsider issue. Generally, project leaders on FPS projects are outsiders. In addition to the outsider project leader, subject matter experts are often outsiders. Us and them becomes a sticking point. The outsiders are standing on the insider's turf. The insiders often think, "Why did management think that we (insiders) would fail? Obviously, that would be the ONLY reason for bringing in people from corporate (outsiders)."

It takes time to break the ice and have an outsider become an insider. Many FPS projects have very little time. Running teams effectively under these conditions will be difficult. Remember, some members of your team may have been working on the problem hard and for a possibly very long time. They may not welcome help.

> ***Wisdom Tidbit – Lone ranger rescues only work in the movies, the work of a project takes a team; so build one.***

Another team problem on FPS projects occurs when there is a high burn rate. High burn rates will create a high-stress team environment. Often high-level stakeholders are pushing the team. When your boss's boss's boss's boss is up in arms about the consequences of the problem, there will be stress and worry on your team. Team members ask themselves, "What is this going to do to my career?"

Crisis projects also mean personal life interruption. It is very difficult to get up in the morning thinking about a normal workday and going to your kid's soccer game that night and instead a crisis erupts and causes you to end up on a plane heading into the thick of the plant problem. The mental shifting of gears is taxing.

Fundamentals of Industrial Problem Solving: A Practitioner's Guide, First Edition.
Zdravko I. Stefanov, Eldad Herceg, Carla Schmidt, David M. Jacobson, Dana Livingston, J.P. Chauvel, Sunil Kumar Chaudhary, and Christopher Paul Christenson.

A nice day becomes a bad day quickly. Team members who are suddenly traveling toward the problem will feel this shift. Making the travel arrangements will cause team stress. Ticket prices will be high due to short notice. Team members worry about those costs. The duration of the stay can often be unknown. How long will it take to get control of the crisis? How can you plan your life when you have no idea how long it will be interrupted. Long hours will ratchet up the stress as well. Existing work will be superseded by the crisis work. Team members will ask themselves, "How can I take on more workload? How can I do my regular job AND this problem?"

Crises also causes confusion. In the military, they call it the fog of war. Although obviously not as serious as war, the fog of a plant crisis is present. What is happening, who is doing what, and what are the next steps are often a mess in the beginning of a crisis. Without strong leadership, the mess can continue well into the crisis. The regular running of the plant and the managing of the crisis mix in a chaos brew. The bottom line is that the very nature of crises will hobble teamwork efforts.

The life of the plant type projects are problems that have persisted for years without being solved. These problems will also inherently induce team stress. Working a problem that has been around a long time and has yet to be fixed bubbles with undercurrents of doubts of success. Because success has been elusive in the past, it will be the tribal belief that future efforts will fail. Because the FPS project is a new effort, work will be clouded with these doubts. Although the pace of life of the plant problems are often slower than a crisis, there are forces pulling the team apart.

Also, FPS teams are using the rigors of the scientific method and will be killing some hypothesis that have been thought to be true. Tribally accepted process myths are going down for what they truly are . . . myths. This will induce more stress as local experts are proven wrong. Being an expert and also being wrong about something is tough to weather. Some folks are open to this situation. Others will not be. Creating and maximizing the openness of the team will be the project manager's challenge.

The headwinds against teamwork on FPS projects are strong. This chapter deals with team dynamics and how they evolve over the course of the project. Particular attention is given to the various roles (project leader, subject matter experts, etc.) within the team to optimize the team's productivity.

An adept project leader can utilize the tailwinds of FPS projects to overcome the many headwinds that are inherently baked into FPS projects. This chapter will show how. Enabling the teamwork needed to be successful at delivering the project goal is the main aim of this chapter. The bookshelves are full of books on teamwork and how to run a team. This chapter will be specific to running FPS teams and will focus on those specific skills. The advice may at times run counter current with tried and true team development techniques. This is unavoidable. Running a crisis team takes a different path when compared to typical projects. The goal of the chapter is for the reader to understand that special team management path and be able to act on it to lead or participate in successful FPS projects.

This chapter will begin by showing how to unite and develop the team. Team building will take different directions depending on if it is a highly time-driven project in crisis mode or if it is less time-driven and more results-driven project. These two specialized FPS projects - Life of the plant and Crisis will be covered giving the reader a "How to" guide to follow for both cases.

Clearly established roles prevent conflict and maximize productivity. This chapter will also share essential team roles and the importance of establishing those roles.

Managing FPS projects is in many ways identical to any other project, but does have some specialized challenge in maintaining parallel processing within the team. This chapter covers how to meet that challenge.

Larger projects may require a hierarchical structure of a steering team with various sub teams that focus on specific hypotheses, fact-gathering efforts, or specific needs of the project. This chapter will share best practices in managing larger projects with sub teams.

Running FPS teams is a difficult job for the project manager. Hopefully, this chapter will make that job easier for the reader.

7.1 Crisis Project Team Building

In most projects, getting to know your team members is a helpful and useful team building process. Maybe work history, hobbies, and a little about everyone's family is discussed and shared. Often personality types or team roles are explored to deepen the understanding of how team members will prefer to work together. Communication styles are discussed to help set up communication strategies. Sometimes whole meetings are devoted to just understanding the people involved in the team, their motivation in working on the project, and all around getting to know each other. Team building will take place, but first things must come first. To illustrate this concept, here is an example of a crisis project situation:

7.1.1 Team Building in a Crisis Project

Imagine you are a young engineer (Sarah) assigned to a critical process that makes an essential component for the military. The component is made about once a quarter. The run typically takes seven or eight days to complete and then is shipped to the military contractor and will not start again until the next order is received in a few months. The warehouse is contractually required to hold six months supply or about two runs of material. The component goes into a new plane being produced for the Air Force. The material you are making is a complex compound designed for the extreme environments in which the military operates. Because of the rare nature of the material being used, a single run costs over $100,000 in raw materials. Sales of the material are worth over a $1 million per run and margins are over 90%. Once the military contractor goes into full production, sales are projected to go from the current 1 run per quarter to 10 runs per quarter. Because of this potential, you have been assigned full time to work on ensuring that production goes smoothly. The material was piloted and produced for over a year meeting specifications every time. No problems. Your job felt kind of boring. Because you are just starting out as an engineer, you do not mind because it is job experience and when production ramps up, you will be in an essential position.

And then batch run 7 happened. The process is pretty simple, the operators add all the compounds that are the raw materials by weighing and adding them to a batch tank. The tank is then mixed over 100 hours while the heating and cooling system on the tank maintains a precise 50 °C. The material is then cooled and down packaged into syringes on a syringe-filling line. Batch run 7 started out like all the other batches. Raw materials were carefully measured and loaded. This can be confirmed because inventory checks afterward showed the correct amounts were left in the raw material containers used in the batch. The batch hit the targeted 50 °C and ran smoothly for 23 hours. Then at hour 23, a temperature spike shot the temperature probes readings up 60–70°: clearly hotter than set point. Immediately afterward, the system returned to the set point and the remaining run went smoothly. Packaging did not go smoothly. The operators called you in the middle of the night and told you that you might want to come into the plant to see the problem. As you meet with the lead operator, he shows you a sample, the normally viscous product that is thick like ketchup was thin as water. Obviously, this wasn't going to cut it. You tell the lead operator to load the batch into scrap totes, do a complete clean out, and then for the next batch use only raw material lots that were not used in the bad batch. The most likely cause was probably a bad lot of raw material, so avoiding the bad batch raw materials would make sense.

Batches 8–10 continued to make water consistency material despite a number of iterations of troubleshooting such as testing the heating and cooling system for process leaks; calibrating the temperature probes and other process instrumentation; dozens of standard testing and sampling throughout the runs looking for contaminants. The scrap totes are stacking up and the attention to the problem has quickly blown up into a nightmare.

The nightmare comes in the nature of this product. The product has contractual requirements to supply and maintain a safety stock. Because it is a military product, there are also legal implications if your company fails to supply. When the supply problem became clear, your boss escalated the issue up the management chain. The next thing you know, your are being copied on notes hitting as high as CEO level. Your manager (Roger) is visibly under pressure and is transferring some of that stress on your back. For the first time in your career, you are actually worried about getting fired.

During the flurry of notes, a series of notes appears making it clear that an outside team was coming to address this issue. You were named as a team member. The team is tasked to do something called Fundamental Problem Solving to address the problem. An FPS project leader has been named. You only know a few of the team members. Your boss tells you to clear out your calendar for the next two weeks and to do your plant work before and after the FPS meetings that will be held. You feel a little worried that you might have done something wrong and that this is a corporate witch hunt. You have been in charge of making the four bad batches that will have to be scrapped, that is over $400,000 of raw material and you know that waste disposal is not going to be cheap. Nobody is going to be happy with this.

The conference room is full. Many of the folks in the room you do not recognize and the few that you do are from research and you have not seen them very often. The project leader kicks on the overheads projector and pulls up the first page of an agenda.

8:00 a.m.–10:00 a.m.
Meet and greet:
 Purpose: Get to know the team members
 Desired result: We know each other.
 Share-
 Your company history
 Hobbies
 One interesting fact about yourself

10:00 a.m.–12:00 p.m.
Team personality testing:
 Purpose: Understand our strengths and weakness
 Desired result: We can work as a high-performing team
 Take personality test

12:00 p.m.–1:00 p.m.
Team lunch at Bruno's diner

What do you think of the project leader? How do you think your boss would feel at this moment? What about the CEO? Do you feel like this project is going to be helpful and do you feel like the project leader understands the urgency?

Hopefully, this example has helped the reader understand that although getting to know your team is important, there are more important things on crisis projects. A project leader's objective is to keep the main thing . . . the main thing. Knowing what is THE most important next step for a team is always a balancing act and requires laser focus and situational awareness.

> ***Wisdom Tidbit – NEVER start a crisis project with team building exercises.***

This poses a dilemma. How do you build a high-performing team in the middle of a crisis? To answer this question, we will follow this crisis example in a much better way and share some of the most useful techniques that are used to build teams in the heat of a crisis. Most useful techniques:

1) Build understanding of the consequences of the crisis:
2) Using the three best questions.
3) WIIFM (What's In It For Me) question.
4) Leverage situational interaction opportunities to build connections.
5) Utilize low tolerance of second guessing.
6) Take action at every opportunity and engage the team.
7) Practice regular progress reminders.
8) Build respect for team members and lead by example.

To illustrate these techniques, we will follow this current military product example and show how the techniques can be used.

7.1.2 Build Understanding of the Consequences of the Crisis

Let us examine the military product crisis example with a different direction. Instead of starting with team building, the project leader, Jane Fundamental, decided to help the team understand the crisis they were facing.

Jane was asked to lead this project from the Divisional Vice President, Bob Executive. Bob had called Jane because Jane had gotten him out of a couple of tough scrapes in the past. When Bob called Jane, he immediately started talking to her about the company's reputation and the importance of the military contract business. In the middle of his talk, Jane interrupted Bob and said "This is really important to understand the impact of if this thing goes bad Bob, but right now it is MORE important that the team hears this. Can you spare a half an hour or so to tell them?" Jane arranged for Bob to tell the team in person at the kickoff project meeting.

During the kickoff meeting with the team, Bob outlined the importance of the project. Including these key points:

- Failure on this project would tarnish our reputation in the military contract space and would jeopardize future contracts that are currently over $200 million in sales.
- In the past, when our competitors were in situations like this, they ended up in litigation for years. Litigation must be avoided.
- Delays in the contract will attract media attention and our company's reputation would be tarnished.
- The military is counting on this technology to keep our edge in the world stage.
- The clock is ticking because our safety stock is running out.
- Our customer's CEO is calling our CEO about this problem; the importance of this project is understood at the highest levels in the company.

Jane captured the key points and then asked Bob if he could describe the situation in one sentence.

Bob says, "Stop making runny product or we let our country down."

Key points from the example on technique of building understanding of the consequences of the crisis.

Go to the Source

In the continuation of the example, the project leader did something that is one of the most important components of building a team. She brought in someone that truly understood the big picture and had them pitch the project importance to the team. Teams need to form around an outcome that is important. The corollary is that teams will rip apart at the seams when pursuing the trivial or unimportant. Strangers in the middle of nowhere will pull together in a crisis such as a major traffic accident or a flood. When a problem is truly important and teams understand that importance, it is easy to develop solid teams. Project leaders that understand this concept will try and clearly understand the relative importance of the project and insure the team understands that importance.

While it is possible for the project leader to pitch the importance of the project, in general that should be a last resort. When the project leader pitches the importance, the authority they have is diluted. It is second hand or even third hand. Go to the source of the importance and the drive whenever possible and have that person explain project importance to the team.

Make Sure It Is a Real Crisis

All projects that need to be done will not be a crisis type; but when they truly are a crisis situation, it is important to recognize them for what they are. Recognizing a crisis is in the authors experience a skill that must be practiced. This is counterintuitive. Everybody understands when it is a crisis, right? Many "crisis" projects are nothing but fabricated excitement. For example, a manager is about to miss his budget because the maintenance costs missed the communicated estimate due to higher than average damage discovered during the annual maintenance. The manager is not going to make his annual targets and is going to communicate and try and drive something.

Any situation where a planned estimated sales or maintenance costs or capital costs that are suddenly projected to miss the plan will repeatedly generate fabricated excitement. This is not to say that action to fix the situation is unimportant; it is to say that these kinds of situations are rarely a crisis. Project leaders can take a project seriously whether it is a crisis or not; but managing noncrisis projects as if they are a crisis can wear team members out and quickly undermine any credibility the project leader has.

Many companies have bad business practices that are clear signs of mediocrity such as shifting costs from the current quarter to the next to make the current one look better. These types of moves fabricate far-reaching panic and a pseudo crisis.

To understand if it is truly a crisis vs. a squeaky wheel from an aggressive stakeholder progressing a personal agenda or a fabricated crisis, there are a few clues that if some of these things are true will tell you that there is a crisis going on:

1) Will the impact of not doing the project significantly affect systems and people outside of the company ($, environmental impact, harm to humans, regulatory, etc.)? For example, a crisis project is done because the feedstock for a major project line is contaminating the product making it harmful to humans. Unless the product is fixed, a key intermediate in heart medication will no longer be available. Crisis.
2) Is the deadline for getting the project done a hard deadline? A hard deadline means that after X date, the consequences absolutely will happen. For example, the plant is producing vents that do not meet a new vent regulation that come into law in three months. Failure to meet the new regulations will immediately trigger fines and eventually will lead to plant shutdown. Crisis.
3) Is it a situation that will likely kill, maim, or seriously injure one or more people if not resolved? For example, a shipment of a chemical at sea is building pressure and efforts to reduce the pressure have failed. Crisis.
4) If the order of magnitude of the financial impact is two or more times higher in size that projects typically done can also signal a crisis. For example, most projects clock in at around a million at ACME X company. A project to solve a problem that would cost the company $100 million dollars (two orders of magnitude higher than the typical $1 million) can be a crisis. Big numbers can crash a business, impact investors, and form a crisis situation. Crisis.

Getting the team on board with the knowledge that they are fixing a crisis will clear the path on team building as a diverse group of people gets focus. Individual team member will own the priority.

Summarize the Crisis into Just a Sentence or Two

In the example, Jane asked the business leader most impacted by the crisis, Bob, to summarize his concerns into just one sentence. Situations are complicated and there will always be many factors that can be consequences of a crisis. The team will need to repeatedly explain the urgency of the project to others. Complex explanations are not useful in quick communications. When the project leader encouraged the business leader to tease the most important reason from the many reasons that really helped to get the organization to understand the priority. It will make a difference as the team members go out and use their network to help solve the problem.

7.1.3 Using the Three Best Questions (The Three Main Pillars of the Charter/DQO Process)

Continuing the military product example, lets see how Jane builds her team.

Immediately after Bob finished his talk to the team on why the project was important, Jane then had the team tell the story of what happened and what is happening now, the background of what happened. Jane had a coworker capture the discussion using a mind map software, but she could have also used a word processor program. This is a get it all out moment. It took a couple of hours for everyone to talk. Jane rarely talked, the few times she did was with clarifying questions to make sure everyone understood the key facts and technologies being discussed. Jane also had the team take short breaks when she could see they were losing focus and getting tired. After the background to the project was done, Jane took over typing in the mind map software and she typed three questions.

What is success?
What is necessary for success?
What is rate limiting?

Over the course of a half an hour or so the team gained agreement and came up with answers to these powerful questions.

They decided success would be to make good (in-specification) product and get it to the customer before the customer uses the last of the safety stock. They decided that understanding the chemistry and physics that were causing the product failures were what was necessary for success. They also agreed that what was currently rate limiting for solving the problem was getting good analytical testing that clearly shows the chemical and physical differences in the good and bad product.

Key Points from the Example on Technique of Using the Three Best Questions

At first blush, the three best question may not seem like a team building exercise. Do not be fooled. They are helping advance the project by gaining agreement on the goal of the project (What is success?). But the secondary goal of building a team is happening at this time. The team is discussing and deciding what goal is the most important to accomplish. They are figuring out what to work on first as the understanding of what is needed to be successful and what is holding them back now from being successful (What is rate limiting?). The building of agreement and discussions with a successful outcome of answering those questions will build the team as well. Answering these questions will give the team it's first small victory and is the first opportunity for the team to work together. It will also serve the project leader well to help keep the team on task. Whenever the team seems to be slipping off target, Jane will ask this: "Is this discussion and the work we are doing now helping us to make good in product and get it to the customer before the customer uses the last of the safety stock?" If the team decides, it is not helping move the ball toward success, they can move in more productive directions. Teams build best when they are moving forward. Use the three best questions to start team building.

7.1.4 Using the WIIFM Question (What's in It for Me)

During a break, Jane started chatting with the chemist who developed the product originally, Mike. After some small talk about the weather, Jane asked Mike why he wanted to be on the project. What was in it for Mike? Mike responded by saying that the product was one of the ones he was most proud of developing. He wanted it to be really successful. He is thinking about retiring soon and felt like this would be a nice way to cap off his career. Throughout the day during breaks or when there was a little time she asked every team member this question. She also told the team why she wanted to be on this project. She said that she loves the challenge of helping teams solve really tough problems.

Key Points from the Example on Technique of Using the WIIFM Question

Aligns Tasks with Motivation Understanding what drives each team members can be crucial to best fit task selection; who is going to be motivated to do the task best. For example, Mike the development chemist is likely to be great at solution development. His motives suggest he is going to drive hard to find answers and make the product successful.

Identifies Motivational Issue Do enough projects and project leaders will see that although most team members want to work on the project, but there will be the occasional team member that absolutely does not want to be on the project. They will answer the WIIFM question in ways that will indicate the issue. For example, if a team member says the only reason they want to do the project is because their boss is making them, the project leader has a motivational issue with that team member.

Asking the WIIFM Question Will Help Team Members Understand Their Own Motives It is strange how we dive into activities without asking why we want to do it. When a project leader asks a team member the WIIFM that will initiate the team member to think about why they want to be involved. Otherwise, they may or may not consider it deeply.

With team member understanding the very basic reasons for why they are here on this project, they will be a leg up on performing as a team.

7.1.5 Leverage Situational Interaction Opportunities to Build Connections

> ***Wisdom Tidbit – To work as a team, the team needs to do work together.***

This simple, but true concept is really important. It is possible to assign tasks, have team members run out and do those tasks, get back together, report back, and run the entire project like this. Essentially, a loose group of individuals that work apart individually and only are a "team" for brief moments of time during meetings. Many projects are done in this manner. It will work when deep teamwork is not needed. Crisis projects typically need deep teamwork. Let's see how Jane, the project leader, uses this technique to build her team.

The team was in the middle of building their hypotheses when Chris, the analytical chemist, started shaking his head. Jane stopped the team and said, "Can we hold on a minute? Chris, you look like you have some concerns. Can we hear them?" Chris took a deep breath and said, "I know we need to form these hypotheses to understand the science behind our problem, but we also said our rate determining step was to be able to analyze the differences between good and bad material. To really do that, we need a good representative uncontaminated sample of both good and bad material. We also will need to be able to get a good representative sample during the run at intervals of time such as hourly. If that is our rate-determining step, why aren't we working on that?" Jane's response surprised Chris. "You're right!," Jane said.

Sarah the young engineer in charge of the process said, "I can go ask the operators to get some bad samples from the bad lots we toted off, go to QA (quality assurance department) to get some retained good samples, and also write a procedure to sample during the run." Jane said, "Sounds great, but I would like more people to be involved in this as well. We need those samples, but we also need them taken in the way Chris needs them. He may have concerns about air or water contamination. Witnessing how those samples are taken will be important to getting it right. We also need someone who runs the process to help get the best sample locations. Simon is the lead operator on the project and knows the process best. I propose we form a small sub team to make this happen and that the sub team goes to do that work right now."

Jane then put the three questions on the overhead and asked what success was for this sub team. The team agreed the first part was to obtain at least five good and five bad uncontaminated representative samples. The second part was to have a safe and effective sample procedure to collect uncontaminated representative run samples hourly. They believed some safety department approvals for the new sampling procedures was the rate-limiting step. Simon, the lead operator, said he would work with the safety folks to make sure they understood the urgency and would let the team know if there were any issues. The team also decided on some supplies such as the right sampling containers were necessary for success. Jane asked Sarah to take the lead for this sub team. Sarah, Simon, and Chris set off to get those samples. The remaining team continued the work of forming good hypothesis.

Key Points from the Example on Technique of Leveraging Situational Interaction Opportunities to Build Connections

Set the Sub Team Up for Success Jane could have let Sarah go and do those tasks by herself. It might have worked out fine, but instead she used the opportunity to both further team building AND put the best experts on the job to do the job right. Sarah was not an analytical chemist, the right sampling technique may or may not have been done, by using this as an opportunity to get more experts involved, Jane is setting Sarah up for success.

Think Before Acting By stopping and having the team answer the three best questions, the scope and what the goal of that sub team was made clear to everyone. It only took a few minutes, but clarity is priceless. The sub team had a very specific scope and goal for the task directly aligned to the project goal.

Team Connections Will Be Built During this work, Sarah, Simon, and Chris will get to know each other through work. Much more than if Sarah had done this task by herself.

7.1.6 Utilize Low Tolerance of Second Guessing

Monday morning quarterbacks are people that say what the quarterback should have done on the Sunday game. They are second guessers. This can be a fun discussion over coffee, but it is pure poison for teams. Team leaders who exhibit a low tolerance for second guessing keep the poison out of the team and prevent a major root cause of team breakdowns. Let's see how Jane handles second guessing in her team.

As the team was working on their hypothesis, Bob, the vice president, walks into the meeting. He says, "Jane, if you don't mind I have some important news." Jane said, "OK Bob, let me finish writing down this hypothesis and then you can tell us." Bob said, "Our customer's CEO just got off the phone with our CEO, he wants a project update presentation Tuesday at 4 pm in a virtual meeting." Jane smiled and said, "OK Bob, I will add that to the communication plan and make sure we are ready for it. When you get a chance, send me the planned meeting duration, the main purpose of the meeting, and the virtual meeting details. Also, please ask the CEO and the head of sales if they want a preview of the presentation before the virtual meeting. I would need help to arrange that if it is needed. Anything else Bob?" Bob said, "No, I guess not. I will send you the details."

Jane returned the team to the task of forming good hypothesis. About 10 minutes into this activity, Mike, the developmental chemist, says, "Is anybody besides me worried about this CEO thing? It feels like we are already so far behind, this team should have been formed when the first batch went bad." Jane looked Mike straight in the eye and said, "I will talk about the CEO update in a minute, but first I need to address something you just said. You said that the team should of been formed earlier, that may be true; but it does not help us in any way achieve our goal to second guess history. Odds are the best decisions were being made at the time with the available information. Second guessing by people who were not there at the time is non productive." Jane stood up an added a "Ground rule" for the team on their ground rules flip chart: No Second Guessing (Monday morning quarterbacks). She then said, "Mike, I know that the pressure of CEO involvement makes us nervous, but we need to put that energy into tasks that are aligned to the goal. Unless we had a time machine, how would revisiting past decisions or mistakes help us start making good product? Mike, let's move forward and stop second guessing. Thanks."

Key Points from the Example on Using a Low Tolerance of Second Guessing

Jane took the poison of second-guessing head on in a respectful way. She made the ground rule clear and shifted focus onto more productive ground. Second guessing in this situation could quickly turn a team against itself. The folks that made the original decisions are going to be resentful for the "Monday morning quarterbacking." The whole team needs to be focused on activities that move the project forward. Revisiting past decisions or possible mistakes will never move the ball forward. At best, revisiting these decisions may be something that happens at the end of the project to capture lessons learned that might show up in the control plan on the project. For example, on the issue of escalating the problem in this example, a control plan might include criteria on when to escalate that is tighter than the current practice. Again, that would be at the end of the project and not disrupting the flow during the project.

Crisis projects are likely full of questionable decisions that are made during the thick of the fight. People are making decisions with a fraction of the information needed and are bound to make a mix of good decisions and bad decisions. This is how it will be. Rubbing folks noses in their bad decisions is VERY counterproductive . . . pure poison. Team leaders and team members need to be vigilant and quickly shut down second guessing.

7.1.7 Take Action at Every Opportunity and Engage the Team

Productive teams are strong teams. Folks need to feel like they are making progress to really feel a part of a winning team. The problem with that on FPS projects the team will not ultimately win until they figure out what science is driving the problem and then figure out a solution. Obviously, this happens when it happens. If they figure it out in a day, it takes a day. If they figure it out in three months, it takes three months. Teams need more feedback than just the ultimate win to stay motivated along the way. Small wins need to be happening along the way. Let's see how Jane uses this technique.

After putting up the new ground rule, Jane moved on to the next problem. She said "I can see from everyone's faces that we are not going to make much progress until we deal with this CEO meeting, so that is what we are going to do for no more than the next 1/2 hour. The meeting is not going to help in any way make progress on our goal, but it is creating a distraction that is making us less efficient. So, put yourself in the customer's shoes: What is the most important information they need?" Jane pulled up the mind map software and captured the team's input. The input looked something like this:

- When will the problem be solved?
- Are you folks putting enough of the right resources on fixing it?
- How much time do we have to solve this problem before the safety stock runs out?
- How will we know the material we already have is good?
- What exactly is the problem?
- Are there any roadblocks in the way that are preventing fixing the problem?
- They will want regular communications and will want a certain frequency.
- Other issues we have not thought of . . . customer may have some unexpected questions.

Jane then set the team to answering these questions. The three best questions already answered by the team helped with several, the make-up of the team addressed the resource questions. Within a short time, the team was satisfied they had good answers to all but two. Mike said, "We can't really tell them how long this problem is going to take to solve, that is an impossible question." Jane said,

"That makes it easy, we just tell them we do not know and cannot know until we get more data. The only thing we can give them is our best estimate on when our first round of data results will come in; we can get that once we finish our hypothesis testing plan. We will have that before the meeting."

The other question that the team could answer but did not have the answer for was about the safety stock timeline. Latonia the production scheduler said "I can get that information and calculate out the timeline for running out; it will take an hour or so." Jane replied, "It is really important information, go do that now and report back to the team. Thank you! Also, Craig (from Quality Assurance), please join Latonia and take the mind map we just made and convert it to a no more than 3 page presentation for the CEO. Please, do not spend more than 1/2 hour on that task because it is not moving us toward our goal; Thanks! Any clarifying questions?"

Jane and the team quickly answered the three best questions for Craig and Latonia's tasks. Craig and Latonia then headed off to get their work done. The remaining team moved on to working on generating hypotheses . . . the really important work!

Key Points from the Example on Taking Action at Every Opportunity and Engage the Team

Jane lived the take action at every opportunity. She alleviated their fears that were distracting them from the most important tasks by tackling the presentation but limiting the time to work on it. This got the team rolling. Taking action fires people up. The team now knows what will be covered with the two CEOs and much of their distractions are gone. The action of calculating the safety stock duration was critical and addressing it NOW also exemplified taking action philosophy. Jane selected a couple of team members that may or may not contribute to the hypothesis generation. She is taking a risk that they might come up with the most important hypothesis, but the risk is low because the entire team is thinking about the problem. If they come up with something, that hypothesis can be added later.

What if instead of taking action, Jane had just captured the tasks in a task list? The energy in the room would have been completely different. The team needs to see progress. Not all actions need to be done NOW, but the important ones that are of a reasonable duration (one or two hours) should be done NOW. It will gear the team up to be can-do, jump on it, wildly energetic maniacs solving the problem.

7.1.8 Practice Regular Progress Reminders

For some reason, the human mind on the average spends more time worrying about what is to come instead of thinking of all the progress that has been made. Imagine you are driving across the country. You are on the last 100 miles of your journey. Where does your mind spend its time? Odds are it is dwelling on the last 100 miles. Occasionally, it may wander back to think of events that happened on the journey; but most of the focus is likely to be the next steps. This tendency to think forward is good, because the next steps are the most important ones; but this can have some negative consequences on teams . . . especially teams working on a crisis. The main consequence is motivation. If that next 100 miles is straight up a mountain, the team can lose confidence. Reminding a team of all of the rivers they traversed and the many difficult miles they conquered before they got to the mountain can really help them be motivated for the next challenge. Team members and team leaders should make a regular practice of reminding the entire team, as well as the stakeholders, of the progress made.

> ***Wisdom Tidbit – Reminding a team of what they have accomplished is a pure adrenaline injection to help overcome the next obstacle.***

Lets see how Jane uses the technique of using regular progress reminders to build and motivate her team.

After two hours of highly technical discussions and arguments while fleshing out the main hypothesis to the problem, the team was getting tired . . . maybe a little cranky as the day moved toward evening. Simon (lead operator) said, "I don't know about anybody else, but I kind of feel like we are spinning our wheels. It doesn't feel like we are getting anywhere!" Jane looked over at Simon and said, "Fair comment, let's take stock. What have we gotten done since this team formed this morning?"

The team though about it and over the course of a few minutes came up with a long list of things they had gotten done so far that day. Smiles spread around the room and Simon said, "I guess we are spinning our wheels in the right direction!"

Key Points from the Example on Practicing Regular Progress Reminders

Ticking off the accomplishments like . . . CEO presentation-done, sample plan for good vs. bad-done, calculation on safety stock duration-done, most of the hypothesis needed to understand the problem-in progress . . . will light a fire in the team generating enthusiasm and energy. Jane knew two things, first that the team was making progress and second that they needed to be reminded of that progress.

Teams do sometimes spin their wheels. If the team is not making progress, it is equally important to remind them of this. Usually, the root cause of this situation is that they are not working on the main thing. The critical to success items can drift away into an abyss while the trivial many that may actually be either easier to do or are more fun to do can sidetrack the team. Team leaders and team members that ask the big question of what has actually happened to move toward the goals can correct these situations.

Wisdom Tidbit – Forward progress toward to goal is NOT the only vector a team may follow, checking for progress periodically can get teams back on track.

7.1.9 Build Respect for Team Members and Lead by Example

Respect. That one word is the mantra for minimizing team conflict and maximizing team performance. It is easy to slip into bad habits of pouncing on mistakes and making the person feel bad. Professional mistakes slip into personal attacks. It seems to be a human condition to prey on those that trip. Addressing problems while avoiding personal attacks is a key balancing act in maintaining an environment of respect. Disrespect is poison on projects. Team leaders that carefully exhibit respect for team members can lead by example.

Respect is also deeply about recognizing efforts as they occur. Noticing good work is as important in developing a culture of respect. Let us see how Jane builds a culture of respect on her team.

Craig and Latonia rejoined the team after completing their tasks of getting the first draft of the CEO presentation, as well as calculating the duration of how long the safety stock would last. Jane welcomed them back to the team meeting and then asked, "How did it go? Were you able to be successful per our 'What is success?' question?" Seeing their nods Jane lighted up "Wow! We gave you a couple of really tough tasks and I half expected they would run well into tomorrow, but here you are! Let's start by reviewing that first draft CEO presentation. We are looking for feedback to improve the presentation to take it to the next level, but we will only spend 15 minutes on this . . . still solving the problem is the most important thing."

Craig began going through the presentation when about halfway through Chris said, "Why did you have to show off so much with all those slide builds and special effects, it is hard to watch!" Jane said, "Chris, Craig worked hard on this; I appreciate that work. I gave him a tight timeline and he delivered. This is just our first draft. We want constructive feedback. I personally do agree we need to simplify this some because our time for the meeting is short. What does the rest of the team think?" Looking around, Jane could see agreement on the team. Constructively gathering feedback, the team had Craig fix the presentation as agreement on the feedback was done. The 15-minute review ended up taking 25 minutes; but in the end, the work was done.

Latonia then reviewed her methodology and the report that was used to do her calculations. Jane said after she finished "Wow! You thought of everything. I for one feel a great deal of confidence in these numbers. Nice job!"

Later that evening, Jane wrote a quick email to Latonia's and Craig's bosses to let them know about the good job they had done on the critical tasks. She gave specific examples about what was done well, as well as the tasks importance to the project. She copied both Latonia and Craig on their individual emails. She knows it is a best practice to send the feedback to both the boss and the employee when it is for a job well done.

Key Points from the Example on Building Respect for Team Members and Leading by Example

Hopefully, it is pretty clear that Jane does not tolerate team members showing anything less than respect for each other. She also notices when good work is happening. She follows up with noticing it by documenting it. Respecting others builds teams. Building that culture by leading by example will build teams working on a crisis.

Crisis Project Team Building: Summary

Traditional team building techniques may not always work on crisis project. The time driver, the fear, and the pressure make team building difficult. Savvy project leaders will keep the main thing (the goal of the project) the main thing. But, there will be ample opportunity to use the work within the project to also have a side benefit of team building. Capitalizing on these opportunities can be the difference between a dysfunctional team not solving the problem wasting energy and a highly functional team that nails solving the problem in a short of time as possible.

7.2 Life of the Plant Project Type Team Building

Life of the plant projects are based around long-standing problems. For example, a major heat exchanger fouls every 2 months taking the plant down for 4 days costing millions of dollars in lost margin and it has been happening for 20 years. These are sticky problems that have proved to be persistent.

Life of the plant are a special case for team building. These types of projects follow the pattern of a costly problem that has been attempted to fix for many years to no avail. The plant folks figure out ways to work around the problem and minimize the overall impact, but it is still a problem. FPS techniques are more powerful that typical troubleshooting, so often an outside team packed with corporate experts is pulled into making another attempt at fixing the problem. In the authors experiences, the batting average of these techniques are quite good and more often than not these long-standing problems get solved. Digging deep into the science turns out to be very productive. Unfortunately, it means that outsiders are coming into a long-standing situation and this can create an "Us" versus "Them" team environment and therefore presents a specialized approach to team building. Here we will cover the following approaches:

1) Crisis project techniques
2) Critical nature of a face to face kickoff meeting

3) Generalized/typical team building approach
4) Asymmetric workload balance for outsiders versus insiders
5) Opportunity pairing of outsiders with insiders
6) Carefully revisit past
7) Something new

7.2.1 Crisis Project Techniques Also Work for Life of the Plant Project Work

The techniques covered during the crisis project section do actually apply to life of the plant project work as well. These techniques can supplement the other techniques covered. Here is reminder of the crisis project techniques and a few clarifying comments about how they are modified to life of the plant problems.

- Build understanding of the consequences of the life of the plant problem. This technique will differ from the crisis in that usually, you are looking for not only a business leader to pitch the big picture, it is also best practice to have a process expert who understands the long-standing history of the problem to pitch the problem from a historical point of view on all of the previous work.
- Using the three best questions. This technique is done the same as in the crisis projects.
- WIIFM question. This technique is done the same as in the crisis projects.
- Leverage situational interaction opportunities to build connections. This technique is done the same as in the crisis projects.
- Utilize low tolerance of second guessing. This technique is done the same as in the crisis projects.
- Take action at every opportunity and engage the team. This technique is slightly different from when done with a crisis project. Because the timeline is stretched out, immediate action may or may not make sense. Default whenever possible to taking action, but when that is not possible the action timeline should be negotiated with the action owner. Give and take is possible with problems that have been around for a long time.
- Practice regular progress reminders. This technique is done the same as in the crisis projects.

The techniques used to build teams during a crisis project can clearly be used to supplement life of the plant problems as well as normal/typical problem projects.

7.2.2 Critical Nature of a Face-to-Face Kickoff Meeting

For life of the plant kinds of problems, there can be a pull to do the project remotely. Flying experts in from around the world can be expensive and project sponsors will be less willing to foot the bill for a problem that has been around a long time. Often outside consultants are engaged, the costs increase as those consultants are brought into the picture. Sponsors will often be reluctant at this stage.

The issue is that this front-end cost control is setting the project up for either failure or for slow progress. In the authors experiences, the likelihood of solving a problem in which the experts have not seen the equipment goes down dramatically when compared to when the experts are given a chance to develop a deep first-hand knowledge of the equipment experiencing the problem or in developmental problems, seeing the material first hand is critical.

People also develop better/deeper relationships when they actually meet in person. In team building, this is a critical to success factor.

> ***Wisdom Tidbit – Bring the team together early on in the project, build the team initially then, and build a deep FIRST-HAND understanding of the problem to set it up for success.***

To counter the sponsors desire to reduce costs, there are some tactics a wise project leader can use.

- Review all of the good reasons listed above with the sponsor.
- Deeply discuss the business case and calculate the face-to-face meeting cost as a fraction of the case. For example, if it will cost \$10,000 to have a face-to-face meeting to try and solve a \$10 million/year problem . . . that cost is only 0.1% of the yearly cost. Hopefully, the ridiculous nature of that fraction will encourage the sponsor to do what is right for the project.
- Make it a condition of doing the project. FPS project leaders are in high demand and rarely directly work for the sponsor. Letting the sponsor know how serious this issue is to the success of the project can be the nudge to move them to do what is right.

Face-to-face meeting trumps all other methods and unless the project business case is low; the project leader and sponsor need to push for this in at least the first meeting.

7.2.3 Generalized/Typical Team Building Approach

In the crisis example of doing it wrong, the project leader put forward the following agenda.

8:00 a.m. – 10:00 a.m.
Meet and greet:
  Purpose: Get to know the team members
  Desired result: We know each other.

Share:
  Your company history
  Hobbies
  One interesting fact about yourself

10:00 a.m. – 12:00 p.m.
Team personality testing:
  Purpose: Understand our strengths and weakness
  Desired result: We can work as a high performing team
  Take personality test

12:00 p.m. – 1:00 p.m.
Team lunch at Bruno's diner

This agenda was wrong for a crisis because of the urgent nature of the crisis problem type. For a life of the plant problem, this agenda would be fine. Generally, in the kickoff meeting, the team will meet for a couple to a few days. For a few day meeting, it would be better to spread the individual team building pieces throughout the meeting, sprinkling them among the project work.

There are many great books and trade articles on building teams for typical projects. We are not going to cover those because there is excellent material elsewhere. As stated before, we are focusing on the unique aspects of an FPS project. Using generally accepted team building techniques is encouraged for the life of the plant problem sets.

7.2.4 Asymmetric Workload Balance for Outsiders Versus Insiders

When outside experts and outside project leadership are brought into a plant to solve a plant problem, an "Us" vs. "Them" or "Insider" vs. "Outsider" split within the team can be common. One of the best techniques to combat that is to create an asymmetric workload where the outsiders are set up to do MORE than their share of work.

The nine words most feared in manufacturing plants across the world are "We are from corporate, we are hear to help." A typical pattern is an overloaded/overworked manufacturing site has some corporate "help" show up because of a problem. The "help" that shows up and adds to the workload of the local manufacturing folks. After enough experience with this, us versus them walls show up from day one with the locals when any outsiders show up. The pattern of the asymmetry in work being that the locals DO nearly all the work and the outsiders direct the work and often take credit for the work is a very disruptive pattern. To break this pattern, set the project up to be the opposite pattern. Negotiate more outsider team member resource time up front when setting up the project. Project leaders need to pay attention to where the work is flowing and do their best to create the right asymmetry. Outsiders that come in and roll up their sleeves and work their butts off are appreciated.

Outsiders working harder and longer than the local folks on the project sets up a favorable pattern. When the outsider is the project leader, this is even more important. Leading by example is critical. There will be tasks that need to be done by a certain person on the project because of their role, this is unavoidable. For example, an operator on the team may be the only person qualified to take the samples needed. That work has to go to them. But there are typically as many or more other tasks that are either up for grabs between any team member such as taking minutes for team meetings that allow for creating this desired asymmetry.

Creating a bias toward outsiders working more is one of the most useful techniques in preventing outsider versus insiders team dynamics that are often present in life of the plant problem types.

7.2.5 Opportunity Pairing of Outsiders with Insiders

The technique covered in crisis project team building: "Leverage situational interaction opportunities to build connections" is the same for life of the plant type problems. Having team members do work together for sub tasks aligned to the project will build a strong team. One useful spin on this is for the project leader to consciously pair outsiders with insiders on these chartered sub tasks. Working together allows people to get to know each other and learn to respect each other's talents. Cultivate these opportunities.

> ***Wisdom Tidbit – Purposefully pair outsiders with insiders on tasks to prevent the outsider versus insider team rifts.***

7.2.6 Carefully Revisit Past

Life of the plant problems by definition have been around for a while. Many folks have worked on the problem. Teams need to RESPECT that work. That means holding fast to no second guessing culture. It means that a deep review of what is gone before needs to be done. Delving into the history part of life of the plant should be the main thrust in initially understanding the problem. What has been tried? What were the results? What technical reports were written? What were their conclusions? Teams should dive deep. During this dive, clearly spell out the hypothesis that the previous work was pursuing. Carefully collect facts and data from those historical efforts. The quickest way to lose respect for the team is to try something that has been tried every year for the last 20 years. Understanding what has been done is the best initial focus for life of the plant problems.

Does this mean that work done before will not be repeated? No! Often previous work does need to be repeated because things have changed. Maybe new analytical techniques are available, or perhaps different raw materials are in use today versus the past. The key is to go into these situations with your eyes wide open with a deep understanding of the work that has gone before.

Often in life of the plant problem earlier attempts at fixing did not have the rigors of clear hypothesis testing and did not use the hypothesis onion. Putting the historical work into the hypothesis onion with clearly stated hypotheses is a highly recommended activity. This will clearly show the gaps in knowledge.

Equally important is pushing the team to generate new hypotheses that have never been tested historically. Often, the real problem lies in these new directions.

Treat the past with respect and there will be a better team cooperation helping to build a high-performing team with great results.

7.2.7 Something New

Life of the plant problems will result in a cast of skeptics. "This problem has been worked on for the last 20 years and nobody has fixed it; this team is a waste of time" are words that may not be spoken, but folks will be thinking this or something like it. One of the biggest ways to build a strong team that minimizes the skeptical thinking is to drive to do something new. A new way to test a hypothesis, a new hypothesis, a new technique for measuring, or ANYTHING helpful that is new.

Maybe the team contracts the industry expert from outside the company or goes to a major university to contract some new research. With life of the plant problems, focus on new is often the key to success. But, in addition to helping the project, being successful will bring an important element to the team ... HOPE. The counter to the skeptics will be there. Why do you think you folks can solve this when nobody else has? Because we have hired a world-renowned expert in the field to help us understand the science. New is an important team building technique because of the hope it injects.

Life of the Plant Project Type Team Building: Summary

Building a team around a long-standing life of the plant problem is challenging. Old problems bring skeptics. But, if team leaders kick the project off right, respect the past, bring in something new, and use solid team building techniques these problems can be solved.

7.3 Roles and Responsibilities for an FPS Project

Typically, an FPS project will contain some or all of the following roles:

- Leadership in the business: With problems, there is money flowing out the door. This leadership cares about that aspect of the project. It can be one person or many. The number of layers of leadership that are stakeholders is dependent on the impact of the project. Shutting down all sales in Asia is going to go further up in the organization than impacting a minor customer in one location. This leadership also will fund the solution costs and needs to understand that as well.
- Team members: The people who are giving up their normal work to move into work on this crisis or critical work are important in communications. Good communication flow will help them be more successful.
- Leadership of the team members: The team member's leadership is critical as well because they can help control the flow of other competing work that will draw your team away from the project and its goals.
- Project leader: The project leader's primary roles are: The leadership of the project team and the delivery of the "What is success?" question in the DQO (Charter). They will need team members and key stakeholders communicating with them to be successful. They will work closely with the project sponsor in project decisions. For this reason, the project leader is also a key stakeholder. They will be in the communication plan as a focal point person on information traffic.
- Project sponsor: Typically, the project sponsor is accountable for the process that has a problem and will control most of the resources needed to solve the problem. The sponsor is the person who has the obligation to ensure a project's successful execution. They will secure the team members and other resources with their influence and their position/role within the organization. Outside of the team, they will be the person that needs to know the most about the project. Ultimately, the sponsor will be the key person in solution approval. Project sponsors have a lot of "skin in the game." Their personal success is closely linked to the project being successful. The sponsor is a key communicator to the project leader, team, and other stakeholders.
- Process owner: Somebody oversees the process that is broken. They will need to be involved because of the impact to their work. They will also be involved in finding and solving the problem. Sometimes the process owner and sponsor are the same person. Not always. The process owner needs to be in the communication loop because it IS their process!
- Customer/customer interface: Depending on the project, there almost always is a downstream customer that is being impacted by the problem. They will need to know about supply or quality interruptions being caused by the problem. The

team may be working with someone in the sales hierarchy and relying on them for doing the communication, but ultimately these communications need to be happening. Like all key stakeholders, the level within the organization is set by the problem impact. If the project is impacting the entire building industry market, then the customer interface people will have broad scope responsibility for that market. If the problem is impacting one customer with a specialized need and a product customized to them, then the interface might just be the sales or technical sales person who works with that customer. Please also note that the term customer in this context could be many things. It certainly could be a direct customer of the company if the problem was affecting the product the customer was buying, but customer could, for example, mean a regulatory agency when the problem was impacting the environment. In that case, the interface might be an environmental point person who interfaces with the government agencies instead of someone from sales.

- Subject matter experts: Some of these will likely be team members, but many may still be outside of the team. These are the people that know the stuff that is essential to solving the problem. Maybe they are a heat exchanger design expert or a chemist specializing in the reaction the FPS team is working on fixing. Whatever their expertise, if they are important to the project, they are a stakeholder. Being on the communication plan may help solve the problem because it gives the experts a chance to influence the outcome of the project. They might steer the team in a new direction by adding a hypothesis different from the ones developed by the team. Subject matter experts might hand the team a solution that they have seen used before to the same problem in some other location. Keep these folks in the communication loop, their expertise can be leveraged.
- Others: There will be folks outside of the list above because each project is unique. The right people will be who they are, and do not constrain the team to typical when thinking about key stakeholders.

These roles will be within the team and adjacent to the team. The core team will contain the folks doing the work. Related to those folks will be supervision that are guiding the goals of the work and providing the resources. A common root cause of team conflict and dysfunction is not knowing who is doing what. That is where roles come into play. Clearly establishing roles and responsibilities up front will help prevent team conflict and dysfunction. Let's look at an example of a team conflict caused by roles and responsibilities.

The story: An outside expert in chemometrics, Nathan, was brought into the team to provide his expertise at analyzing complex data sets. Nathan had been doing this type of work for the past 7 years and was much respected in the company. The project leader Don was a certified Lean Six Sigma Master Blackbelt.

George was the project leader, but had secondary skills in analyzing data. George was also much respected in the company and took great pride in his skills.

The team had just finished a major data collection plan and had tens of thousands of lines of data that needed to be analyzed to try and kill a handful of hypotheses. The team planned to meet the following week to look at the results of the analysis. The meeting to look at the data was a DISASTER! Nathan and George both prepared presentations with their analysis of the data. The meeting quickly spiraled out of control into arguments about techniques and results between George and Nathan. The remaining team was confused and frustrated by the waste of time. Nathan felt that George did not respect him and George was just confused about what happened.

This example is illustrating the root cause of not understanding roles and responsibilities causing team conflict. This root cause will result in either multiple people doing the same work and fighting over it, the wrong person doing work not suited to them, or the work falling in between people and not being done.

To prevent this root cause, establish roles and responsibilities. This is best done in advance of any work being done, but it can be done after a conflict occurs to fix the root cause for it the future.

One of the best techniques is to do a roles and responsibilities exercise with the team or with the individuals on the team that are experiencing conflict. The exercise is to first spend some time answering the following question.

What Kinds of Work do we Expect to be doing on this Project?

The team or the individuals with the conflict then brainstorm this and write one item per sticky note. Then circles are drawn on the wall with team member names over them. The team then takes the sticky notes and put them in the circles. If the work does not fit any of the people, it is put outside of all the circles. The team then reads what is in their circle and negotiate if the work should be in their circle or someone else's circle. When one or more of the sticky notes are outside of everyone's circle, this says that a role is missing. Ad-hoc or new team members will need to be found to fulfill those roles. The outcome is that the roles get defined and the responsibility to complete the work within an individual's scope is developed. It is also useful to come up within a one-line statement that defines what the role within the project will be.

Let's see how George and Nathan complete this exercise for the work of data analysis.

After the meeting, George and Nathan met. George apologized to Nathan and said that he should have worked harder at establishing the roles and responsibilities for data analysis work better. No time like the present! George and Nathan sat down and brainstormed the work they expected to occur. This was the list they came up with:

- Putting the data into a format that works with the analysis software.
- Using advanced chemometrics techniques to analyze the data to kill hypotheses.
- Using standard techniques such as ANOVA, SPC charts, and t-tests to kill hypotheses.
- Preparing presentation to summarize results for the team.
- Independently reviewing analysis results prior to sharing to insure correct interpretation of results.
- Presenting analysis to the team.
- Presenting analysis during stakeholder reviews.
- Making sure the analysis presentation is understandable to a lay person.

After George and Nathan finished with this list. George put two circles on the board: one was labeled Nathan and the other was labeled George. They then put the work into the circles. Here is how the work landed:

George:

- Using standard techniques such as ANOVA, SPC charts, and t-tests to kill hypothesis.
- Independently reviewing analysis results prior to sharing to insure correct interpretation of results.
- Presenting analysis during stakeholder reviews.

Nathan:

- Putting the data into a format the works with the analysis software.
- Independently reviewing analysis results prior to sharing to insure correct interpretation of results.
- Preparing presentation to summarize results for the team.
- Using advanced chemometrics techniques to analyze the data to kill hypothesis.
- Presenting analysis to the team.

Nathan then summarized his role. "My role on this project is to provide chemometrics analysis to kill hypotheses and present those analyses to the team and I will also include any new insights from the traditional data analysis techniques you do."

George then summarized his role with respect to data analysis. "My role is to gain additional insights to the data and pass those on to you for incorporation into the full analysis and to present the overall conclusions during stakeholder review."

George and Nathan had one item that was not in either circle.

- Making sure the analysis presentation is understandable to a lay person.

They enlisted another team member to help with this. The team member was a lay person not familiar with the data analysis techniques and could provide the insight needed to make the presentations understandable to the masses.

Once roles and responsibilities were established, George and Nathan worked well together gaining respect for each other's contribution.

7.3.1 Roles and Responsibilities for an FPS Project: Summary

Roles and responsibility issues are a root cause of conflict on a project. Savvy project leaders spend some time establishing those up front and also during any conflicts. This establishing of roles and responsibilities can be done using standard lists for example: Standard roles and responsibility of a project leader, or can be established using a team exercise. The same is true for the stakeholders for the project. They sometimes also need to understand their roles and responsibilities.

7.4 The Challenge in Maintaining Parallel Processing Within the Team

At the core of the fundamental problem-solving method is parallel processing. The team is typically working on 4–6 hypotheses at the SAME time. The work is divided, but the progress is on multiple fronts. Often the team is working toward killing between 20 and 30 hypotheses for every project. The strategy being used is to focus on the most likely 4–6 hypotheses and then as soon as one of those hypotheses is killed, the next most likely one takes its place. The hypotheses will be characterized as either Must do now, Easy to do, Low priority, or Dead (MELD). The must do now and the easy to do ones are the first ones to be worked. The later ones get worked on after some of the initially prioritized ones are killed.

This parallel processing is much faster that a one at a time – serial approach. Often traditional problem-solving ends up following a serial approach because there is usually a theory about what is going wrong that seems much more compelling than the others. Problem solvers gravitate to working on that one. This works for many cases because often the most compelling theory is exactly what is going wrong. The team focuses its time and energy on where the problem is located. In FPS, we are not generally aimed at the relatively easy or obvious problem. The nature of the problems we recommend using FPS for is the rock hard challenging one. In these, the first guess on the hypothesis of what is going wrong is rarely the right guess. That means that we are slowing the project down if we use serial processing. Working on multiple hypothesis at the same time is the most effective approach for these problems.

Attempting to parallel process more than 4–6 hypotheses is generally not recommended. Teams lose focus when they try and do 10, 20, 30 things at once. The trivial gets mixed in with the important. Figuring out which set of hypotheses are easy to kill or must do now because they are the most likely will bring that focus to the team.

Unfortunately, guiding a team to maintain parallel processing is quite difficult. Here is an example to illustrate the issue.

A compressor essential to operating the production line was failing every month. The same key part was repeatedly breaking. A team was formed to fix the problem. Over 47 times, this had happened, but luckily the maintenance folks had kept the old broken parts. The team had formed 32 hypotheses and prioritized 5 to initially work on. One of the hypotheses was that the part was undergoing chloride-ion stress corrosion cracking. This specialized failure mechanism was well known in industry and was the most likely candidate. The team's metallurgist specialist was tasked with testing for this mechanism. The first part the metallurgist tested showed clear signs of chloride-ion stress corrosion cracking. The team was ecstatic! They immediately dropped all other work and sent the remaining parts off site for testing. The team then focused efforts on finding the source of chloride ion to eliminate this problem. For two weeks, they worked on this and then the result of the testing came back. Of the 47 parts that had failed, only one . . . the first one tested exhibited chloride-ion stress corrosion cracking. One of the possible sources of chloride ion the team had been investigating was from a special clean out that was done in an upset early in the year. The part that failed form chloride stress corrosion cracking failed right after that special clean out. Although the team understood one unique failure mechanism, they had wasted the better part of two weeks chasing something that would not do much at fixing the bigger problem.

In the example, the team did explain 1/47 failures. It is common for FPS projects to have more than one thing working to cause the failures. Sometimes two or more hypotheses cannot be killed.

As the example illustrates, serial processing is a high-risk business on FPS projects. FPS project leaders should make it a regular practice of maintaining parallel processing. This can be accomplished by:

1) Every-time a hypothesis is killed to make it an agenda item to select the next one to take its place.
2) Regularly remind the team of the consequences of serial processing.
3) Require updates every team meeting on ALL of the active hypotheses.
4) Stop working on the other hypotheses ONLY when the propensity of evidence fails to kill a hypothesis or hypotheses that explain the problem.

7.4.1 The Challenge in Maintaining Parallel Processing Within the Team: Summary

Parallel processing is one of the core critical to success factors within the FPS methodology. Project leaders and team members are wise to avoid the trap of dropping hypotheses when one looks promising. Stop working on the other hypotheses ONLY when the propensity of evidence fails to kill a hypothesis or hypotheses that explain the problem.

7.5 Strategies in Team Management for Very Large Projects

FPS projects can sometime be quite large. For example, a problem was costing a company over $100 million/year, the company was in danger of losing money and reporting the losses to shareholders. If the problem was not fixed, the company could go bankrupt. A team was formed and 53 experts and additional resources were tapped to work on the problem.

When the teams get this big, they can quickly become unwieldy and unproductive. Managing these teams takes a specialized approach. First, usually a steering team should be formed.

The role of the steering team is:

1) High-level coordination of resources including sub teams.
2) Makes the decisions needed by the project within a broad sphere of influence.
3) Gives guidance and advice to the project team as needed.
4) Gives directional input to the project team.
5) Ensures resources are appropriate for the problem. Note: Too much or too little can both be bad for projects.
6) Trusts the team to do the detailed within project work.

Steering teams are made up of the project leader and key stakeholders who will make the major decisions about resourcing, solution selection, communication, and any other critical to success decision.

7.5.1 When to Form a Steering Team

Some rules of thumb that indicate that a steering team is needed are:

1) The project burn rate is high. On the order of magnitude, hundreds of thousands of dollars lost each day or higher.
2) The number of people that will be spending most of their time on the project till the problem is solved exceeds 10.

3) The most important two to three stakeholders have interests and motivations that are in opposing directions, and therefore the steering team is needed to provide unity in the team's direction.
4) Swift decisions are needed on actions that are: 1) beyond the team's control and 2) within the steering team's control.

If these conditions hold true, think steering team. Get the right people on that team and schedule regular meetings at a frequency that makes sense for the problem at hand.

The project can then be run under a structure of a main steering team setting direction and writing charters/DQOs for sub teams doing the work of the project. The project leader will coordinate the sub projects and likely run one of the sub projects. The sub projects will have leaders and team members. They are chartered with the best three questions – What is success? What is necessary for success? and What is rate limiting? This controls the scope of the work for each sub team and keep teams from duplicating work or missing work.

Sub team structure will depend on the nature of the problem. Here are some examples of sub team structures to illustrate the variety of ways sub teams can be structured. Steering teams will need to decide how best they want to structure sub teams for their problem.

1) A plant failed with all of the liquid chemicals inside turning to a polymer of such a high viscosity it was essentially a solid. The team was split in half. One sub team was charged with doing the short term work to get the plant cleaned up, repaired, and ready to start up. The other sub team was charged at finding the root cause and science behind why the polymer was formed.
2) A highly complex formulation failed causing an extremely high burn rate, dozens of people were assigned to figure out the problem; 35 hypotheses were developed and 7 sub teams were formed and assigned 5 hypotheses each to kill.
3) A product was failing as it aged. There was no predictive test to determine if the product would fail, leaving a situation of having to age the product for months to find out if the problem existed. One sub team was charged with developing a predictive test that would quickly tell if the product was going to have the problem. This team was only developing how to measure the problem. Another sub team was formed to solve the problem.

As you can see, there are many ways to structure sub teams. The nature of the problem will dictate how to do this. Best practice is to charter the sub teams and regularly meet to apprise the different teams of the progress being made.

7.5.2 Strategies in Team Management for Very Large Projects: Summary

When projects get big, managing them is essentially the same as small project with the exception of dividing the work between sub teams. A steering team coordinates the sub teams and makes the critical decisions.

7.5.3 Team Management and Roles: Summary

An adept project leader can utilize the tailwinds of FPS projects to overcome the many headwinds that are inherently baked into FPS projects. Enabling the teamwork needed to be successful at delivering the project goal is a balancing act of using effective techniques to build and manage teams. The main thing is to keep the main thing the main thing.

8

Stakeholder Identification, Analysis, and Communication Plan

Stakeholder communications are critically important within any project. The team's ability to get resources and implement interventions to fix the problem hinge on the ability to get buy-in from the stakeholders. The critical nature of stakeholder management is unfortunately often learned the hard way.

An example of the critical nature of stakeholder communication is one of a failed solution selection presentation. The FPS team was working on a life of the plant problem. Since inception, a critical heat exchanger in the plant had consistently fouled after a few months of operation. The plant folks had never figured out the cause but had worked at methods of cleaning out the unit quickly. They could turn around from a fully fouled heat exchanger to a clean one in an amazing process that over the years had gotten better and better. Better and better was still not good enough and the cost of the fouled heat exchanger in lost production made it worthwhile to charter the problem-solving team to figure out fundamentally what was happening and develop a solution. The team did exactly that. It was determined that running the shell side temperature higher would prevent the chemistry that was causing the fouling. To achieve the higher temperatures, some improvements needed to happen to the heat transfer fluid furnace and system to allow it to run at higher temperatures.

The cost of the improvements was far less than the cost of the fouling impact on production, so the team felt like the presentation to the stakeholders to get buy-in on the solution was going to be a rubber stamp of their recommended solution. The team went into the stakeholder presentation with a high degree of confidence that the presentation and approval for spending the money to solve the problem was the easy part of the project. These feelings of confidence were quickly destroyed as the meeting progressed. It was a hostile room with aggressive

Fundamentals of Industrial Problem Solving: A Practitioner's Guide, First Edition.
Zdravko I. Stefanov, Eldad Herceg, Carla Schmidt, David M. Jacobson, Dana Livingston, J.P. Chauvel, Sunil Kumar Chaudhary, and Christopher Paul Christenson.

questions and body language that said that the approval for funds was going very badly. The team almost did not get through their first slide of the presentation by the time the meeting was over. A shocked team promised to schedule another follow-up meeting to finish and reach decisions.

After the meeting, the team reconvened to try and understand what happened. After much discussion, one of the more experienced team members pointed out that most of the approving body had in fact worked on this problem during their career. They had worked hard to try and figure out the problem and had essentially given up and instead worked on minimizing the impact of the fouled heat exchanger. Their work to minimize the impact had been important at that stage of the plant life. They gained production and likely had been rewarded for their hard efforts. This problem-solving team was essentially causing two things that cascaded into a bad stakeholder review. First, the solution would make the work to speed up the heat exchange cleaning process obsolete or at least less important. Nobody likes to see their hard work erased. This is a difficult pill to swallow. Second, the nature of the solution was relatively simple. This problem was one that these engineers who are now key stakeholders had worked on solving. They had concluded that it was not solvable. Then, an upstart team concluded that it in fact was not that hard to solve. It was almost certainly a blow to the stakeholders' egos and implied that the problem could have been solved years ago if they had been doing their job. No wonder the stakeholder review went badly.

The team decided that it was time to communicate better. First, they worked individually in one-on-one meetings with the key stakeholders to explain the complex chemistry that was going on in the fouling process. Second, they knew that the fouling would not go to zero. It would be greatly reduced, but zero was not achievable. With this knowledge, they estimated the fouling frequency and pulled together the business case of keeping the fast cleaning process intact. It made good business sense and it was a nod to those that had done the work to develop the procedures. Armed with better stakeholder understanding, the team was then able to move forward and get approval to implement the solution to the fundamental problem that was causing fouling.

Another example of stakeholder communication going bad is one where a process improvement team was working to understand why a compressor was failing almost as soon as it was started up. The cost was enormous with the site production at near zero and customers screaming for material. The team was in crisis mode. The FPS team figured out that it was a materials of construction issue with a part of the compressor. Purchasing was brought into the picture and the part made from a material more robust in the application was ordered and expedited. A chartered dedicated airplane was chosen to get the part to a nearby city as quickly as possible and a dedicated delivery truck was arranged to get the part from the airport to the plant. All was set to bring the part in as quickly as was physically possible. Maintenance and production folks were set up to be there and install the part the second it arrived. The part did not show up. Despite calls and other efforts to find out what was happening, it could only be confirmed that the plane had delivered its leg of the parts journey and what happened after that was a mystery. Ultimately, the part was found, installed, and did the job; albeit much later than planned and it cost the company a significant pile of money. The story of what happened to the part is an illustration of why knowing all the stakeholders in a system is critical to a project's ability to deliver results in a timely fashion.

The part's journey had a hiccup; the truck experienced a flat on the way between the airport and the plant. Once that was fixed and the driver was underway, they were later than the original schedule. The plant had a dedicated delivery entrance. The truck driver followed procedures and came in through the delivery entrance. The plant had a policy and signs at the delivery entrance stating that it did not accept deliveries after 4 p.m. The team that received deliveries worked days and after-hours deliveries were done by exception only. The guard stationed at the gate did his job and turned the truck around and gave instructions to come back when the entrance officially opened in the morning. The part was now delayed.

Should the guard be punished for doing his job? Should the truck driver be punished for following instructions and procedures? People impacted by a project can span a broad spectrum. Failure to communicate with them is clearly a problem. Better communication with the people involved in the delivery workflow would have prevented the problem. The fault of why this cost happened is really a story of dropping the project management communication ball; not a failure of the delivery process.

These examples illustrate the critical nature of stakeholder communication. Stakeholders big and small can make or break your project. This chapter discusses how to identify stakeholders, when to use a steering team, how to assess the importance of the stakeholders, how to understand stakeholder needs, how to develop content for the communication, the best ways to communicate, how to manage antagonistic stakeholders, and the changing nature of the team's communication plans. The goal of the chapter is to arm the reader with the skills to prevent the most common communication failures and have successful stakeholder communication.

8.1 Stakeholder Identification

8.1.1 The First Step, Stakeholder Identification-make a List

During the initial stages of the project, chartering and early launch key stakeholders should be identified. Identifying stake holders is about answering three key questions.

1) Who will be impacted by this project?
2) Who can make or break this project?
3) Who holds the resources that we may need to complete this project?

To identify stakeholders, it is helpful to ask these three questions of the team, of the sponsor, and to repeat the questions as the project moves and changes. Capture their answers and include them in a list. The key stakeholders will be a part of your charter. All the stakeholders will eventually be added to your communication plan. At its heart, stakeholder identification produces a list of names and reasons why they are important to the project.

8.1.2 Typical Stakeholder List Information

1) Name.
2) Role in the organization.
3) Role in the project.
4) E-mail.
5) Location.

8.1.3 Technique for Generating a Stakeholder List Within a Project Meeting

One useful technique for generating a stakeholder list within a team meeting is to hand out yellow sticky notes. Then ask the team big three questions:

1) Who will be impacted by this project?
2) Who can make or break this project?
3) Who holds the resources that we may need to complete this project?

On a white board or flip chart, draw three large circles in a bull's eye labeled as shown in Figure 8.1.

Then, have team members write name of potential stakeholders on their yellow stickies and then put them on the bull's eye where they feel they belong. Simple, but effective.

8.1.4 Where to Store Stakeholder Lists

The stakeholder list can be kept in a database, a spreadsheet, series of route lists, a document, web-based project software, and many other ways. The way chosen is ultimately just a decision with a relatively meaningless outcome. So, personal preference trumps all else. Keep the list where and how you want to do it.

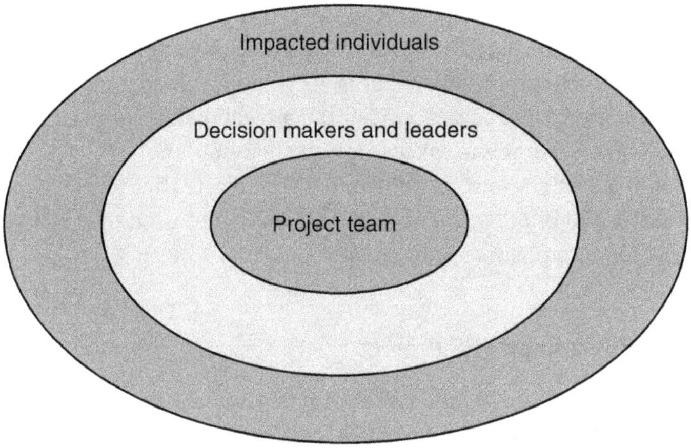

Figure 8.1 Stakeholder identification chart.

> ***Wisdom tidbit: The person who cares the most about any team-shared document is the one who typically updates that document.***

In the case of key stakeholders, the person that cares the most about the stakeholder list is almost always the project leader. As a project leader, if you spent your life in spreadsheets and that is your most comfortable go-to tool, it is fine to use a spreadsheet to store your stakeholder work. If you are awesome with web-based project collaboration tools and love that, do that to keep your stakeholder lists. Content is what counts.

8.1.5 Tips on Making Stakeholder Lists Effective

Using Route Lists

A simple way to make a stakeholder list practical and useful is to set up a series of route lists that can then be used in future e-mails and meeting notices. For example, a route list of the team can be used for team meeting minutes. A separate route list for those that should be copied with the minutes, but who do not attend team meetings can be on the .cc part of meeting minutes. As another example, a route list of the most critical people responsible for making decisions can be used when setting up stakeholder review. Because much of the project communication is likely via e-mail, this method has an advantage in simplifying future communications.

Shared Systems Approach

When information can be accessed by the entire team, it can be helpful when team members are updating information together in a collaborative fashion. For example, a team member thinks of a new stakeholder and goes into a file share area and updates the stakeholder list. This is a shared systems approach. As an example of how this works, a team member that is covering for the project leader while on vacation uses a shared route list to send out the project update. In principle, make shared needs a shared system with access for all stakeholders.

The stakeholder list is about content, not format. Use the one that works with your preferences. Make it a living document that does really get used to improve communication and accelerate positive outcomes.

8.1.6 Typical Stakeholders in an FPS Project

In a typical FPS project, there are many stakeholders your team will identify during the initial analysis. In general, the key ones they will identify are:

- **Leadership in the business:** With problems, there is money flowing out the door. This leadership cares about that aspect of the project. It can be one person or many. The number of layers of leadership that are stakeholders is dependent on the impact of the project. Shutting down all sales in Asia is going to go further up in the organization than impacting a minor customer in one location. This leadership also will fund the solution costs and needs to understand that as well.
- **Team members:** The people who are giving up their normal work to move into work on this crisis or critical work are important in communications. Good communication flow will help them be more successful.
- **Leadership of the team members:** The team member's leadership is critical as well because they can help control the flow of other competing work that will draw your team away from the project and its goals.
- **Project leader:** The project leader's primary roles are: The leadership of the project team and the delivery of the "What is success?" question in the DQO (Charter). They will need team members and key stakeholders communicating with them to be successful. They will work closely with the project sponsor in project decisions. For this reason, the project leader is also a key stakeholder. They will be in the communication plan as a focal point person on information traffic.
- **Project sponsor:** Typically, the project sponsor is accountable for the process that has problem and will control most of the resources needed to solve the problem. The sponsor is the person who has the obligation to ensure a project's successful execution. They will secure the team members and other resources with their influence and their position/role within the organization. Outside of the team, they will be the person that needs to know the most about the project. Ultimately, the sponsor will be the key person in solution approval. Project sponsors have a lot of "skin in the game." Their personal success is closely

linked to the project being successful. The sponsor is a key communicator to the project leader, team, and other stakeholders.

- **Process owner:** Somebody oversees the process that is broken. They will need to be involved because of the impact to their work. They will also be involved in finding and solving the problem. Sometimes, the process owner and sponsor are the same person. Not always. The process owner needs to be in the communication loop because it IS their process!
- **Customer/Customer interface:** Depending on the project, there almost always is a downstream customer that is being impacted by the problem. They will need to know about supply or quality interruptions being caused by the problem. The team may be working with someone in the sales hierarchy and relying on them for doing the communication, but ultimately these communications need to be happening. Like all key stakeholders, the level within the organization is set by the problem impact. If the project is impacting the entire building industry market, then the customer interface people will have broad scope responsibility for that market. If the problem is impacting one customer with a specialized need and a product customized to them, then the interface might just be the sales or technical sales person who works with that customer. Please also note that the term customer in this context could be many things. It certainly could be a direct customer of the company if the problem was affecting the product the customer was buying, but customer could for example mean a regulatory agency when problem was impacting the environment. In that case, the interface might be an environmental point person who interfaces with the government agencies instead of someone from sales.
- **Subject Matter Experts:** Some of these will likely be team members, but many may still be outside of the team. These are the people that know the stuff that is essential to solving the problem. Maybe they are a heat exchanger design expert or a chemist specializing in the reaction the FPS team is working on fixing. Whatever their expertise, if they are important to the project, they are a stakeholder. Being on the communication plan may help solve the problem because it gives the experts a chance to influence the outcome of the project. They might steer the team in a new direction by adding a hypothesis different than the ones developed by the team. Subject matter experts might hand the team a solution that they have seen used before to the same problem in some other location. Keep these folks in the communication loop then their expertise can be leveraged.
- **Others:** There will be folks outside of the list above because each project is unique. The right people will be who they are; do not constrain the team to be typical when thinking about key stakeholders.

There will be other stakeholders who are not key or critical. That does not mean that these folks are excluded in communication. It just means that the nature and priority of communication is different. For example, a key stakeholder may get a face-to-face meeting with the team to answer questions and concerns. The minor stakeholders may only get an e-mail letting them know who to contact with their questions and concerns.

8.1.7 Special Case of Key Stakeholders-Steering Team

In some specialized cases, where the projects are of high value, are complex, and have a high degree of multifunctional resources at play, it can be easy to lose direction and ultimately control over the people working on the problem. These types of FPS projects can also need quick decision-making capability for tough questions. Decision makers need to be present and involved to keep progress rolling and avoid decision-based project bottle necks. In these cases, it is often helpful to form a steering team for the project. The steering team is usually made up of the project leader, sponsor, business leaders, and customer interface people. The role the steering team plays is:

- High-level coordination of resources including sub teams.
- Makes the decisions needed by the project within a broad sphere of influence.
- Gives guidance and advice to the project team as needed.
- Gives directional input to the project team.
- Insures resources are appropriate for the problem. *Note: Too much or too little can both be bad for projects.*
- Trusts the team to do the detailed within project work.

The steering team will meet at regular and appropriate intervals as well as ad hoc as needed. This strategy is about getting the key stakeholders together on a regular basis to accelerate progress. These are the stakeholders that are jointly setting the direction of the project, making the tough decisions, and coordinating complex multifunctional efforts.

Make the decision on having or not having a steering team during the same timeline as completing stakeholder analysis. That decision should be mostly a sponsor and project leader consensus decision.

Some rules of thumb that indicate that a steering team is needed are:

1) The project burn rate is high. On the order of magnitude, hundreds of thousands of dollars lost each day or higher.
2) The number of people that will be spending most of their time on the project till the problem is solved exceeds 10.
3) The most important two to three stakeholders have interests and motivations that are in opposing directions and therefore the steering team is needed to provide unity in the team's direction.
4) Swift decisions are needed on actions that are: (i) Beyond the team's control and (ii) within the steering team's control.

If these conditions hold true, think steering team. Get the right people on that team and schedule regular meetings at a frequency that makes sense for the problem at hand.

8.1.8 Changing Nature of Stakeholders

Identifying the stakeholders initially is just the beginning of looking at stakeholders. To do this right, more will need to happen.

As the project progresses, the stakeholder analysis will change and therefore needs to be revisited as changes happen. For example, in an FPS project, a team was working on understanding an unwanted side reaction that was occurring and impacting the quality of the product. When the team discovered it was a raw material contaminant causing the reaction, the team had to move to the previous process that was feeding the process with the problem to figure out what was causing the contaminant. There was a reshuffling of stakeholders due to this new information. The team revisited their stakeholders to change the list around and add more stakeholders from that newly in-scope process area. The stakeholder analysis for a project is not and should not be a one-time event. It should be a regular process within running the project that is updated as new stakeholders come on line or as missed stakeholders show themselves.

There are patterns to why a team misses stakeholders or why new stakeholders appear on projects. Understanding these patterns is helpful in maintaining an impactful stakeholder list.

8.1.9 Main Root Causes of Having an Incomplete List of Stakeholders

- Initial stakeholder analysis incomplete.
- Scope change due to problem location understanding.
- Phase changes on project such as moving to solution implementation.
- Company churn as people change jobs.
- Problem severity shifts due to outside forces.

We will cover these root causes in detail.

Initial Stakeholder Analysis Incomplete

> ***Wisdom Tidbit – When teams do stakeholder analysis, their defect rate is rarely zero. Almost always stakeholders get missed.***

Here is an exercise to understand why this tidbit of knowledge is true.

Just think of a simple question. Who are all your friends? Write down the list. Set it aside. Pick it up later today and ask yourself, "Is this list complete?". Odds are, you will think of more names later due to interactions with other people jogging your memory and more time thinking about the question. Did you forget who your friends are? Are your friends unimportant? An incomplete list has nothing to do with importance. It just has to do with the fallible nature of human thought processes. Your team will have the same problem. Your team will leave people out. Project teams need to keep looking. As you seek stakeholders, you and your team will need to look for clues along the way.

Here are some rich sources of information that help complete the stakeholder list.

E-mail Trails The project leader, sponsor, and team will be communicating with e-mails. New names will pop up as the e-mail communication trail gets forwarded to those outside of the stakeholder list and replies are made. These are rich ground for noticing who the stakeholders really are and can help build communication bridges. It helps in completing the list. The trick is to notice these additions to the e-mail chain. Ask questions and be curious as a project leader or team member. Why did that individual forward the e-mail to someone not on our stakeholder list? This will help understanding the growing list of stakeholders.

Hallway/Breakroom Interactions When a high burn rate project is underway, project leaders and team members will be getting plenty of unsolicited advice, solution proposals, criticism, and other forms of feedback throughout their work days. These events can be annoying and detract from project progress but are also an indication of communication issues. Paying attention to who is offering their feedback and the motivations around the feedback are very helpful in identifying new stakeholders or new concerns. Be mindful of these interactions and try and understand them deep enough to make the call on if this is just casual interaction or is this something worth following up on. Is this someone just being nosy or is it a stakeholder that was missed? Often subject matter experts will bring their thoughts to the team this way. For example, on one FPS project, it was the person who fought and solved this same problem over a decade ago who approached an FPS team member during a coffee break. The right science and the right solution were known by the individual, but not documented in a lasting way. Following up on this information resulted in fixing the problem. Pay attention. Old operators who used to run the problem plants often approach team members with these casual interactions.

Always ask yourself and the team, was this information request something that we missed out in communicating or is this just vocalization of the concerns being handled by the project team? Is there value embedded in these interactions? It is a balancing act of listening and not listening; of paying attention and ignoring. Project leaders play an important role in striking the balance by sorting through trivial interactions and noticing the vital interactions.

Meeting Attendance Adjustments When people invited to a meeting bring uninvited attendees to a meeting the project team is running, it is usually a signal that there may be a stakeholder missed. Often, the addition is another process expert or an individual that has a role that would be impacted by the outcome of the meeting. For example, on an FPS project, the team was meeting to understand the results of a rheological study. The person who was going to review the results forwarded the meeting to two other process experts in running these tests. The results were complex, and the person felt like they needed help; so, they reached out to other experts. The project leader paid attention and added those new experts to future meetings in that subject matter. When running a meeting and a new attendee shows up, it is best practice to spend a few minutes at the beginning of the meeting doing introductions and some understanding of the role the individual plays. This allows the team and project leader to clarify the individuals stake in the game and often will result in growing the stakeholder list.

Phone Calls or Instant Messages During a project, the project leader and team members will be calling or instant messaging folks outside the team. Those interactions will be made while doing the business of a project. Like all networking, one thing will lead to another. For example, on an FPS project, an instant message discussion about possible chemistries at play with a chemist lead to discovering a person in purchasing who was asking about the project wondering if they should bring the raw material vendor into solving the problem. These interactions will grow your stakeholder list. It is helpful to pay attention.

Accepting that your initial stakeholder analysis will always be incomplete and imperfect is a useful approach for project leaders and project teams. It is not about doing a sloppy job on that initial analysis, it is about recognizing that perfection is not going to happen unless future thoughts and interactions are captured and used. Make your stakeholder lists evergreen.

Scope Change Due to Problem Location Understanding

It seems at first that it would be easy to know the location of the problem. Unfortunately, this is not always the case. We have complex systems with complex interactions. FPS projects are typically unusually complex otherwise the front-line folks would have solved it. FPS projects tend to start in the location that the process experts believe is the most likely location of the problem. Choosing a starting point is an educated guess, but nothing more. For example, an FPS project team found that a trace contaminant in a raw material that was causing an unwanted reaction in the process displaying the problem was introduced seven processing steps earlier. Discovery and new information changed the project scope. A new team had to be formed around the processing step that was causing the problem; not the first guess at what was causing the problem. In another example, an FPS team was trying to understand why the product being produced was unstable and was reacting prematurely. It was supposed to react within the customer's process and instead it was completing the reaction in the package during shipping to the customer. The team formed around the production unit thinking the problem was in how the product was produced resulted in unstable product. After measuring the system in detail to kill hypotheses, they found that product that was not packaged was always stable. The team eliminated all the hypotheses tied to product production and shifted their focus to packaging. Air contamination in packaging was ultimately found to initiate the premature reaction. The team had to change to focus on packaging. New people to work on the problem and new stakeholders had to be identified.

As hypotheses are killed, many projects will change. Project leaders and teams need to be prepared to change with new information and new discoveries. When these changes happen, the stakeholder list will need to be updated.

Phase Changes on Project such as Moving to Solution Implementation

This root cause is hopefully obvious. The folks forming the hypotheses can be different than those testing them. The folks implementing the solutions can be different than those who are troubleshooting the problem. Each phase of an FPS project or any project for that matter will have new and different stakeholders.

Company Churn as People Change Jobs

It is just math to understand that during the life of a project, people are likely to change roles. People move jobs, quit, or retire. Thinking about this from a probability standpoint is useful. Imagine a project that is going to last four months. This imaginary project has 25 key people that are impacted as stakeholders or are working on the project as team members. A useful rule of thumb is that people will change jobs, quit, or retire an average of three years into a job. What are the odds that 1 of these 25 people will do that during the life of the project?

There are several ways to do the math around this question. The best way is to use the actual distribution of the time in jobs and run a Monte Carlo simulation of the question. But, for the purposes of making this clear, we will just do some simplistic probability math. The odds that a person will leave the job during any given month is 1/36 or 2.8%. In 1 month out of 36 months, somebody will leave (3 years × 12 months/year). The project duration is 4 months, so 4 × 2.8% = 11.2%. There is over an 11% chance that any one person will leave their job during the project! But there are 25 people, so we need to account for more people in the math $(1 - 0.112)$ ^25=0.05. There is only a 5% chance that all those 25 people will stay in the same job for the duration of the four-month project. So, as a project leader, you are 95% certain that at least one of these people will change. Pretty strong odds that there will be churn.

Your project will have different math around this question, but odds are there will be people turnover. Project leaders need to be prepared for this. Each new person that gets involved in the project will need to catch up to the problem and the project status. When project team members or sponsors change jobs, then the pain can cause delays in project execution. Add churn to your reasons for stakeholder changes and plan on it.

Problem Severity Shifts Due to Outside Forces

Projects are not stable animals. They change as the world interacts with them. A project can start out as medium impact and burn rate, but quickly turn into the most important project of your career. For example, an FPS project team was working on a problem that was restricting a process capacity. The process was meeting customer demand, so it was important; but not critical that the problem get resolved. The project leader was surprised when the head of manufacturing called him directly about the problem. A competitor had an explosion at their facility effectively shutting down production. A major customer had called asking for more product willing to pay a premium. A relatively sleepy project got escalated to weekly board level updates.

On the flip side, critical projects with high burn rates can cool quickly. As an example, an FPS team was working on a product quality problem that was shutting down a major customer causing panic throughout the corporate structure. Daily updates were being requested by senior vice presidents and other senior executives. The customer was driving daily project updates as well. The customer stood to lose a significant amount of money. During one of the updates to the customer, suddenly the tone changed. The problem was no longer a crisis and the customer became willing to take the product with a quality problem on waver. Panic shifted to a strange calm. Eventually, the customer explained the change. They had been working on the problem from their perspective. They had discovered that changing a simple setpoint on their process enabled them to use the product with a quality problem and still produce good material in their process. This "quick win" enabled the customer to start back up and avert the major impacts of the crisis. The problem shifted from one that had to be fixed today to one that could be fixed in the coming weeks without costing more money. The burn rate dropped, and the FPS team had to adjust.

The communication plan for a project that heats up its burn rate usually needs to increase frequency and generally will need to move up the chain of command in stakeholders. Communication needs to match any project priority shift, or the project and project team will suffer. A project leader should help the team shift with those changes.

8.1.10 Last Thoughts on Stakeholder Identification

Stakeholder identification is an imperfect process that can get tough to do to perfection. Project leaders and their teams should make thinking about this list a regular part of project meetings and interactions with folks outside of the project. The goal is not perfection because on the seriously important problems, perfection seekers on the average slow down progress. Instead, seek to do good enough with robust systems that pick up any stakeholders that were missed or pop up during changing conditions. These initial identification efforts and systems will increase your probability of project success.

8.2 Stakeholder Analysis – Deepen Your Understanding of the Stakeholder List

Once you have an evergreen list of stakeholders residing in a spreadsheet, route list, word document, or whatever your preference of location and method, the next step is to gain a deeper understanding of the stakeholder's needs as well as their current bias toward the project. The analysis phase is really all about understanding the stakeholders and how best to deliver on their needs.

8.2.1 Initial (Current) Bias Toward Project

The initial bias for stakeholders happens the second a project is announced. Think about any significant announcement in your work life or home life. If it was announced a new boss is taking over your group, do you think the group has the same opinion of how good or bad that news is? Likely, some folks will think it is good news and there will be good changes coming. Some folks will think that bad things are about to happen. Everyone is going to have different reactions in dramatic or small ways. When there is a significant problem, a project leader announced, and a team is formed, that fact will cause waves of opinions within the stakeholders. Some stakeholders may think that another team should be working on the problem or that another project leader should have been assigned. Some stakeholders may be assigning blame for the problem and now feel like something is finally going to happen. Some stakeholders will be thinking about how much resources this problem is going to pull from their group and are worried about all the other work on their backs. There will be reactions. Many different reactions. How do you gain understanding of what stakeholders think?

> ***Wisdom Tidbit – To understand what the stakeholder's concerns are with your project, ask them. Strangely, most project leaders will default to guessing. Guessing is ineffective. Stop guessing and start asking.***

Why should the team or project leader care about that initial bias a Key stakeholder have toward the project? An example helps. An FPS team had formed and was working on a high burn rate problem, and the project leader started the calls to key stakeholders about understanding their concerns. Research and development (R&D) resources were almost certainly needed for the project because the problem involved some complicated chemistry and all the world's experts in that chemistry were in the R&D department. The project leader picked up the phone and called the R&D key stakeholder to level set expectations about resource needs and understand concerns. The call started off very difficult. The R&D stakeholder could care less that manufacturing was down, and that cash was flowing out the door. His group was in the middle of a reorganization and might be having layoffs. He needed his top people to help with those problems. The stakeholder had their hands full and the project just added to the department's workload. The top people R&D needed to handle their problems were the same people the FPS team needed for technical help. The project leader understood those new concerns and discussed ways to minimize impact. The project leader adjusted the plan to shift the folks that the R&D stakeholder needed to an important, but less resource-intensive consulting role. A more junior R&D expert would help with the other project needs such as literature searches. The project was able to progress at the fast pace it needed while still minimizing the impact to the special situation R&D was experiencing.

For this example, think about if things happened differently. Imagine that the project leader did not show the leadership it takes to make the phone call. Suppose the project leader guessed that R&D would be fully supportive of the project because of its tremendous impact to the company. The issues with the R&D would continue with or without that phone call. Odds are that symptoms of the R&D issues would pop up in the FPS project. Maybe the R&D folks would fail to show up for meetings. Or maybe they would get behind on their tasks. Clearly a full plate with another scoop on top topples in some way. Or worse, the R&D folks would try and accomplish everything and do one job during the day and the overflow work at night. Short term that may work, but ultimately people break down under high levels of stressful work. This key stakeholder problem would show up and could directly impact project completion and success. Disengaged experts are less helpful than fully engaged experts. This example shows the value of talking to your key stakeholders.

Here are some tips to help with this key stakeholder discussion:

- One person talking to another person face to face will be more successful than a broader meeting, e-mail, instant messaging, or other methods. If possible, try a conversation first. If face to face is not possible, a phone call works almost as well.
- Share the project's "What is success?" first to get them on the right page. Having the key stakeholders know what is at stake helps them form an informed first impression bias.
- During the conversation mostly listen, write things down, and ask clarifying questions.
- Target 5–10 minutes for the time spent on this conversation. A five-minute conversation on the most important issues is focused. Running over time because the stakeholder has more to say is fine; but setting a meeting for one hour on a short topic will just fill the time allotted without necessarily delivering additional value. Time is critical, and you are burning time.
- Lesser stakeholders with less project impact can be reached with lesser methods like e-mail, group meeting, or instant messaging because the risk is lower, and everything is a risk reward balance.
- A good question to ask is "What are your concerns or thoughts about the (project name) project?". The question is open ended and typically gets great high-value answers.
- Divide and conquer while leveraging relationships. If one of your team members knows a stakeholder, they can be the one who asks that stakeholder about their concerns and thoughts. Team members hearing firsthand the thoughts of stakeholders is also helpful in keeping them on target and in scope for the project. Leveraging existing relationships helps deliver open and honest answers. Dividing up this work helps speed up project execution.

8.2.2 Needs vs. Initial Bias

With key stakeholders, you will get feedback that is beyond just the initial bias; you will also get the needs of the stakeholders. Needs are linked to the stakeholder's role and interest. Needs are about why the stakeholder cares. The other feedback category are statements of initial bias. Initial bias is more complex because it is a question of perception and history. Initial bias is what a stakeholder thinks at that moment in time. A person's preconceived notions and perceptions are difficult to predict. Needs are different. Needs are more predictable, yet are often not directly stated. Needs take deeper analysis. With practice, project leaders can use their understanding of both initial bias and needs to help with how and what to communicate.

8.2.3 Needs

The stakeholder may or may not say that they need something, but they will indicate that they have needs. The technique of gathering initial bias will also gather stakeholder needs. Project leaders and teams just need to think hard about what the stakeholders are saying.

For example, suppose a project leader asked a key stakeholder "What are your concerns or thoughts about the (project name) project?". The stakeholder responded by saying, "I am worried about how long this is going to take and if we are going to shut our customers down. If that is going to happen, we had better start letting them know now."

Although the statement does not say that the stakeholder needs anything specifically, it does really speak volumes on a need.

The stakeholder needs an estimated project timeline and an initial prediction of risk of running out of product while the issue is being resolved. Filling this need will enable the stakeholder to make the decision on if and how to communicate this problem to the customer.

Listening to stakeholders is hard, but essential. It will take practice. Stakeholders do form patterns. There are typical stakeholder needs that will show up time and again after listening long enough. These needs will prioritize differently depending on the nature of the FPS project. For example, a problem that is shutting down an intermediate that will impact hundreds of downstream customers has a different feel than a project that impacts one customer on one product. What is important will be different. That is why project leaders and teams will need to talk to their stakeholders vs. just trying to generically address needs. The needs priorities will change and be different depending on project and stakeholder.

8.2.4 Typical Stakeholder's Needs

- ***Know that Core Team has acceptable competence*** – The first concern which may go unspoken, does the Core Team or Project Leader have the required ability to successfully fix the problem.

- ***Know that progress is being made – What is the progress?***

Stakeholders want to know that there is energy being expended on the problem and that the energy being expended is making progress.

Filling this need can be difficult because early on the only progress being made is often that the team is gaining deeper understanding of the problem. Address this need by counting hypotheses generated and describing the science around the hypotheses. Also, use quick wins and moving in the direction of good to move from crisis to less crisis. Highlighting these mitigation activities will show progress. Stakeholders need to know this is hard work and the team is working hard on it.

In the middle of the project, the team is killing hypotheses without solving the problem. From an outsider perception, it can appear like nothing is happening. In fact, the most important work is happening. Address this need by highlighting the work that killed the hypotheses and counting the kills to date.

Toward the end of the project, the team is working on solutions. Again, the problem still exists. From an outsider's perspective, nothing is happening. To fix this problem, put the outsiders to work by asking them for solutions to the hypothesis proposed. Give the stakeholders a decision timing. They will need to help with the work on deciding what solution needs to be implemented.

During solution implementation phase, the progress is the execution plan on building the solution. Communicate % complete and other indicators showing progress.

Project leaders and teams need to construct updates that reflect the movement of the team toward problem resolution. Tell the story.

Rate-determining Step – What Is Slowing Down Progress?

Rate-determining step sometimes also called critical path is of interest to the stakeholders because it is what is holding up the project. Stakeholders want to understand what is slowing things down.

Making the rate-determining step go faster will make the project go faster. Communicating what that step is and what resources are needed to make it go faster is a helpful part of stakeholder communication. Often, stakeholders can intervene and remove roadblocks if they know what road is blocked. Stakeholders who do not know the rate-determining step will often try and drive teams to activities that from an outsider's perspective need to happen. Work on things that are outside of the rate-determining steps will often slow progress. Communicating the rate-determining step will help fulfill the stakeholder need to solve the problem as soon as possible.

> ***Wisdom Tidbit – If your team needs help with road blocks that are impeding progress, let stakeholders know the roadblock and the team will have a better chance of knocking down those barriers. Teams that keep roadblocks quiet get what they deserve, slow progress.***

Timeline – When Is This Going to Be Done?

The problem you are working on has ripple effects on customers, coworkers, sales, business, and many other downstream impacts. Your company is adjusting for these impacts. Folks are suffering at work and want to know when that suffering will end. Your stakeholders want to know when it will end.

As a project leader or a project team, this can feel like an overwhelming question. When is this going to be done? The pressure will be EXTREME on high burn rate projects to state the date this project will get solved. If your company is losing $100,000/hour on a problem, you will get asked "When is this is going to be solved?" On all projects big and small, there will be some pressure. Stakeholders will want the team to commit to finishing by a selected date. They will often want the team to pick that date. Early in a project, this is like selecting, cutting a stick, and handing it to stakeholder to beat you with it. Giving the date early on in a project can only go well if that date is right. The likelihood it will be the right day is approaching zero. From a stakeholder perspective, if the problem is solved after the date, the team failed. If the problem is solved earlier than the date given, then it can also be a problem. Obviously, the problem was easy, and the team was sandbagging. Completion dates are land mines waiting to blow up.

> ***Wisdom Tidbit – If the project leader and team give a date at which the problem will be solved when they do not know the root cause of the problem, they will get what they deserve – angry stakeholders. Avoid this.***

So, what does the project leader and team communicate? FPS projects do not follow traditional project planning. Building a house for example, is something that can be planned with traditional planning method. Timelines can be estimated based on history and the steps it takes. Solving problems does not flow like that. It must flow by a nontraditional path. To communicate the FPS project timeline means communicating what is known and what is unknown. Use typical project planning for the next step known phases. Typical project planning is identifying the next known steps, estimating a timing for each step, and sequencing the steps in a logical plan. This means following the FPS process and communicating those milestones. As your team passes milestones, you will need to get a good assessment of time to the next milestone. For milestones further out, you will need to schedule and use check-in points to manage risks that get generated due to an unknown completion date.

As an example of a milestone that is more known, an early step in all FPS projects is hypothesis generation. How long is that going to take? The team can come up with a traditional plan for this because it is mostly a known. The team might plan to get all the process experts together for a three-day summit and plan to have almost all hypotheses generated by the end of that summit. That piece of the problem of hypothesis generation is something that can be communicated with conviction. It falls in the realm of traditional project planning. Once the hypotheses are generated, the project leader and team can then communicate that the desired outcome was achieved and form the plans on killing the most important hypothesis. The plan for the next section can then be communicated with conviction. In this manner, the team can step through timing communication.

The work that is totally known (done) and the work that is known enough to achieve a solid plan is where the focus needs to be on communication of timing. As a project leader, establishing credibility among stakeholders is important. Your credibility will climb if you deliver on your commitments in the space where you have a respectable understanding of timing and by letting stakeholders know what is currently unknowable.

In the uncertain unknowable space, the communication should be about setting check-in milestones that match external expectations. For example, although the team is rolling through killing hypotheses efficiently at a point in time, committing to when an unknown solution for a yet unknown root cause would be silly and high-risk communication for a team to make. Instead, the team can commit to a stakeholder review of progress that includes a review of the current project plans. The team can also commit to risk mitigation planning such as what to do if the root cause is not found by the date of the review.

Another example of using a review milestone for an unknown total project duration would be to set up a decision milestone. Suppose your customers may or may not be impacted by the problem your team is working to solve. When do you tell your customers? The team and stakeholders may not know when the problem will be fixed, but they do know when the problem will start to effect customers assuming it continues. The project leader can set up a decision meeting to decide to notify the customer. That date can be communicated to stakeholders. Using future meetings to sort out directional issues is one great way of navigating the sea of unknowable answers. It is a hard pill to swallow, but solving problems is as ambiguous as it gets. How long will it take you to solve something nobody has solved so far despite their best efforts? The authors have seen extremely difficult, technically challenging, multi-factor, complex problems solved on the same day as the team first met. At the extreme end, we have also given up on problems because teams could not figure them out within the timing stakeholders allowed the team to work on the problem. Typical problem-solving timing is the order of magnitude between hours to months, but years to solve sometimes happens as well. Typical is a WIDE, but not very useful range of timing.

As a project leader and a team, think about your next steps and ask yourself and your team if they are knowable with reasonable estimates. If it is knowable, then have your team and process experts estimate the timing with traditional project management tools. Communicate that timing to stakeholders and stick to that. Escalate communications if the timing is going off course and course correct. Use traditional project schedules methods.

Unknowable things such as how long is it going to take to install the solution to a problem your team does not understand, is best kept out of the planning space. Communicate that these activity durations are unknowable and use scheduled risk mitigation on those impacts.

Ambiguous finish timing will be a source of tension with your stakeholders. What holds it together is credibility. Commit to what can be committed and deliver on that over and over. Build your reputation as a project leader that delivers results and that credibility will help with the stakeholders worries. Focus on communicating that the right people are on the job and that they are giving it their best. When they know the best team is on the job, your stakeholders will understand the uncertainty side.

Problem $ Cost $ Benefit Analysis – What Is This Problem Costing and What Is It Worth to Us?

The language of management contains dollar signs. In many ways, if the team and the project leader are speaking to management in terms that do contain dollar signs, then management will be listening. The reverse is also true. We must become fluent in financial terms and use that knowledge to understand the right responses for a problem. Flying an expert in from Japan tonight to work on a $50,000/month burn rate problem feels extreme and can get the team into hot water. With the same situation of needing to quickly fly in an expert with short notice for a $1,000,000/day burn rate problem would be appropriate and could even be considered under-kill. Clearly knowing something about the costs and benefits can help with within project decisions. Costs and benefits are also a challenging communication topic. If you put 10 accountants in a room and asked them the value of the project but add the rule that they need to calculate it independent of the other accountants, odds are there will be 10 different answers with possibly significantly different answers. With this kind of ambiguity, what should be communicated?

One of the first principles of financial communication it to minimize the work on the financial side to fit the purpose. For example, if the question is, "Is this problem worth a team spending time on fixing?", does it matter if it is costing over $100,000/day or $115,456.25/day from lost sales margin? Eventually it might matter, but just knowing the order of magnitude of the problem as being hundreds of thousands of dollars per day is good enough to decide to fix the problem or not to fix the problem. Following this line of thought, use Pareto 80/20 principle when initially finding burn rates. Figure out the few big costs that will result in 80% of the total dollars. Focus on those. State values during communication by knowing those few costs and that there is more to be tracked later. For the previous example, the stakeholder communication would say this problem is costing us at least $100,000/day in lost sales margin. Are there other costs? Likely, maybe there are higher maintenance expenses or overtime getting worked. Ultimately, these will need to be calculated and communicated, but early in the project and for SIMPLICITY of communication it is important to use the main costs and benefits in your communication about the dollars.

A technique for money communication is to let the folks that want to talk dollars during communication to supply the estimates for the dollars being discussed. As project managers, the authors have estimated and calculated many costs and benefits for projects. Just because you can do the cost calculations, does not mean that you should be dragged into the debate about costs. When your senior managers speaking dollars say it is costing $50,000/day and your math says it is $55,000, think about the purpose. The purpose is for the folks that are worried about the dollars of the problem buy into the dollars of the problem. Obviously, gross differences need to be discussed. But, the best way to achieve buy in is with their own numbers.

Another principle on money communication is to defer to a money process expert. For example, if your sponsor believes that this problem is costing us over $50,000/day and another stakeholder disputes that number and thinks it is $4000/day, it will be a problem. Project leaders could get in between and try and figure out who is right. That would be a bad choice. A better choice would be to pull in someone from finance to get the best number and then the stakeholders are discussing this with an outside credible expert. The war is being fought where it needs to be. The project leader can watch the process and proceed with a good direction when the dispute is settled. Sometimes, companies have done a fantastic job of translating what the process experts know into standardized workflows and reports that can be used by people who are not money experts. If this is the case, the standardized workflows and reports can be circulated and that can be the higher authority referee and be used in lieu of a financial process expert.

Another principle on money communication is that the bigger the costs needed to make a change to fix a problem, the more scrutiny the costs will receive. Increasing remedy costs means that there will need to be alternatives considered and more detailed analysis. Communication will be about numbers credibility at these stages of the project, so the work that needs to be done by the team needs to reflect that need. Use facts and focus communicating the facts. Rough estimates are only appropriate when the remedy costs are disproportionately low when compared to the benefits.

Another principle on money communication is that not all stakeholders care about this part of the communication. For example, you were told that you need to work this Saturday because it is costing your company money. The first thought might be something along the lines of "If it is costing my company money, I am all-in! I will be there Saturday!" or it might be something like "Saturday! I promised my daughter I would go to her soccer match. What am I going to tell her?".

Just because some of the stakeholders speak in the language of money does not mean that all stakeholders are motivated by it. The key on communication is to have something for multiple motivations. An operator may not be motivated by the problem costing the company money but might completely back the team's solution if payout is that the process will run better and make their workday saner. Communicating about money is about knowing when to talk about it and when to shut up about it.

Communicate the money first during presentations, verbal discussions, and written communications with money-driven stakeholders. These stakeholders have many demands on their time. They need help prioritizing. If you come to them with a $10,000/day issue and they are currently dealing with $1,000,000/day problems, then knowing that will help them prioritize. In typical FPS projects, the burn rates are high, so leading with the money involved helps the team get the full attention of the stakeholders. Leading with the money side helps the big question that folks will have when they meet or read something which is "Why do I care about this?". Consider these two quotes:

1) "We are working on a problem and need some help. We need some analytical testing and would like to have your team work this weekend to get the results in a timely fashion. Thanks in advance for your help."
2) "We are working on a problem that is costing our company $1,000,000/day that is getting upper management's full attention and we need some help. We need some analytical testing and would like to have your team work this weekend to get the results in a timely fashion. Thanks in advance for your help."

The second quote gets the "Why do I care about this?" question out of the way. It gets the attention needed and helps orient the persons involved in knowing the magnitude of the crisis. The project likely just jumped to the front of a long line of folks trying to get help. It is the right thing to do. The first quote does the stakeholder a disservice. What if the stakeholder deprioritizes the work and decides to push the work till Monday? The stakeholder is right to do that for statement (1) because there is some critical information missing. Put in your statements the most important information and lead with it. Give your stakeholders a chance to do the right thing.

The last principle about communicating the dollar side of the equation is choosing how the number is stated makes an impact. Think about the message. For example, if the burn rate of your project is $250,000/day, it could be communicated as that. For more urgency, try saying it is a least $10,000/hour. It is the same number but putting it on a per hour basis ups the urgency. If that is what needs to happen, use that. As another example, if the project is costing $500/hour, it could get marginalized. Stating it as over $1 million every three month moves it away from trivial to significant. Think about the time and amount that will have the most appropriate impact. Another piece is use of significant digits. Saying the problem is costing \$12,652.12/hour is going to imply extreme precision where there is likely none. It also will dilute the message as the reader or listener gets lost in long list of numbers. This is not helpful. \$300,000/day is a useful statement. It anchors the listener and the readers' thoughts. Communicate effectively the money involved.

> ***Wisdom Tidbit – Simplify money communication and talk money only to those that care about money, think about the message and you and your team will be money ahead.***

- Impact to customer – Will our customers be affected by the problem? Customers pay the bills and are the only reason a business exists. Minimizing the impact to this critical group of stakeholders will weigh heavily on the minds of your sponsor and other stakeholders inside the company. When communicating about customer impact, your team needs to understand what the product is doing for the customer and how the problem is impacting that outcome. For example, a problem had shut down an operation that was providing product for an automotive company. The product was one of a kind and was needed to make a critical part for trucks. There were no competitors and if the problem was not solved, truck manufacturing would grind to a halt. The communication focused on "If we don't fix this, they can't make trucks." Keeping that in front of stakeholders enabled the team to make the problem concrete and real for the folks involved.

To clarify the impact to customers, work with the sales and technical sales folks. It is their job to understand customers and what is going to happen moving forward with a problem. That clarification will help with resource alignment and urgency. For example, on an FPS project, the sales person clarified the customer impact. They did this by pointing out that the impact of the quality problem was that windows would fall from skyscrapers. The quality problem was a technical physical property parameter that was not exciting to talk about. Translating the physical property to customer impact engaged the team and their stakeholders. Ho hum became an urgent issue.

The opposite reaction can also occur. For example, a problem had prevented the manufacturing site from making material that would pass quality testing. Talking to the sales person, the impact to the customer was almost nil because the customer primarily bought from other suppliers and the market had excess capacity. Internally, lost sales were costly making working on the problem worthwhile; but from a customer point of view, the impact was small. Communicating with the customer was about letting them know that the supply would need to come from competition for a couple of months while the problem got sorted out. The team could then primarily think in terms of lost sales knowing that the customer relationship would only suffer a relatively small amount.

Direct communication between the FPS team and the customer should rely heavily on the folks whose jobs depend on those communications. Sales, marketing, and service representatives will need to help tailor the communications to meet customer needs. At times, legal will also need to be involved to keep the language of the customer communication in line with the contracts and other legal issues. Customers typically have one big question. When will the problem be solved and back to normal? Usually, that is the one question that the FPS team cannot answer or make solid commitments about. FPS teams should avoid the trap to committing to an unknown. Instead, the communications need to be about what the team can commit to doing. Progress updates at an agreed upon frequency would be an example of something within the team's control that can form a commitment. An estimate of completion timing can occur after the team has a hypothesis that it has failed to kill, and the solution has been selected. Then, a problem solution timing commitment can be made.

Building credibility with a customer is about making commitments that are possible and delivering on those commitments. For example, an FPS team committed to test all bad lots by next week with an agreed upon analytical test. Communicating the results of that test the following week, built confidence with the customer.

Successful FPS teams communicate consistently with customers and deeply understand the impact to customers. Managing this stakeholder group takes working with the customer facing people and thinking about the issues from the customer's viewpoint.

Impact to Employees – Will Our Employees Be Affected by the Problem?

Big problems are going to impact the folks associated with the work that is in the area of the problem. People may need to work longer. In the chemical industry, sometimes a big problem can be a dangerous problem. Stakeholders need to understand how the problem and how the solution will impact employees.

The communication strategy for understanding this is twofold. First, asking impacted employees about their personal impact will help to understand their views. Knock-on effects of problems are hard to guess at, so do not guess. Ask. Second, making those effects clear to the wider group of stakeholders. Everyone knows and understands a problem within their circles of influence or circles of work. What is missing is knowing the people impacts beyond those circles. FPS project leaders and team members will need to understand that broader impact to bring understanding to all stakeholders during the communication plan.

For example, an FPS team was working on a serious plant problem. The mechanics and operators working on the issues were working overtime and long hours. By communicating that impact, the analytical folks who were doing the tests to help understand and ultimately resolve the problem realized that they needed to up their game and help solve this as soon as possible. They ramped up to do extra coverage and ran the analytical testing round the clock because they understood what others were having to do while the problem existed.

Communicating about the problem's impact on employees will be a typical need that stakeholders have, and teams need to be prepared. To address this need, talk to people. Instant messaging or e-mails can also be done, but on the average are less effective than simple conversations. One of the FPS team's roles is to communicate impact across and in-between functions. Doing this effectively will help accelerate the project.

8.2.5 Last Thoughts on Needs and Bias Gathering

The process of understanding the stakeholders may feel a little overwhelming. There is a crisis going on! Why would we waste time on this? This kind of push-back is typical of newer project leaders and teams. Urgency and quality feel they are working against each other. Experienced project leaders understand that parallel processing is possible with projects and that this piece is essential to project success. An experienced project leader can get both fast and good stakeholder understanding.

For example, an experienced project leader was called in to lead a crisis FPS project. All hands were on deck. While the team was deep in the process of forming hypotheses, the project leader, with the sponsors and process owners' help, prepared the list of stakeholders. The sponsor then set up a meeting to get the most critical ones in a room all at once. The project leader and a good notetaker accomplished all the bias and needs gathering during that meeting in a little over an hour. They asked the group one at a time "What are your concerns or thoughts about the (project name) project?". The notetaker typed the responses in a mind map with the key stakeholder's information. Once all the folks had their say, the meeting ended. The project leader and notetaker identified the key needs and bias during the next half an hour from the notes. They rejoined the project team with that job done before the project team was done with hypothesis generation. During the subsequent communications, the key stakeholders felt the team did excellent clear communications and met their needs. Great stakeholder understanding was achieved with zero delay to the project through parallel processing and efficient use of time.

> ***Wisdom Tidbit – Taking the time to understand stakeholders does not have to impede progress. It will accelerate progress as communications about the project begin in earnest.***

8.3 Acting on Your Stakeholder Analysis – The Communication Plan

Once the stakeholders are identified, agreed upon, and their needs are understood, the next step is to develop communication plan. The communication plan is just that-a plan to communicate. Fortunately, there are many ways to communicate. Unfortunately, some are better than others in specific situations. This section will show the 5Ws and an H method of constructing communication plans and some examples of when communication has gone bad. Using this technique will help readers communicate better.

As an example of a communication plan going wrong, a project team decided to have an informal web-based review of the project with the sponsor. The team was spread out and this made sense. Because the team could not see the sponsor, they could not understand the facial expressions and body language of the sponsor. The sponsor was very pleased with the team's progress, but the questions being asked made the team second guess their work. The sponsor challenged them on their sampling plan. The sponsor's purpose was to gain understanding. The team's second guessing led to a post-meeting revision of the sampling plan. This caused rework of the sampling plan and ultimately led to project delay. In a subsequent weekly sponsor/project leader meeting, the mistake was discovered. The sponsor was surprised that the sampling plan was changing without a clear technical reason for the change. The project leader sorted out the misunderstanding and stopped the sample plan revision and minimized the delay this communication error caused. The original plan was executed. Not being in the same room was a problem. Maybe if the project leader or any of the team members would have overcome the location issue and sought to understand and asked more questions of the sponsor during the meeting, then the extra wasted work and delay could have been avoided. As this example shows, HOW the communication is done matters as does WHERE the communication takes place.

> ***Wisdom Tidbit – Communication is both verbal and nonverbal. When the nonverbal piece is absent, both parties involved in the discussion must work harder by asking more clarifying questions.***

As another example of a communication plan failure, a key stakeholder meeting was scheduled to be done monthly. The project leader knew that the key stakeholders had FULL calendars. To try and fit the reviews into those full schedules, the project leader tried to act proactively and get the reviews on the calendars so that the team had the times reserved. The proactive approach had served the project leader well on past projects, so it was a habit. The difficulty arose as the project leader and sponsor pulled together the burn rate, the business case for doing the work. It was a VERY large burn rate because there were significant sales and margin from those sales at risk. It was a high-profile problem. Because the project leader had scheduled the meetings before understanding the burn rate, problems occurred. The key stakeholders paid attention to the meeting timing and many of them became very worried that the project leader and team did not appropriately understand the urgency of the situation. A simple proactive act by the project leader quickly became a crisis for the project. Subsequent discussions fixed the issue. As this example shows, the WHEN of communication matters.

> ***Wisdom Tidbit – Communication timing is a balancing act. Too late, there will be uninformed guesses by stakeholders leading to trouble and miscommunication. Too early, the communication can fail due to insufficient information for a productive communication.***

These communication failures show how there are many aspects of a communication plan that need to be done right. The main tool for getting it right is the 5Ws and an H technique.

8.3.1 5Ws and an H. Who, What, Why, Where, When, and How?

Previously, we discussed how to find WHO the key stakeholders are on the project. We also found out their needs and biases to understand WHY they are interested. We will get more specific or granular on the WHY question by generating a purpose and desired result for each specific communication. We will also cover the WHAT, WHERE, WHEN, and HOW parts of a communication plan. These together these make up the 5Ws and an H.

8.3.2 WHY – Purpose and Desired Result

In all forms of communication, starting out with understanding the purpose of the communication will help clarify and make communication effective.

The purpose of a communication is really "Why am I bothering to call, or meet, or e-mail, etc." It is the essential reason for doing the communication. Remember you are taking somebody's time, if you do not know why you are doing that, then you are a time waster.

Typical reasons for communicating are to:

1) Gather information
2) Answer questions
3) Solicit input
4) Decide

5) Brainstorm ideas
6) Plan activity or strategy
7) Train or coach
8) Solve a problem
9) Kill a hypothesis
10) Gain agreement

A good practice is to write down or think through your purpose for communicating before doing the communication. Sometimes, there are multiple purposes but writing them down helps clarify and clear the mind. Telling the person or persons you are communicating with will help them help you be successful in that purpose.

> ***Wisdom Tidbit – If you cannot write down what your purpose is for communicating, then odds are it will wander around and get lost.***

The desired result is just how you will know that the purpose has been achieved or not achieved. For example, if your purpose was to understand when sample results will be available, your desired result is a date for completion. If you do not understand that date, the purpose was not achieved. If your purpose was to reach a decision to support or not to support the implementation of the solution your team recommended, then the desired result is stakeholders reaching a decision of supporting or not.

Tell the folks with either written or verbal communication your purpose and desired result and your communication will be more successful. For example, compare these two meeting notification possibilities for the same situation.

- **Meeting 1**

Subject – Catalyst Failure Fundamental Problem – Solving Team meeting
8:00–9:00 a.m. on 1 January 2001

- **Meeting 2**

Subject – Catalyst Failure Hypothesis Generation
8:00–9:00 a.m. on 1 January 2001

Purpose:
Develop hypothesis on why 1234 process catalyst is prematurely failing.

Desired result:
Generate at least five hypotheses with researched references on why the catalyst is failing so soon.

Imagine you just received the first notification. Do you know why you are going to this meeting? Do you know what to do to prepare for it? Contrast that to the second meeting for the same thing. Which meeting is likely going to be more productive? Purpose and desired result help focus and drive results.

Agendas for effective meetings can be thought of as the steps planned to achieve the desired result. For example, the following could be the full meeting notification for the previous example.

Meeting
Subject – Catalyst Failure Hypothesis Generation
8:00–9:00 a.m. on 1 January 2001

Purpose:
Develop hypothesis on why 1234 process catalyst is prematurely failing.

Desired result:
Generate least five hypotheses with researched references on why the catalyst is failing so soon.

Agenda:
- Review history and background of the problem (Joe)
- Share the main chemical reactions thought to be happening (Mary)
- Research and generate hypotheses (All)
- Review and select most likely hypothesis to develop (All)
- Document hypotheses (All)
- Did we meet the desired result? (All)

The agenda in this example shows a sort of roadmap of how the purpose is going to be achieved and the desired result accomplished. Thinking through planning meetings with purpose and desired result will make more productive meetings.

Purpose and desired result go beyond just meetings. For example, phone calls benefit from thinking through purpose and desired result. Compare these two conversations on the same topic.

- Hi Joe, I was wondering how things are going on our catalyst testing?
- Hi Joe, the purpose of my call is to get an estimated completion date for the catalyst testing. The team needs it to schedule the next plant trial.

The first one just feels like fishing for information. Clarifying purpose on the second one helps everyone get what they need to do their jobs. Without purpose, meandering prevails.

Written communication benefits as well with purpose and desired result thinking. Compare these two opening lines from two e-mails on the same topic.

- The FPS team has found nitrogen purged catalyst to be more effective in maintaining catalyst life. (remaining e-mail text)
- The purpose of this communication is to update our catalyst consultant network on our recent catalyst trials aimed at extending catalyst life. We are seeking feedback from you experts on potential next step trials to further extend catalyst life.

The FPS team has found nitrogen purged catalyst to be more effective in maintaining catalyst life. (remaining e-mail text)

People need to understand why they are reading something. It changes how the text is read and the thoughts that are in the heads of the folks reading it. With a clear purpose and desired result stated in the second case, the experts are thinking about ways to extend life. Clearly stating what is needed has a higher chance of meeting that need.

Purpose and desired result are very helpful in answering the specific WHY question of each communication. This specific WHY is used in conjunction with the bigger high-level WHY questions of the stakeholder needs.

8.3.3 WHAT

The WHAT part of communication can be practically anything, even a text about WHAT we are having for dinner. The what is the content of communication. As project leaders and team members of an FPS project, we will have many WHATs. There is however a pattern and like many things those patterns follow the Pareto (80/20) rule. A very few WHATs will constitute most communication plans. The remaining many WHATs, will be infrequent and difficult to predict; and therefore, should not be in the communication plan. They will be handled and thought through as one-offs as they occur. The vital few WHATs within a typical FPS project are:

- In-progress project work
- Team working meetings
- Team working meeting minutes
- Standardized stakeholder project updates
- Formal stakeholder project milestone review

This list is sorted by approximate frequency of occurrence and will vary by each project. Each project will have other specific communications needs unique to the project that are not in the list. For example, maybe there is a regular sales team meeting that occurs every two weeks. An FPS project closely tied to the number one selling product may need to update that group of stakeholders once a month.

That meeting would get put on the communication plan and be geared toward the needs of that audience. Communication plans are not a formula, but they do follow patterns. Fit your communication plan to the project needs, not the other way around.

Now, let us examine the vital few WHATs that typically occur on projects and the communication tools to use within those WHATs.

8.3.4 WHAT 1: In Progress Project Work

(HOW: e-mails, war room walls, instant messages, phone calls, person to person, or web-based project tools)

Project work will create information traffic between anybody working on the project. In the communication plan, it is useful to pick a way to broadcast critical outcomes of those communications to the project leader and team members and to set up a ground rule of always using that way.

Ground rules are rules used by the team to shape team behavior. Most teams use e-mails on a standard team route list as their ground rule, but newer cloud-based project tools also accomplish this function as can many other means. Standardizing using a ground rule to communicate with the entire team any critical information via e-mail for example, does not mean that only e-mails initiate that communication. For example, a team member could go talk to an analytical expert about sample results that are important to the project; but to follow the ground rule the team member would then summarize the key points of that discussion and send an e-mail to the entire team.

The mode of how that information transfer happens will vary, so e-mails are not always the right answer. Any mode can be the right answer. For example, an FPS team that was working on a high burn rate project set up a war room. War rooms are a designated space, usually a conference room, that is then used exclusively by the project team for the duration of the project. Key learnings such as hypothesis being killed were put on large pieces of paper that were stuck to the walls of the war room. Team members that were in and out of the war room as they completed project work would read other team member updates posted on the walls. Team members breaking that ground rule who just used verbal updates did not keep all team members or the project leader in the information loop. When this occurred, the project leader had to step in and enforce the ground rule to keep free communication flowing.

> ***Wisdom Tidbit – If communications are given by verbal modes, always follow-up with written communication. What was said and what was heard are often different things. Written confirmation prevents this communication failure mode.***

8.3.5 WHAT 2: Team Working Meetings

(HOW: person to person, telecom, or web-based meeting)

WORKING team meetings are critical to FPS project success. These usually are aimed at the main project steps such as generating hypotheses. They are called for when it makes sense for the team to work together on some aspect of the project. For example, a team meeting might be called to develop an action plan to kill three of the Must-do-now hypotheses. Remember the high-level purpose of a team working meeting before calling the meeting.

Purpose: Complete necessary project work to move toward the project goal that depends on all or most of the team interacting to be successful.

Project leaders should be mindful of deciding when the entire team should be involved or if a smaller subset should contribute. It is a trade-off decision. In general, smaller groups act swifter on tasks and can get more things done due to less discussion and more focused expertise; but trade-off is creativity and information transfer. Larger groups tend to enhance creativity and transfer information faster, but are slower in completing tasks and are often less focused. Team meetings are overused. The best project leaders clearly understand the purpose of the meeting and bring the right number of people together to deliver that desired outcome. Despite the cautions around team working meetings, they are the most common and usually the best way to keep the team moving forward and delivering on a successful project. Use them wisely.

8.3.6 WHAT 3: Team Working Meeting Minutes

(HOW: e-mail or web-based project tools)

Writing meeting minutes can be a complete waste of time or the most valuable thing that can be done for a project. The extremes in the value of meeting minutes can and do occur leading to doubt on value. It is simply hard to tell the value upfront. In addition, most folks are reluctant to type in minutes due to the influx of other work and other distractions. Sometimes, this task can take longer than the meeting itself. Because of this, often teams will get lazy and will fail to publish minutes. This then becomes a root cause of duplicate communications, dropped tasks, misunderstandings, and a whole host of gremlins that haunt a project. Thinking through the two main purposes of meeting minutes helps engage the team to doing them to the appropriate level of detail.

Purpose:

- Communicate the IMPORTANT work that occurred during the meeting to team members and select stakeholders that were not at the meeting.
- Document new task assignments generated during the meeting.

To achieve this purpose does NOT require perfection. Most of the procrastination and foot dragging on writing up minutes has to do with striving for perfection. Perfection is the enemy of good. High performing teams will have somebody who is the least involved in the current meeting typing the minutes in a mind map or in a document file as the meeting is happening. The meeting main points can then be quickly checked prior to ending the meeting and the minutes can be issued at the end of the meeting. This is the best case. An okay compromise is to get the minutes typed up during the 10 minutes after the meeting. Strive to meet the purpose of meeting minutes, do the minimum to meet that need. Strive to capture only what matters.

> ***Wisdom Tidbit: Ideally, the meeting minutes should be taken real time in the meeting and projected on the wall or shared in virtual meetings. Taking real time minutes is hard work. It can inhibit the participation of the notetaker. The notetaker needs to not worry about spelling, etc. The goal is to remember the key things. Minimal cleanup after the meeting is sufficient. Asking the question: Is this the way you want this remembered is an enabling question for building trust. Change the notes as needed by the team. Real-time note-taking is particularly helpful if there are language challenges. Notes can be read and reread. The verbal stream only goes by once. Remember to share the note-taking between members comfortable with this challenging task.***

Make the title simple, include the date of the meeting, and in the title call out if there are new action items. For example:

Minutes for 1 January 2001 Catalyst Failure Hypothesis Generation meeting. Please read action items.

Copy the purpose, desired result, and agenda into the team minutes and capture the notes below the agenda items to add context to the notes. Knowing the context helps the reader understand the minutes.

Put new action items on the top of the minutes, call out who is responsible, and a planned due date. If any of those are missing, put who and when you think should have the task done, but ask the readers to send you the corrections. Bold or color the peoples name assigned to tasks, so that the task's responsible person jumps out to the reader. People who attended the meeting read tasks but are unlikely to read the remaining content. They were at the meeting, why read the minutes beyond the task list? The task list can be tracked by many means, but even if the team is using a shared spreadsheet to track tasks for example, still put the new tasks in the meeting minutes and then just copy paste the action items into the spreadsheet. Tasks posted in the minutes cause readers to put those tasks into their personal task tracking systems which seem to be as varied as humans are.

Here is an example of a simple spreadsheet-based action tracker:

Meeting	Action item	Responsible party	Date due	Date completed	Status

This rolling action tracker can then be used in future meetings and be cut and pasted into the meeting minutes.

> ***Wisdom Tidbit – If you fail to track actions, you are asking to fail in getting actions done.***

Attach any presentations, data, charts, or whatever content was shared during meetings. Screen captures can be used during web-based meetings, but let the person sharing know that screen capturing is happening.

Document any decisions made. Leaders often forget they decided something. Documenting the decision saves rework later. With consensus-based decisions, document the names of who supported the decision and any detractors.

Here is an example of how simple team minutes can be done based on the previous meeting agenda example:

Subject – Minutes for 1 January 2001 Catalyst Failure Hypothesis Generation meeting. **Please read action items.**

Action items:

Mary – Send team references for chemical reactions reviewed (Due 5 January 2001)
Joe – Schedule a hypothesis review with our analytical team (Due 9 January 2001)
Team – Document and circulate to the rest of the team any additional hypothesis concepts you may think about after the meeting (Due before next meeting)

Meeting minutes:
Note: Original agenda in ***bold italics.***

Purpose:
Develop hypotheses on why 1234 process catalyst is prematurely failing.

Desired result:
Generate at least five hypotheses with researched references on why the catalyst is failing so soon.

Agenda:

- ***Review history and background of the problem (Joe)***

See attached presentation.

- ***Share the main chemical reactions thought to be happening (Mary)***

See attached presentation.

- ***Research and generate hypothesis (All)***

Details are in the hypothesis onion spreadsheet. Click here for link to spreadsheet.

- ***Review and select most likely hypotheses to develop (All)***

Team selected five hypotheses to initially pursue. Air poisoning, catalyst crushing, catalyst polymer coating, bed bypassing, and active species leaching.

- ***Document hypotheses (All)***

Click here for link to document in file share.

- ***Did we meet the desired result? (All)***

The team agreed it was a productive meeting. The team decided that they may have some more hypotheses pop in their heads later and should document those as well.

This example is far from perfect, but does it meet the purpose of meeting minutes? Overprocessing is considered a waste in lean production systems. Overprocessing of meeting minutes is a waste as well. Minutes for meetings do create value, so best practice is to make them happen with the least amount of work possible.

8.3.7 WHAT 4: Standardized Stakeholder Project Updates

(HOW: e-mails, war room walls, instant messages, phone calls, person to person, or web-based project tools)

High burn rate projects especially need standardized stakeholder updates. Information control becomes a major problem when there is a crisis. For example, on an FPS crisis project, the team was trying to kill the leading hypothesis. They had devised a DQO that was a lab experiment. The team had multiple runs planned because the team knew that this was the first time to ever do this experiment and there was a significant risk of a false positive. The first test failed to kill the hypothesis. The team got excited about the results but knew to wait for the remaining lab runs. Unfortunately, word got out through back channels and senior management got word. Next thing the team knew, each team member and the project leader were getting questions from all sorts of folks from high up the corporate ladder to the operators running the process. Communication was chaotic and could have caused wasted time. Fortunately, the project leader had set up a standardized project update e-mail. The project leader added the details around the experiment and clearly communicated that the results needed to be repeated prior to being conclusive. The team and sponsor then used this standard communication whenever folks outside the team asked about what was going on. The misinformation was quickly set right. The subsequent experiments ultimately killed the hypothesis, so the false positive and the stir it caused were quickly forgotten. This event could have set the project back significantly; instead, standardized project update saved the day.

In some form at some selected cadence, stakeholders will need to hear from the project team. These updates are the heartbeat of the communication plan. They will include the "vital" few pieces of information that stakeholders need to understand the current project status. Content will vary depending on the project stage and activities, but in general will have:

- Key background to the problem
- Reminder of what is success (project goal)
- Current impact to X (Customers, employees, business, etc.)
- Progress on killing hypotheses
- Progress on any quick wins or impact mitigation efforts
- Progress on any solution pilots or solutions
- Next steps plans
- Contact method for if stakeholders need more information

These can be done many ways, for example on paper on a war room wall or a standardized e-mail or a phone call verbally covering the key points. When making the communication plan, decide which way best fits the situation (see the HOW question). Standardize the format and most of the content. Then, send the update at the selected cadence (see the WHEN question). This will save time and keep the communication flowing.

8.3.8 WHAT 5: Formal Stakeholder Project Milestone Review

(HOW: person to person, telecom, or web-based meeting)

Periodically, your key stakeholders will need to make decisions about the project. For example, they may need to decide to fund the more expensive preferred solution or maybe a second choice less-expensive solution. There will be decisions that are outside the scope of the team. These decisions typically occur at project milestones such as when the hypotheses have narrowed down to one that will not die, and the preponderance of the evidence supports it. The team will then need support for solutions the team generates. These important milestones need a formal stakeholder review. The need for a formal review can just naturally occur at critical milestones or it could suddenly occur with just one piece of new information. For example, an FPS project team was working on a project with a critical customer. The team had gotten far enough to understand the science around the problem but realized that there were quick win fixes at the customer's process that could solve the problem fast. The in-house solutions looked like they would take a significant amount of time. Because the customer relationship situation was complex and at risk, the team realized it was out of their scope to decide if they should approach the customer; that was a decision for the key stakeholders to make. A formal review was called.

Calling them formal is intentional. Decisions need to be documented and confirmed. They need authority behind them. Otherwise, it can be like they never happened. This is a weird but true piece of running high burn rate projects. The blame game can be played at very high stakes. Decisions are often full of knowns, but they are also full of unknowns. In the previous example, nobody knew what the customers' response would be. The relationship was at risk. Asking the customer to change their process could be very bad for business or it could be the best thing to happen because maybe solving the problem anyway possible was fine by the customer. The decision had an unknown. The outcome could go either way. Imagine if asking the customer to solve the problem went very badly and the customer chose to take their business elsewhere. As a project leader, would you prefer meeting minutes from the stakeholder review with a well-documented decision supporting going to the customer that included who supported that decisions? Or would you prefer a verbal decision from a single key stakeholder with no one else but you in the room? When the stakes are high, clear communication is critical.

The formal stakeholder reviews can be either technical or business reviews. The technical reviews are essentially peer reviewing the science being done. They are there to get folks outside of the team to try and poke holes in the science and ensure that high-quality conclusions are being reached. Business reviews are aimed at resourcing and business decisions. They are there to get key decision makers up to speed on the project and then to do their jobs on making critical decisions needed by project teams.

> ***Wisdom Tidbit – One rule always applies "Never surprise stakeholders," new information needs to be digested, give them a heads up prior to any stakeholder review.***

Some projects fly through all the stages of the FPS project workflow and reach the end without a formal stakeholder review. Even these fast-paced projects need a formal stakeholder review. For example, FPS team formed and did a blitz on their problem. The existing data from the timeline and troubleshooting data collected earlier by the folks working the problem prior to the FPS team formation was enough to kill all but one hypothesis. The hypothesis onion did its job. The remaining hypothesis was a root cause that could be fixed by a procedural change within the control of the team members. The change worked and the problem was solved in a little over a day. The project was essentially finished. Documentation needed to be done and some control plan work, but there were no decisions to put in front of the stakeholders.

The project leader set up a key stakeholder review anyway. Some would question this move. Why bother with a stakeholder review? The project leader knew there were decisions to be made. The stakeholders needed to see for themselves that the project was successful. They needed to decide if the project was done or was there follow-up work to solidify the science. They also needed to support the change for the long term and buy into the control plan to the project. Every project needs at least one formal stakeholder review.

The content for stakeholder reviews is like stakeholder updates or storyboards and will follow that same pattern. The key piece that is different is that stakeholder reviews will also include any decisions that the team needs stakeholders to make. They will provide enough background to make those decisions.

Formal stakeholder reviews are an essential piece of running an FPS project. Be flexible in how you do the review but be rigid in the decision to do them.

8.3.9 Communication Tools to Use

Within these WHATs, there are some communication tools worth knowing about. These can be used within any of the WHATs as needed. For example, a storyboard can be used within a formal stakeholder review or it could be used within a standard stakeholder update as an attachment to the update e-mail. Here are some effective communication tools to consider using:

- Storyboard
- Elevator speech
- Standardized e-mail project updates

8.3.10 FPS Storyboard

Storyboards have an origin from animation moviemaking. Because there were many cartoonists on a project, there needed to be a way to link all the work of drawing together. The animators took representative drawings that tied the scenes from the movie together and told the story of the film.

Six Sigma practitioners took that concept and applied it to projects. With just a few slides or pieces of paper on the wall, the story of the project is told. In FPS, we often will extend and modify this concept for FPS projects. The genius of storyboards is that with a few conceptual charts, pictures, or words, the story of very complex concepts like troubleshooting a multivariate complex problem can be told in 5–10 slides, or posters, or easel papers.

FPS story boards are like Six Sigma storyboards, but the methodology story being told will follow the FPS methodology instead of DMAIC. Because of the complexity of FPS problems, there are often multiple stories being told and therefore there can be as many sub storyboards about each hypothesis as well as the overall project storyboard. The overall project storyboard will typically flow like the typical standardized updates and have panels or slides that follow that same story. The overall project storyboard can be used instead of a written update or can be used in conjunction with a written update. Here are the typical components of the project story.

- Key background to the problem
- Reminder of what is success (project goal)
- Current impact to X (Customers, employees, business, etc.)
- Progress on killing hypothesis (hypothesis onion)
- Progress on any quick wins or impact mitigation efforts
- Progress on any solution pilots or solutions
- Next step plans
- Contact method for if stakeholders need more information

The critical piece that sets storyboards apart from traditional written update are the concepts and pictures. So, for example if the current story was about looking for signs of stress corrosion cracking, the words could be something like:

The team discovered stress corrosion cracking in the microscopy analytical results of the valve stem.

The storyboard picture would say the same thing, but hopefully in a more visual impactful way (see Figure 8.2).

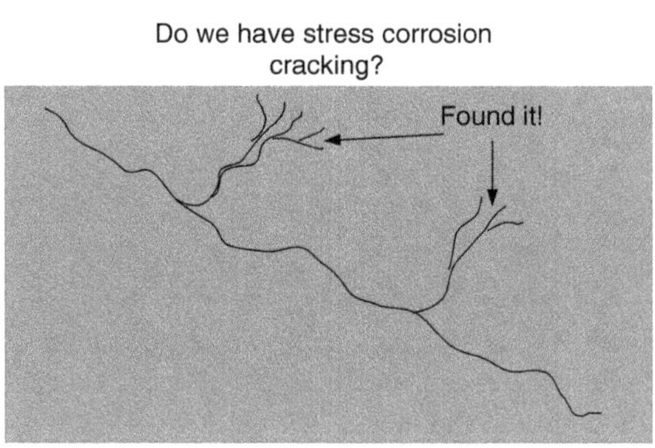

Figure 8.2 Storyboard picture.

Storyboards are not constrained to presentation software. They are often posted in the war room for crisis projects. These storyboards are typically handwritten and updated throughout the troubleshooting effort as new hypothesis are killed.

There is a subset of the storyboard concept that is unique to FPS projects. These are the substories of the hypothesis. Instead of telling the story of the overall project, they tell the story of the quest to kill a hypothesis. Because projects can have dozens of hypotheses, there can be mini storyboards for each one. These mini storyboards are typically kept in individual presentations or an appendix of the main project storyboard, or as tabs on the hypothesis onion spreadsheet. In the case of tabs on a spreadsheet, the spreadsheet itself is the storyboard. Paging down through the spreadsheet will tell the story of killing or failing to kill that hypothesis. These mini stories will generally have the following content:

- **Title of hypothesis** – It is helpful to use letter number designation for filing structure, hypothesis onion reference, and file names. A common structure is H#, so H1, H2, H3,. . .etc. H is for hypothesis and the 1 is just a reference number. Sub hypotheses will continue as letters. So, there can be an H2a and a H2b hypothesis. Sub hypotheses are related hypotheses. For example, a high-level hypothesis could be "Metal contamination is inhibiting the reaction" and sub hypothesis could be "Iron contamination or Nickel contamination." For the Iron sub hypothesis contamination title, it could be called "H1a – Iron contamination is inhibiting the reaction." For the Nickel sub hypothesis contamination title, it would be called "H1b – Nickel contamination is inhibiting the reaction."
- **Order of magnitude math/science** – For any hypothesis to be plausible, it needs to be in the right ball park. Order of magnitude thinking is about that ballpark sizing. For example, would a 1 part per million iron contamination cause the problem? Or does it have to be tens or hundreds of parts per million? Researching the science can answer this kind of question. So, if researching the science shows problems in the 10 part per million level, for this hypothesis to be plausible, it needs to be in that range or above. If a mass balance around a raw material contaminant calculates that the most iron contamination could be is 1 part per billion, this hypothesis is then dead. Order of magnitude math tests can be quick wins of killing hypotheses and should be done early. As another example, overcooling was a concern for one FPS team. If the reactor got too cold, it would result in unwanted reactions. There was a cold recycle stream that the team formed a hypothesis around that the recycle stream was resulting in overcooling causing side reactions. Order of magnitude math was done in the form of an energy balance. The science had shown 10 °C cooling could cause the problem. The energy balance showed that at most the cold recycle stream could result in 0.1 °C cooling. Not in the right ballpark, so the team could kill that

hypothesis. It would be impossible for the recycle stream to cause overcooling. This first pass first principles math will be a part of your story.

- **Full hypothesis statement** – This is the logic of the hypothesis. It is the full clear statement of the hypothesis that is testable in such a way that it can be killed. A well-constructed hypothesis is essential. For example, we had shared a hypothesis title earlier – H1a-Iron contamination is inhibiting the reaction. This is not a complete hypothesis. It is a title. The hypothesis that is testable and can be killed may be something more like this:

  At least 10 parts per million iron contamination is inhibiting the reaction in 1234 reactor causing the 10% or more reduction in conversion currently being experienced.

  In this case, some pre-existing science would have to be in place to pick the testable level of iron that would cause the problem. See the section on hypothesis forming for writing good hypotheses for more details on a well-constructed hypothesis.

- **References** – Hypotheses for FPS are based on first principles and pre-existing science. Internal or external research reports, articles, book references, formula references, etc., all go here.
- **Plan to kill hypothesis** – This is where the team is planning to pursue a fact to kill the well-constructed hypothesis. A DQO is usually here. The "What is success, what is necessary for success, and what is rate-limiting step" get answered here. For example, in the hypothesis H1a-Iron contamination is inhibiting the reaction, the DQO could be:
  - ***What is success?*** Obtain a sample from the reactor system while the reaction is being inhibited and test it and show single digit or less ppm iron contamination to kill this hypothesis.
  - ***What is necessary for success?*** A good sample plan to collect a representative sample from the reactor.
  - ***What is rate limiting?*** Analytical expertise to run the sample.
  - These DQO questions will form the basis for the fact-seeking plan to kill the hypothesis.
- **Most compelling supporting data/facts/research** – These are all the most important green/supporting crossections in the hypothesis onion around a specific hypothesis. In the storyboard, this can be graphs, pictures, test result, models, research references, etc. This is a hypothesis-specific set of facts supporting that hypothesis.
- **Compelling data/facts/research that kill the hypothesis** – This is the red crosssection in the hypothesis onion. That which killed the hypothesis go in this part of the mini storyboard.

The mini storyboards around each hypothesis are aimed at the technical audience. They are very helpful in getting peer-reviewed science. They are also helpful to manage the complexity of tough problems. The team is typically working four to six hypotheses at any given time and may have dozens of hypotheses ultimately. It is easy to lose track of what is going on and what has been done. Mini storyboards for hypotheses do the job of tracking.

Also, sub teams may be deeply involved in one specific hypothesis and while other team members are working on others. How do you communicate the complex testing and conclusions between those involved with different hypotheses? These mini storyboards fill that communication gap.

They are also very handy as outside technical experts interject their knowledge and thoughts to the team. New hypotheses can be documented if the technical experts have new ideas. Questions of progress on existing hypotheses can also be quickly communicated. These hypothesis storyboards are very effective at complex technical communication on each hypothesis.

8.3.11 Final Thoughts on Storyboards

Storyboard value must be experienced to be appreciated. Be brave and try the storyboards both for the project and the hypotheses to experience the value they bring. Storyboards help teams speak with one clear understandable voice. They control information and drive broad understanding of difficult science. People are programmed to listen to stories. Tell the stories on your FPS project.

8.3.12 Standardized E-mail Project Updates

These are the written form of a storyboard. They have the same typical content as any of the standard project update types:

- Key background to the problem
- Reminder of what is success (project goal)
- Current impact to X (Customers, employees, business, etc.)
- Progress on killing hypotheses
- Progress on any quick wins or impact mitigation efforts
- Progress on any solution pilots or solutions
- Next step plans
- Contact method for if stakeholders need more information

Covering the same information in the written update as the storyboard could just be a short paragraph and then the project storyboard attached to the e-mail. Or they could be the typed form of a storyboard. A typed form of a storyboard should be typical of the regular communication from the project team to the key stakeholders. Here are some tips to make them effective:

- **Put the vital few key pieces of critical new information at the top of the note:** Think what are the one or two (or maybe three) things the stakeholders need to know now. Then, put details below a line under those vital few. We recommend telling the stakeholders that the details are below the line. Busy people often will only read the first couple of sentences in an e-mail. Putting the critical information at the top leverages it. Details are helpful to folks that are focused on the details. So, if you type something like "Read on if you need to know the details" or "Details on project progress below the line," then these people get what they want as well. Big picture readers or detail readers both like this style.
- **One-page updates:** Strive to keep the content to under a page. This can be difficult to do, but in stakeholder communication less IS more.
- **Standardize your format:** Title the sections and essentially create a form that you follow for all updates. This will save time when writing the updates but will also save the readers' time. The readers will know where to look to find the specific information they need.
- **Use route lists developed from communication plan:** Anytime you are regularly sending something to a group of people, the time it takes to prepare a route list is well worth it because it reduces the work to communicate.
- **Always leave the contact information for questions or comments:** If you tell the folks to "Contact John Doe project leader with any questions or comments by replying to this e-mail," then you funnel the communication in that direction. If you tell the folks to "Call John Doe with any questions or comments in the project War room at 123-4567," then obviously you funnel communication a completely different direction. Tell the readers your preferred feedback path in the update or else you will get every direction of contact possible.

As in all things, minimize the work in regular e-mail updates. Only do what it takes to deliver the essential information. Standardizing this work minimizes the work. This tool is very useful in getting essential project progress information out to many people.

8.3.13 Elevator Speech

The concept of an elevator speech is that it is the very high-level summary of the project. It is something that is verbal. The big idea is it is meant to only last as long as an elevator ride. Short, sweet, and to the point. For FPS projects, the elevator speech is going to include some background to the problem, the goal, the impact, the resources deployed, and current progress. Just a few sentences. Use this tool if the team is experiencing or expecting to experience many outsider questions and interactions. To create one, have the team come up with the vital few pieces of information they would tell someone during the time of an elevator ride. Have them practice it with each other. It is not intended to be verbatim, it is intended to be the key points in their own words. This is a useful tool for outsider verbal interactions that seem to happen at higher frequency than you might think.

8.3.14 WHERE

In general, there are only a couple of decisions on the where question for any meetings that are part of a communication plan. They are whether is it going to be:

- **Face to Face** – meeting with everyone in the same room.
- **Remote Web or telecom-based systems** – meeting with everyone in their own locations.
- **Mixed** – meeting with most folks in the same room, but some participants in their own locations connected through web or telecom-based systems.

8.3.15 Face to Face

The most expensive in terms of meeting time and resources are usually the face-to-face meetings. They also have the best interaction and tend to make the most progress. Highly technical or high stakes meetings should be done this way. Project leaders and teams whenever possible should push for this type of interaction; especially on high burn rate projects during the early stages and critical stages of the project.

> ***Wisdom Tidbit – Calculating burn rate helps justify the work that needs to be done. For example, a project costing $1 million/week justifies flying in people from all over the globe for a face-to-face session to solve the problem. Project leaders can use burn rates to make big things happen.***

8.3.16 Mixed

Mixed meetings are a way of enabling interactions, but also being flexible to the needs of people. Sometimes, people's personal schedule can be prohibitive to a face-to-face meeting. Sometimes, a person's importance to the project may mean that they are not essential to being in the room, but still need to interact. These cases mean a combination of face-to-face and web-based meetings will be used.

The trick when facilitating a mixed meeting it to remember to involve the people who are not in the room. The barrier of being out of the room can mean the person gets ignored. To avoid this, set up ground rules. A good rule to use is to do an around the real room and around the virtual room feedback system. Ask for input from each member one at a time while going around the real and virtual room. Do this for critical to success decisions. Another way to ensure input in a mixed environment is to recognize who needs to give input. Everyone may have input on why a heat exchanger is failing, but the project leader needs to focus on getting the heat exchanger subject matter expert's input. Often the smartest, top experts are quiet and need focused questions to have a voice.

Mixed is the second-best way to meet. It can decrease costs of the meeting resources substantially, but the reduced interactions need to be a part of the cost benefit thinking.

Crisis challenges can have a timeline that precludes all the key people from getting to the meeting location. It would just take too long for them to arrive. In these cases, the ideas above in the Mixed discussion are particularly critical. This is also a time to have the meeting notes being taken real time. Written notes provide clarity that the audio/video channels may not provide. A camera controllable by the remote people is particularly helpful.

8.3.17 Remote Web or Telecom-based Systems

Remote web or telecom-based systems have come a long way and are approaching face-to-face interactions in value. As of this writing, they are still not like being in the same room. Having said that, sometimes it is best not to be in the same room. These cases are when there are many people, when only simple people interactions are needed, or when the team has worked together for enough time and can overcome the restrictions inherent to virtual meetings.

Communicating with anything greater than a dozen people usually will have the purpose of information transfer-type purpose. Hundred people in a meeting room for all practical purposes are mostly going to do more information absorption than actual interaction. Web meetings can be a great way to communicate information to a large audience in the comfort of their own office. Everyone can see their private screen and with planning, can get good sound. Chat features can enable interactions that do need to happen with large groups.

Not all meetings are doing complex or difficult things. Sometimes, simple agreement or a quick information review is all that is needed. Being in the same room is simply not that important. Simple meetings can be effective virtual meetings.

When teams work together closely, then they can function better and better. At some point they can become a high-performing team. These high-performing teams seem to be less impeded by not being in the same room. They are calling each other for clarification or speaking out during the meeting and just plain function better despite being away from each other. Virtual meetings can work well with high-performing teams.

The bottom line is that when project leaders are making the WHERE decision for a meeting, thinking about how important or complex the meeting purpose is going to be will help in selecting between face to face, mixed, or remote web or telecom-based systems. In high burn rate projects, project leaders should default toward face to face if possible. It sends a signal that the problem is serious and needs attention. The other types work but take work to make work.

Remember to focus on real-time notes that everyone can see. This is hard work. It is worth the effort.

8.3.18 Other WHERE Decisions

Obviously, things like selecting the right sized room with the correct equipment such as screen projection or white board availability is part of the where decision for choosing a meeting location. But a bigger part is about if you are going to meet near the problem or remote to the problem. Either one can be the right choice depending on the situation.

Near the problem enables teams to go see the problem. This is critical to solving problems. Often, when the right process expert walks down the problem, potential root causes are clear to them. Choose near the problem when these interactions are critical.

Remote from the problem enables teams to get away from the urgency and distractions caused by the problem. This can be important during decision making and for times when clear thought is required. Choose remote from the problem when you want to pull people away from the hustle and bustle to do clear thinking.

Remote or near the problem are important conscious WHERE choices.

8.3.19 WHEN

For any form of communication, be it a meeting, a phone call, an instant message, or an e-mail, there will be a question of how frequent the communication needs to occur.

Frequency or Cadence of any communication can be any regular interval that makes sense. Figuring out what regular interval makes sense is a judgment call. High burn rate problems will tend toward shorter cadence. Life of the plant problems that have been around for many years will tend toward longer cadence. To figure out that frequency, a good technique is to ask the stakeholders or a representative group of folks a simple order of magnitude understanding questions.

"How frequently should we provide (insert specific communication)?"

"Is it Hourly?"
"Is it Daily?"
"Is it Weekly?"

"Is it Monthly?"
"Is it Quarterly?"
"Is it upon major milestone completion?"

The specific communication can be whatever is planned. For example, it could be "How frequently should we provide project updates?" or, it could be "How frequently should we have stakeholder project reviews?" or, it could be "How frequently should we have a sponsor/project leader chat?".

After asking these questions, typically a logical time frame between communications will be selected. Sometimes, folks will choose something between these simplified periods. They may say weekly is too often, but monthly is too long. Then, an in-between period such as every two weeks can be negotiated. As these timings are selected, they get documented in the communication plan. In a later section, an example communication plan will be shown to help clarify this.

8.3.20 HOW

The how question is determining the mode of the communication. The mode can be a phone call, face to face, web-based meeting, an e-mail, normal mail, a website update, or an instant message. All the different modes can be the right choice or the wrong choice depending on the situation. To select the right one, it helps to understand two things about your communication: Purpose and Desired result. Once the purpose and desired result for the communication are known, the best mode of communication becomes a much easier choice.

For example:

Purpose:
Let the team know when and where the celebration lunch will be held.
Desired result:
Team attends the lunch.

Quickly, many modes could be eliminated because they depend on somebody writing the details down correctly.

This just leaves a mode choice of an e-mail, normal mail, and a website update to advertise the details. Because the purpose is a celebration, then thinking through what would be the most effective enables the final mode selection. Maybe the project leader would choose normal mail because they wanted to make it like a formal invitation to a big event. Or maybe the team's main website would have the details because it is something everyone sees, and it helps to further recognize the team because those outside the team will also see that the event is going to happen.

Understanding the purpose of the communication ultimately helps select the mode of communication and prevents mismatches. If in the example above, the project leader made a series of phone calls and called everyone invited to the celebration, it could work but could frustrate those on the phone who are frantically writing down the details and hoping they got them right. The people could be thinking, "Why didn't they just send an e-mail meeting invitation?". This is a mismatch because the mode was a poor choice. Mismatches are frustrating.

How often are meetings called when a simple e-mail could achieve the purpose better? How often do e-mails evolve into back and forth notes that are really extended e-mail conversations resulting in wasted time and energy when a short face-to-face meeting could have achieved the purpose in a fraction of the time? Understand your purpose, then you will arm yourself with the ability of selecting an optimal mode and answer the HOW question.

8.3.21 Pulling It All Together in a Communication Plan

With the 5Ws and an H, then the full communication plan can be put together. Communication plans are almost always put into a grid format, so lend themselves to spreadsheets or in a war room situation could be on paper with hand-drawn grids. As explained earlier, the workhorse from a communication plan are route lists created based on the plan. This allows fast and easy people lists for invitations to project-based meeting, e-mails, IM communications, and many of the cloud-based project tools systems. The communication plan itself has the names and contacts of the stakeholders for the project and that is extremely helpful for any phone calls or single-person e-mails. For projects where the project leader and team members all have relationships with the stakeholders, lookup of phone numbers or e-mails is less useful. But in the authors' experience, the most common case is that the project leader is meeting many of the stakeholders and team member for the first time when the project kicks off. Having a list of who is who on a project is extremely valuable. The communication planning process also allows the project leader some piece of mind. Without the communication plan, the project leader sending out communications would need to worry about each communication about are they sending it to the right people. Because most of the standardized communications are covered in communication plans, the standardized work is simplified and will go much quicker.

When constructing communication plans, people are sometimes grouped due to like needs. For example, if three folks from the sales organization look to have identical or nearly identical needs, then the communication plan may just have a column dedicated to Sales (or row depending on your choice of orientation). Often, the folks impacted, but not deemed a "key" stakeholder in communication plan will show up as groups. As an example, on an FPS project, there was one operator on the team, but the other operator on other shifts and in the same shift were not team members. Those operators needed to see the progress going on, but for the project were not ranked as "key" stakeholders. They were grouped into an Operators column on the communication plan. Compare that to the "key" stakeholders on the communication plan where everyone had a column. Grouping will help developing the plan go faster. Over grouping is a problem if unlike needs are grouped together. Like many things, grouping stakeholders is a balancing act.

In Table 8.1, there is an example communication plan. The example plan will only contain four stakeholders for illustration purposes, but obviously an actual communication plan would contain many stakeholders. The four stakeholders in this example are a business leader, a group of technical chemistry experts supporting the project, the sponsor, and the project team. The example project is a high burn rate crisis causing the company heavy losses and will be run in a war room dedicated to the team solving the crisis.

8.3.22 Final Thoughts on Communication Planning

Poor communication is a root cause within projects that will manifest itself as outcomes of tension, frustration, in-fighting, false starts, slow progress, missteps, and many other value-eroding events. Thinking through a living communication plan and executing on that plan will help maintain the focus on the problem itself instead of people issues. The time it takes to create a good communication plan is made up in the time it saves by faster project results delivery. There are patterns that make communication plans similar, so leveraging those helps minimize the work it takes to execute a plan. Having patterns is different than a recipe to follow. Communication plans need to be tailored to the problem and the people impacted by the problem. Practice will make project leaders and team members more effective in communicating. This improved effectiveness hinges on understanding the purpose and desired result for each communication. Hopefully, the reader will try using communication plans for their projects and will gain deeper understanding of how much better results flow with excellent communication.

8.4 Managing Difficult Stakeholders

There is a proverb that a bad apple will spoil the bunch. We wish we could tell the reader that everyone of your stakeholders will be respectful of other people, support what is best for the company, employees, and customers, will make logical fact-based decisions, will be good listeners, will speak their mind, will expect the scientific method to be used, and will value healthy opposing view discussions. Unfortunately, that world rarely exists. Because there are many stakeholders with

Table 8.1 Example communication plan.

Details				Stakeholders (WHO)			
Name				**John Doe**	**Jane Smith**	**Team**	**Chemists**
Role in organization				VP of the business	Production leader for process	Engineers and operators solving problem	Chemists from R&D
Role in project				Key decision maker for project	Project sponsor	Team members	Technical support
E-mail				j.doe@co.com	j.smith@co.com	Team route list	Technical route list
Location				Texas site	Louisiana site	Louisiana site	UK site
Current bias				Not sure right people are involved	Any help possible is needed now!	Do not need outside help	Too busy for this
Main needs (WHY)				$, timing, customer impact, critical path	Timing, resources	Next steps direction, work scope, technical consulting	Relative priority of project, timing, resource estimates
WHAT	**HOW**	**WHERE**	**WHEN**				
Blitz meeting	F2F	Site	Daily	X	X	√	As needed
Blitz meeting minutes		E-mail	1/day	X	√	√	√
In process work	Post	War room	ASAP	X	X	√	X
Std. update		E-mail	1/day	√	√	√	√

(Continued)

Table 8.1 (Continued)

Details				Stakeholders (WHO)			
Name				John Doe	Jane Smith	Team	Chemists
Key stakeholder review[a]	F2F	War room	1/wk	√	√	√	X
Technical stakeholder review[b]	F2F	War room	1/wk	X	X	√	√
Review minutes		E-mail	1/wk	√	√	√	√
Final technical report		E-mail	When done	X	√	√	√

[a] Project storyboard presentation to be used in review.
[b] Mini hypothesis storyboards to be used in review.

competing agendas and it is a human system, odds are that at least one of the stakeholders for your project will be a bad apple. How to handle those bad apples, prevent them from spoiling other apples, and deliver on project goals is an important skill for project leaders. This section is about stakeholder conflict and attempting the win-win, but also if needed, how to achieve a win for the project and a lose for a bad apple.

Here is a bad apple example story, to understand their destructive nature for a project.

A top-level business manager was very frustrated with a life of the plant problem. Repeated efforts to fix it resulted in no progress and the costs continued to mount as the years passed. The manager attended an FPS presentation from another business reviewing a difficult problem that was resolved and grew hopeful that the problem could be attacked in a new way. The top-level manager told his direct report who was a middle-level manager in charge of the production unit with the long-standing problem that he had to bring in an FPS project manager and fix this problem once and for all. It was a direct order.

The middle-level manager, we will call him Bad D. Apple, had given up on the problem long ago and was frustrated to have to deploy resources on this. He decided to minimize resources, let the project fail, check the box that the task was done, and be able say "The best tried, failed, now we can let this problem go forever." Obviously, this bad apple perspective was never stated or talked about with anyone. It was a hidden agenda of Bad D. Apple. The project leader worked on the problem as normal, but quickly began to understand that there were sponsorship issues. Resources were being redirected, expert team members were replaced with new hires, and stakeholder reviews were not attended by Bad D. Apple who was supposedly sponsoring the effort. Project progress was slow. The project leader had a conflict that was jeopardizing the project or more specifically sabotaging the project.

The project leader scheduled a meeting with Bad D. Apple with a purpose of deciding to cancel the project or to decide to do the project by resourcing it. The project leader presented the facts around resourcing and proposed options. During the meeting, Bad D. Apple privately explained that the project was never going to succeed. The team just needed to endorse that conclusion and move on. Why should Bad D. Apple the project sponsor waste resources on the project? The project leader reviewed the meeting purpose and stated that Bad D. Apple obviously wants to cancel the project, so that is the decision to be documented and will be published per the communication plan. Bad D. Apple fearing his boss's response to a canceled project due to lack of resourcing backed down and agreed to resourcing a set number of hours that was negotiated between him and the project leader from the key people needed. The team went on to solve the problem and Bad D. Apple became a huge proponent of FPS methodology. Bad changed their name to Good D. Apple.

Conflict from difficult stakeholders is just like any other risk to be managed on a project. Mitigation efforts start first. In the case of stakeholders, the mitigation effort is the communication plan. Understanding their bias, needs, and then communicating to them with these in mind will minimize the impact. There will be fewer bad apples because of transparent and effective communication channels. If that initial mitigation fails, follow good people interaction principles to confront the bad apple. These are about seeking Win-Win solutions.

- **Confront only things worth pursuing** – Fixing every people problem is out of scope for project managers, instead only focus on behaviors that are jeopardizing the project.
- **Confront early** – Waiting for bad apples to go away is a terrible strategy. Time is money on projects and damage control on bad apples will be smaller if addressed earlier.
- **Confront privately** – Respect the person. Public confrontation is a last resort. Your purpose should never be to attack the person.
- **Use facts** – Facts aim at behaviors, not people; use them. If you cannot list out the facts on the reason that the person's behavior is jeopardizing the project, it may be that you only have a personality conflict. A personality conflict is not a bad apple issue. Those are handled differently and involve learning about each other's personality and finding ways to work better together.
- **Seek agreement that there is a conflict** – Until the key people involved in the conflict can agree on what the conflict is, there can be very little progress.
- **Seek Win-Win solutions to the conflict** – Often, it is best to ask the Bad Apple how best to resolve the conflict. It is surprising how much people want to get out of conflicts and handing them a rescue rope is often all it takes to find solutions that serve the best interests of the project and the people working on the project. What you believe will solve the conflict is often the same as what the Bad Apple believes. Communication will surface this.
- **Listen more that speaking** – Most conflicts are communication issues. Keeping your mouth shut while the issue is discussed and actively listening to the Bad Apple's perspective is critical to moving forward.

90% of bad apple stakeholder issues will be resolved with Win-Win solutions developed using these principles. The remaining 10% are the Win-Lose or sometimes Lose-Win situations. Here are some principles in handling those uglier situations.

- **Bad apples can sometimes be pulled out of the bunch** – The most common root cause of a bad apple on your team or on your extended resources working on the problem is often quick-win fixable. Work with your sponsor to adjust the team or extended team membership. Get the bad apple out of the equation. When doing this action, it is critical to respect the person. The sponsor should let folks know there is a team or extended team membership change for a good reason. Good reasons are typically other projects the Bad Apple should focus on or other workload issues.

- **Walking away from the project is often an option** – Project leaders and team members have more power than they think. FPS project leaders are sometimes tossed into extremely hostile territory where the bad apple population is 100%. In these situations, the root cause is usually that the local team feels that they will beat the problem soon. They just need a little more time. They perceive the FPS effort to be an outsider jumping in and taking the credit. An effective move for an FPS project leader in this situation is to set up a meeting between the local folks working on the problem and the sponsor with the purpose of negotiating how much time the local team needs to solve the problem. If the team can sway the sponsor to delaying the FPS project start to let them solve the problem, then this is a great outcome. The sponsor believes this is the best investment of time and there is now a date at which the team agrees it is time for some help. The project leader can walk away if the team solves it in the allocated time. If the team cannot sway the sponsor, then the sponsor has a chance to fully explain why outside help is needed and local endorsement of the FPS project will shift some or all of the bad apples to new positions.
- **Make your sponsor work** – Your project sponsor is the local leadership with the most credibility and positional power. It is their job to take care of bad apples too. Pull them into solving the issues and stay out of the sponsors way when they are working with the bad apple.
- **Bad apples have bosses** – Bad apples are taking actions that are jeopardizing the project. Those actions are either supported and endorsed by their management on up, or they are not. Sometimes, important for this support or lack of support is to be transparent. Before using this strategy, the project leader and the sponsor have already privately confronted the bad apple. When a project is jeopardized, sometimes takes a more public approach to stop a behavior. A stakeholder review at the next level of management where project roadblocks based on the facts of the issue are discussed is often all it takes for a bad apple to change their behaviors.

8.4.1 Final Thoughts on Managing Difficult Stakeholders

Difficult stakeholders are part of the project management game. If you run enough projects, you will almost certainly be managing difficult stakeholders. Focus on building a solid communication plan to minimize these issues. Confront these issues early and they will stay small. Respect the people and seek win-win solutions. As a last resort, accept that there may be a loser in the game, but do your best to respect the people on the losing side. Managing difficult stakeholders is something that takes practice, lead projects and you will have practice.

9

Generating, Selecting, and Implementing Solutions to the Problem

9.1 Introduction

At first blush, the solution phase would seem to be straightforward. It is much more in the wheelhouse of traditional engineering. It should be nowhere near as challenging and difficult as what has gone before. At this stage, we have killed perhaps dozens of hypotheses and gotten at the science involved in what is causing the problem. We have already by the time we reach this solution generation, selection, and implementation phase gone from an unknown beast of a problem to something known and tangible. The detective work is done. We know who the killer is. The finish line should be in Sight! Unfortunately, in our experience, knowing the science of the problem is still a long way from the finish line of solving the problem. It is just as challenging as figuring out the science of the root cause and needs to be done well. The work continues.

For example, in one real-world problem, a team discovered that trace amounts of weak hydrochloric acid were catalyzing the product making it perform poorly at the customer's application. It was a credible scenario that made sense from known well-researched chemistry. The weak acid was detected in the product. Lab-scale studies and analytical analysis showed trace amounts of weak acid clearly would catalyze the system in the amounts present. Subsequent work showed the phenomenon could be reproduced; just add or remove trace weak acid and the reaction would turn on or turn off. The science around the root cause was solid. The hypothesis understanding lined up with the data. Green on the hypothesis onion for this root cause across the board. Good to go!

A solution was implemented to remove the acid that involved changing process conditions to evaporate off the majority of the acid in an existing process step. The solution did work to remove most of the acid. In fact, more than two orders of

Fundamentals of Industrial Problem Solving: A Practitioner's Guide, First Edition.
Zdravko I. Stefanov, Eldad Herceg, Carla Schmidt, David M. Jacobson, Dana Livingston, J.P. Chauvel, Sunil Kumar Chaudhary, and Christopher Paul Christenson.

magnitude of acid was removed (>99% by mass balance). Time to celebrate! Maybe not. Despite excellent acid removal, the problem got worse. Catalytic activity increased. The selected solution was worse than no solution. What happened?

Ultimately, in the team's second run at the problem, it was discovered that although the first solution was removing acid, it was also concentrating the remaining <1% by mass acid to azeotrope concentrations. The acid strength was effectively as strong as it could get. As acid concentration increased, the catalytic activity grew exponentially. Most of the acid was gone, but what was left was a much more powerful catalyst. The solution needed another step. Trace water was added at the end of the process to decrease the acid concentration and ultimately the problem was solved by the combination of acid removal followed by trace water acid dilution. Less total acid **and** a low acid concentration resulted in very slow reaction rates and the customer was no longer impacted. This demonstrates the crux of what makes the solution phase feel more like navigating a mine field than a short sprint to the finish. ***Knowing what is wrong is not the same as knowing how to fix it.***

After a credible scenario has been developed to explain the root cause(s) of the problem, then interventions or solutions must be selected and implemented to prevent future reoccurrences. In this chapter, we will cover how to navigate the waters of ***successful*** solution generation, selection, and implementation. This chapter describes key considerations to be evaluated by the team when determining which interventions to offer as solutions such as success criteria, mitigating risk, evaluation of capital projects, returning to design intent, developing new key variables, and monitoring systems to forewarn of issues, and other improvements.

9.2 The Joy of a Solution that Just Works

Although there are risks and problems in the solution phase, it can result in some of the most joyful moments you can have (while at work!).

Picture this. You are working on an odor problem that is costing your company its reputation and its sales. The problem is intermittent with no discernible pattern. As project manager, you are getting regular calls from the vice president of manufacturing wanting updates and more specifically. . .. "Are you done yet!?" Keeping the stink out of the product and returning to historical baseline has been what you and your team have been eating, sleeping, and living for the last couple of weeks. These high-flying problems consume the lives of the people on the team and impact everyone tied to the process. The long hours of testing likely hypotheses have resulted in the most likely hypotheses all dying in the hypothesis onion with combinations of theoretical modeling, research, analytical testing, and lab-scale runs.

But, one hypothesis has not died. This living hypothesis centers on a small air leak into the process that forms trace formaldehyde in the heating step of the process. That formaldehyde then slowly leaves the product as the customer uses it. Analytical testing struggled to find formaldehyde in the process, but the product has all kinds of complexity that makes the analytical chemists very uncertain of any testing conclusions from in-process samples. The analytical results and the fact that the process runs **under pressure** made the team initially decide to deprioritize this hypothesis. Because the set of most likely hypotheses had been killed, you and the team decided to make a bold move and try and get a sample from an unopened drum headspace at the customer site. Stop chasing it down in process samples and go to the source of the complaint. Analytical testing of that customer material found formaldehyde. A big green is achieved (evidence supports the hypothesis). This green test result has hit the hypothesis onion to add to all of the other greens already lined up with this hypothesis. Encouraged by the new data, you and the team crawl all over the process to find the air leak. You find that rework drums are being added through an open pan. The rate of pumping was drawing air through the pan drain point when the pan was almost empty. The low level and relatively high viscosity of the fluid was causing bubbles to suck into the pump at the drain point on the suction side of the pump. This was bringing in air that would then be pumped to the rest of the process. Sampling the discharge of the pump confirms air bubbles, another green square for the hypothesis onion. Rework was intermittent and the problem was intermittent. That is a credible scenario!

You and the team decide to change the rework pan design to include a level switch that stopped the pumping before the problem occurred. That would be a simple and very likely to be effective solution. The new design is installed. Piloting demonstrated that it worked and the air bubbles in the pump discharge are gone. Now, it becomes a waiting game. Will the odor go away? That is the question on everyone's mind. Every day first thing in the morning, you check your e-mail looking for customer odor complaints. Eventually, the first record is set and customer complaints that were weekly have now crossed the month mark with no complaints. Odor complaints have dropped off the map. Eventually, odorless is the consistent constant feedback from the customers and the months roll by with happy customers. Obviously, the team is ecstatic. It is hard to think of anything more enjoyable as an engineer or technical professional than solving a tough problem. These "ghost" problems that appear and disappear at seemingly random intervals are ugly, solving one is incredible. Years pass by and then while doing a tour of the new installation in China, you see the solution that was put in place so many years ago by your odor team has been replicated in the latest installation. You think back on the moment that that problem first got solved by your team. You almost laugh about how simple it seems like now. At the time it did not seem simple because then it was a major crisis. For engineers or technical professionals, those moments are what make up our legacy; our designs that live on solving the problems we face for many years. In the broader context, the solutions make these products possible are what makes modern life possible.

The solution phase can be one of the most joyful phases of the project. It is where bad turns to good. If your solution is robust and solid, it can be something that lives on beyond your career. It can be a source of pride.

9.3 What Happens During the Solution Phase?

The solution phase is really made up of three subphases: solution generation, solution selection, and solution implementation. In the solution generation subphase, the main goal is to generate a number of solution options that will all address the problem in different ways. Solution generation includes completing concept or high-level design. In the solution selection subphase, the solution options are ranked and the one(s) thought to be the best one(s) are picked by leadership and the experts. In the solution implementation phase, the selected solution is designed in detail, bought, built, and started up. Almost always, some form of piloting is done to validate the solutions before full commitment is made. **Note:** Most fundamental problem-solving projects do **not** include major capital/new equipment. So, open your mind up when the terms detailed design, buying, building, or starting up is used. For example: Detailed design could just be programming the Distributed Control System (DCS). Or as another example, buying could be buying an alternate raw material. What is done in solution implementation is solution specific. Different solutions required different types of implementation. Adding a filter to the process is a completely different implementation than changing a catalyst pretreatment method to include a longer heating step.

The three phases of solution feed from one to the next. High-quality solution selection depends on high quality generation. High-quality solution implementation depends on high-quality solution selection. Executing with excellence during all three steps of the solution phase is what it takes to finish strong.

9.4 How Do We Generate Solutions?

Solution generation can range from dead easy to be an innovative difficult development. The nature of the solution is really dictated by the nature of the hypothesis that could not be killed. There are several strategies for solution generation. These include bridging, dead easy/obvious, standardized (sometimes called copy*.*), benchmarked, and innovative.

9.4.1 Bridging Solution

Sometimes, we can solve the problem before the science around the root cause validation has been done. This sounds counter to most of what is written in this book, so we need to understand clearly what this type is aimed at accomplishing. To illustrate what a bridging solution really is, let us look at a simple common example. Imagine you are experiencing a slow loss of oil in your car. The root cause could be a gasket leak or something similar. The work needs to happen to find that leak before the problem will be truly fixed. But what do you do if you absolutely have to drive the car? Add oil and monitor the oil level closely is what most folks would do. This is NOT a permanent solution and fails to fix the root cause. All it does is buy time. Sometimes, when the money is flowing down the drain, you need to buy time.

Bridging solutions will often be one of the first steps to be explored in a Fundamental Problem-Solving exercise. We are covering it in this phase; but consider it very early in the game as well. When a problem occurs and money is getting flushed, we try and stop the bleeding. For example, when a plant is losing a million dollars a day, you had better have explored the bridging solution space! Here are some real-world examples of bridging solutions that saved millions of dollars but were later replaced with solutions that addressed the real root cause.

- A critical heat exchanger was failing for unknown reasons every three months costing millions of dollars. The team installed an online spare and bought seven off-line spares spending hundreds of thousands of dollars to keep the plant running. The short heat exchanger life was being managed, not fixed. Independent and parallel to the bridging solutions, the science of understanding the failures in detail was being done. Within a couple of months, the permanent changes to the heat exchanger design were done to solve the problem. Ultimately, the heat exchanger life was driven to years. The bridging solutions were no longer needed, but nevertheless were much appreciated and highly valuable!
- A color-causing particulate was ruining millions of dollars of product. Specialty fine filtration was rented and installed. Ultimately, an old line with incorrect materials of construction was replaced to become the permanent fix for the problem. The old line was introducing the color-causing particles.
- A temporary higher pressure steam generator was hired in to compensate with higher temperatures to make up for fouling in a heat exchanger. Once the root cause of the fouling mechanism was identified, the temporary steam generator was removed, the heat exchanger was cleaned, and distillation operations were changed to prevent fouling species from reaching the heat exchanger.

Bridging solutions are not by nature bad. They can be extremely helpful in mitigating losses. A key strategy to use when doing bridging solutions is to run the solution implementation independent and parallel to the fundamental problem-solving team. In other words, if possible, separate that bridging solution work into a standalone team focused on that implementation. Avoid dragging the folks working on the science into figuring out the logistics and work tied to implementing a bridging solution. If you run them with the same team, then it is highly likely that the timeline will get extended as the team focuses on getting the bridging solution in place instead of the root cause. It almost always slows the progress on the final solution. The case to run both bridging as well as permanent solutions within the same project team is for projects that are highly people-resource constrained or relatively low in value; then timeline extension is tolerable.

To generate bridging solutions, have the team look at the problem and ask the team. How can we get by now? What is something we can do to fix this now? Give them this list of typical bridging solutions to get them thinking of what it would take.

Some typical Bridging solutions:

- Hire in temporary equipment (filters, heat exchangers, tanks, etc.).
- Pay overtime or bring in temporary people to do the extra work the problem is causing.
- Lot select product for a specific application.
- Lot select raw materials for a specific application.
- Buy an abnormally high number of off-line spare equipment to keep plant running while equipment is failing at a high rate (filters, heat exchangers, tanks, etc.).
- Install an online spare equipment to mitigate the impact of failure on production (filters, heat exchangers, tanks, etc.).
- Create rapid cleanout procedures that clean the fouling and minimize down time.
- Create online cleanout procedures.
- Move production to facilities that are not experiencing the problem.

Note that none of these examples of bridging solutions truly treat a root cause. They just help the plant limp along.

Sometimes, bridging strategies become the permanent solution and should in fact be selected as the best solution. This is counter current to much of what this book is about, we are trying to understand the science of the problems we are fixing. Having said that, there are special cases where understanding the science is not the best path forward. As a project leader, it is important to understand when you should make a bridging solution into the permanent solution. Here are some criteria to help make that call.

If the bridging solution is extremely cheap or free of cost and is also very effective, then the business case for developing the science around the root cause can fall apart. The operational definition of cheap in this case is: It is considered cheap if it is likely cheaper than the cost of doing the work to understand the science of the problem.

If both the cheap and effective boxes are checked, then lead your team to assess the remaining risks. Think about questions like: How likely is this band aid going to hold? Will we be putting on a bigger band aid in a couple of months? Is there more to be gained by understanding the science deeper? Will this come back to bite us next year? Is this bridging solution increasing the risk of injury for our fellow employees doing the procedures? The bottom line is that as long as the risks tied to remaining ignorant of the root cause are low and the solution is cheap and the solution is effective, then plow on with implementing the bridging solution permanently. Otherwise, follow the normal fundamental problem-solving process and get to the understanding. To put in perspective, bridging solutions becoming permanent solutions are order of magnitude of ~single digit percent of all solutions. So, if you tracked all of the solutions within your organization, it is likely that between 1 and 9% of all solutions will be bridging solutions that rolled forward. It is rare that a bridging solution is the right answer; but it does happen.

9.4.2 Dead Easy/Obvious Solutions

Sometimes, when the root cause is understood and a credible scenario is spelled out, the solution is obvious. This type of solution does not initially require much effort to generate the solution. Here are some real-world examples:

a) Air leak into the process is causing trace chemistry corrosion problems. Solution – Fix the air leaks.
b) Small ~ 50-μm catalyst particulate is escaping the current 200-μm filters and causing side reactions in a downstream heating step. Solution – Switch to 20-μm filters.
c) Heat exchanger is intermittently leaking trace water into the process causing a reactive chemical problem of a runaway reaction. Solution – Fix the water leak.

The trap of dead easy solutions is teams often rush to the obvious, but with a little more thought they can drive to a more robust solution. For example, in (a) above, fixing the air leak does not lend itself to a robust control plan. It would be more of a onetime fix event. For example, it would be much better to institute a shutdown air leakage measurement that gets institutionalized into the way things are done. Then, future air leaks that are bound to happen would be detected as well and could be fixed before causing the problem. Or, in example (c) above, fixing the water leak does not give the most robust answer. Elimination of water from the cooling system by using a compatible buffer fluid could be a better answer.

As a project leader or team member, one of the best strategies for dealing with obvious solutions is to challenge the obvious answer with alternatives. Ultimately, the obvious answer may still be the selected solution; but it should still be challenged. When only one solution is available, it is by definition your best and your worst solution. Challenge the team to do better and develop alternatives. Sometimes, your challenge will mean the difference between a solution that stands the test of time and one that ultimately fails to permanently solve the problem.

9.4.3 Standardized Solutions

Sometimes, when the credible scenario is defined, it follows a path that is very well traveled and therefore the solution needed is in the standards. Each company has a set of standards that are followed in specific cases. For example: When installing a pump, your company likely has a standard for suction line design, materials of construction, pressure gauge installation, vent piping, drain piping, etc. The project team does not need to reinvent these. In addition, there are also external standards that drive specific solutions to problems. For example, ASME codes cover heat exchanger design. Here are some real-world examples of where teams picked standardized solutions.

- Trace acid was discovered to be a by-product of the intended reaction through extensive analytical work. The acid was eating away a downstream heat exchanger that was not designed for acid service. Once the corrosion mechanism was understood, the company standard for materials of construction for that acid service (carbon block heat exchanger) was implemented.
- Insufficient lubricant cooling was found to be destroying compressor bearings. The vendor's standard design that included better coolant flow was implemented.
- A valve that was only used during start up and was normally closed was found to be the problem. It was leaking during normal operations and was determined to be introducing low levels of contaminant into the product. The company's valve standard for highly reliable shut off configuration was implemented. Two block valves with a vent and drain on the piping between the two block valves; to ensure the valves hold and are not leaking, the space between is drained and is empty during normal operation. Configuration includes pressure monitoring to validate that the space between the two valves is not seeing leakage.

The standardized or copy*.* solutions are about having the experts on these standards. Usually, these experts are part of the new plant or capital design teams. As a project leader, your job for these types of solutions is all about networking. Bring in the folks that know and live these standards every day to help your team copy*.*. Standardized solutions are great because they have history. As long as they are made to address the root cause your team is facing, they are often the best way to go. Why fight a standard when it has a track record of working? Also, if your team drives to a similar, but nonstandard solution, you will find out that a nonstandard solution is hard to adopt. When the project team hands the work off to the folks that are doing the detailed design and installation, they will work HARD to drive you to follow the standards. Getting an exemption from a standard is in general an uphill battle. Only fight that battle in cases where the standards fail to address the root cause.

9.4.4 Benchmarking Solutions

This solution type seems like but is not really a case of standardized solution. It is looking within your company, within your industry, and outside industry for specific solutions to your problem. This is different from using a standard because most often these are much less frequent and therefore do not get written into the standards. For example: If you are trying to remove trace acid from an oil like stream, with a little searching, you will find a ½ dozen ways to solve this problem by looking within your company or within the petrochemical industry. It is being done. You may also find that none of these ways appear in your company's standards. If your company is not doing this type of unit operation, it is unlikely anybody bothers with a standard. Or as another example, if we are having mechanical failures due to a thermal expansion differential between two metals, the team can look for systems that have addressed this.

A benchmarking solution within your company has a fantastic advantage over other benchmarking forms - you are almost always able to go, look, and see the solution up and running. You can read the procedures and gain or lose confidence in the ability of that solution to address your root cause. The trick to within company benchmarking is to find it. Research reports and key word searches are a starting point. Books on the topic are also valuable. Always start with researching.

Finding hubs (people who know who the experts are) and process experts is the next step. This must be done well to be successful. We can say with certainty that one way to irritate an internal consultant or process expert is to have done zip first. Research sets you up to approach the right people with more than hope and guesses. Consider these two possible approaches for a benchmarking question for a process expert:

1) Do you know of any good ways to separate water from oil?

   Or

2) I read the research report that you wrote in 1995 in which you selected coalescers for our 1998 plant expansion's water removal service. I have what I think is a similar application, would you mind having a look at it? I could use your help.

These examples may seem a bit extreme but we have encountered both directional extremes. Our instant response to questions like (1) are . . . "go read a book on it and come back later when you have better questions" . . . or . . . "delete that e-mail" . . . or "if I am feeling like actually responding, I might ask the questions the person should have been asking. What density of oil? How much water? What flow rate? What temperature?" etc.

We love questions and requests like (2). Spending the time to at least understand the parameters involved, the work that has gone before, and to have read enough to know the words and language of the solution are very productive. Be that. Lead your benchmarking that way.

Going outside of your company, but within your industry can be insightful and worth the time. Unfortunately, there are barriers. Agreements must be made about the intellectual property. The motivation for an external company to work with you needs to be in place. For example, we have approached competitors about safety practices/solutions. This was successful because in general everybody wants people to be safe. Who cares if they are the competitor's people or our people? They are people, so safety solutions are often quite shareable within an industry even with competitors within that industry. There can be other motivations. For example, if you are benchmarking with somebody supplying product raw materials or is a downstream customer the business relationship and mutual benefit can be the right motivation. Again, intellectual property agreements will need to be in place, but this benchmarking can work. Additionally, sometimes there is information in the public domain that can be used. Protected solutions that are under patents can also be used by reaching licensing or other agreements with the folks that own the intellectual property. Go for outside benchmarking in cases where your internal solution options are struggling. This can be hard to do, so pursue it in cases where it is worth the effort.

Benchmarking outside of your industry is another place to look for solutions. These are nearly identical from an intellectual property and motivation standpoint as the within-industry benchmarking. The big difference is that it takes some more creativity to understand the link between your root cause with a credible scenario and the solution. For example, suppose the problem is that the microbes in your wastewater treatment plant are dying and no longer eating the alcohols that you need them to consume to purify the water. You might find solutions in the brewing industry or in biomedical. It is also similar to the within-industry benchmarking, but only pursue this direction where it is needed and necessary for success.

9.4.5 Innovative Solutions

Innovative solutions are perhaps the most fun. By nature, they are often the riskiest. In this solution space, you are pushing boundaries. Your team is solving problems in new ways. It is used usually as a last resort. Cases to be innovative are:

1) Conventional solutions are not available or are unlikely to work. This happens when the problem is truly unique, and solutions are just not there for the specific problem.

   or

2) When conventional solutions are prohibitively expensive due to cost to implement or time to implement.

The trick with innovative solutions is to try and push a team to de-constrain themselves, open up the possibilities, keep creative juices flowing, and yet at the same time stay on target aimed at the root cause; the credible scenario.

There are many books on brainstorming techniques; so, it is not our intent to cover all aspect of idea generation. Instead, we will go through what experience has shown to be the most effective innovative solution generating techniques for these multifactor complex problems that fundamental problem-solving techniques are best suited to attack.

Channeled Brainstorming

In channeled brainstorming, the folks doing this work are VERY focused on the problem itself. As a project leader, you will prepare a one pager that clearly shows the credible scenario for the root cause with as clear as possible explanation of what is going on. One pager just means the explanation only consumes one page. Because of this, pictures or graphs are preferred ways to communicate. One page of nothing but writing will ultimately struggle. A few clear pictures and short explanations on the page will explain it. A diverse team is then brought in that should include the original project team plus outsiders that have expertise in the solution space or have a reputation for being creative forces. The goal of the exercise is to generate as many possible solutions to the problem as possible. Later, you will sort wheat and chaff. As a team leader (or facilitator for the exercise), your job will be to consistently and periodically asking them to fix the problem. Reminding them of what problem they are solving keeps them in the "channel" part of channeled brainstorming. It is counterintuitive that creativity can stem from focus; but it does. To understand this, do this exercise. Write in big letters: "**Problem** – my ping pong ball is floating to the top of the pond. To be successful, it needs to sink." This may seem to be a strange problem, but it will help you to understand why channeled brainstorming works. Now, write as many ways to solve this problem as you can. Whenever you get stuck, read the problem again. I think you will find that going back to the problem drives a next level of creativity. Humans do not like to focus, but when they do, great things happen. Channeled brainstorming can be a very effective tool in generating innovative solutions to problems.

Functional Analysis

Keep It Simple. Engineers (and other technical folks) like to build monuments to themselves. There is a strange sometimes hidden desire to develop complex systems that impress fellow engineers. It is an ego thing. Simple solutions just are not that sexy.

Fight complexity wherever possible. Solutions should be complex enough to solve the problem at hand; but never more complex than that. The way to think of it is that each layer of complexity results in layers of maintenance to keep each complex item running and also requires maintaining expertise to understand how to make changes to the system. In short, the more complex solution, the more work it is to maintain. Simple elegant solutions tend to last a very long time because they just do the job. Think about the humble water tower that maintains our water pressure precisely everyday of our lives. Open the faucet and water flows. There are much more complex answers that would also work at maintaining that pressure. The complex answers would be less perfect.

To get into this frame of mind, value engineering techniques can be used. They help us only solve the problem in hand and yet decouple the system to allow for very innovative solutions. Value engineering works on thinking only of the essence of the problem. The technique for making innovative solutions is to translate the problem into a verb-noun. This is a very strange process for those who have not tried it before. Here is a simple example to wrap your head around the power of the verb-noun. We use pumps everywhere in the petrochemical industry. Imagine we need to pump liquid from one tank to a distillation column. But, what is the function of that pump? What is a verb-noun that would describe what it is doing? Pump-liquid could be used, but pump-liquid is really using one fixed solution. It is static not dynamic. We really want to use unconstrained verb-noun combinations. Use combinations that do not lock and close down solution options. Does the pump transfer liquid? It does. Verb-Noun. Transfer-Liquid. That verb now phrasing is not specific to a solution and opens up possibilities. Generate solutions that transfer-liquids.

To see how that opens the mind up to innovative solutions, start thinking about other ways to do the transfer. We could use gravity. Think about water towers. Gravity works all of the time and is really reliable. Pumps have failure rates. So far gravity does not have a failure rate. Alternatively, we could apply pressure on the tank and create a differential pressure between the tank and the distillation column. No pump needed. The tank is likely inerted with nitrogen anyway. We could physically bucket the liquid and carry it to the top. Sounds like work, but it might be the best answer. The possibilities just keep opening. Understanding the function of something is the essence of what it takes to be innovative.

To apply this to your fundamental problem, translate the root cause and credible scenario solution possibilities into a verb-noun description (unconstrained) that your team can then brainstorm around. To practice this, look at everyday objects and think of the verb-noun of what they do. For example: What does a light bulb do? With practice, it becomes second nature and can really help going the extra innovative mile.

This verb-noun thinking has been very well elaborated in the book Innovation on Demand by Victor Fey. This methodology has been very effective in creating and documenting creative ideas in very difficult problems.

How Do You Mitigate the Risks Created During Solution Generation?

There are risks that pop up at every phase of a project. Savvy project leaders understand this and work to mitigate these risks throughout the life of the project. The solution generation stage is no different. It has risks that must be mitigated. Here are some of the ones that pose the greatest hazards.

a) **Excessive solution ownership by an individual (My baby will never be ugly)** – We all like our ideas to be the one that saves the day. It is often why we got into the business in the first place. Most folks like being the hero. The issue with this desire is that it can lead to a few dominant personalities driving their solution to a finish line. That is awesome if those solutions are equally awesome at solving the problem. The reality is science is what dictates outcomes. If the science behind the solution works, the solution works. If a person loves or hates a solution, it has no bearing whatsoever on if a solution works or not. Believing in a solution or not believing in a solution also has no bearing whatsoever on if a solution works or not. It is the way the world works. Our feelings have zip to do with the formula for gravity. Our beliefs have zip to do with if the world is flat or spherical. The issue is that strong personalities can drive bad science forward, spend a lot of money, and never solve a problem. As a project leader, we need to lead with science and ***follow*** with harnessing the passions of team members. To do this, it is important to separate the solution from the people. If those strong personalities have a weak association with any individual solution, then this risk is mitigated. That will help enable them to follow the science instead of their passions. To accomplish this weak association, a project leader can creatively force team members to build up several possible solutions. Having them work on multiples helps with this separation process. For example: If a couple of team members have worked to develop an initial concept, consider swapping out and having other team members further the concept. Or try having parallel efforts. Have a couple of teams independently developing the same concept and then bring the teams together to compare notes and get the best of each. Generating many solutions can also help mitigate this risk. When teams press themselves and generate many solutions, it is very hard to remember ownership. Solutions become more of a team ownership. As a project leader, watch for the solution ownership risk and intervene as needed to keep the team on track.

b) **Excessive solution ownership by a team** – Similar to solution ownership by an individual, the team can also take too much ownership. The team ultimately will not be in place years down the road. Often, only a couple of team members are going to be a part of the process that is changing for the short term. Many team members are outside experts or short timers from the process standpoint. If this short-term team overly owns the solution, the pushback outside the team can stop good science from moving forward. For example, on a project, it was determined that more heat was needed to stop a process from undergoing larger energy balance-related cycles. The project team came up with an almost free solution – A direct contact heat exchanger. The liquid would be pumped through and in direct contact with a vapor stream with excess heat. Since both fluids were the same composition, the only thing to exchange was the heat. The team did the math and validated that the science would work. Taking it outside the team was an absolute uphill battle. Heat exchangers do that job, put one of those into the process. Ultimately, the only way the team could get the work approved was to propose a pilot. The pilot worked, so the solution was implemented; but it could have just as easily been never tried. To mitigate this risk, it takes communication and it takes getting some skin in the game. Involve those influencers and stakeholders in the solution generation work. Take the time to show them the science around the credible scenario that is the root cause. Project leaders must think hard about the broader impact of where the project is headed.
c) **Solution concept science does not work** – Science is one of the harshest task masters there is. The cards are dealt before you even begin the process of generating solutions. The outcome is not known; but it ***is*** predetermined by the laws of physics/chemistry/first principles. Solution A will never work, solution B will work if the engineers design it correctly, and solution C will robustly work in a broad operating range. If we could only know outcomes before investing time and money into our ideas, this phase would be much simpler. We have improved our odds of getting the right solution by deeply understanding the science of the problem, so in some ways this risk has been mitigated. But solutions that fail tend to always cost time and money.

This risk is usually worth extra effort. This is a great time to pilot the concepts on the small scale. Is it possible to test concepts quickly? Cheaply? Can you try the solution in a beaker? Could you build a small-scale mock-up? Can you quickly check the math? Can you build a preliminary model? During the concept-generation phase, we will start to transition into the concept selection. Think about what work would help to pull the chaff from your concepts. Is there any small-scale work that can help understand which of the generated solutions could work? For example, a team has determined that the current sterically hindering nonionic surfactant will not stabilize the existing emulsion because it is physically not big enough. Could you put all the ideas to replace it through a simple prescreening exercise? Through this type of work, you will gain dividend in that as the team moves to solution selection, they are selecting from good options. Also, a little bit of work at the concept phase will generate more solutions. The process of moving from an idea on a piece of paper to a simple mass or energy balance around the solution or a preliminary sizing will give the team some "Aha!" moments that lead to new solutions. Deeper understanding of a solution almost always generates more. Project leaders should always have the north star of understanding the science in front of them at all phases including solution generation.

Solution generation is the feedstock for solving the problem. Take the time, do the work, and set the project up for success by creating viable options.

9.4.6 When Do We Select Solutions?

Selecting solutions is a reversible process. We try and generate multiple solutions because project teams need to mitigate the risks tied to solutions and solution implementation. Until a solution is proven in the field, we do not know if it will work or if it is practical. Maintaining multiple concepts will ensure that there will be another one ready to go when one fails. Why bet on one horse? A team's odds of success increase significantly if multiple solutions are pursued. The math on that does have diminishing returns. Going from one to two solutions in the running mitigates a significant amount of risk. Going from 20 solutions to 25 mitigates much less incremental risk.

Ultimately, there will need to be a decision. The word decision has Latin roots meaning "To cut off." We will cut off and not pursue alternate solutions at some point during the select solutions step. When to do it is an important consideration. If we select one solution too soon before we reach a reasonable certainty it will work, we can waste time and cause rework due to finding out this solution did not work and now we are starting over with another one in the pile. If we select a solution too late, then we will waste time and resources pursuing parallel solutions that will just end up on the scrap heap.

Table 9.1 Questions driving premature decisions.

Timing thought process questions	Answer driving premature decision
Is solution easily reversible?	Yes
How certain are we that the solution will work?	Moderate
How complex is the solution?	Simple
How costly is the solution?	Almost free or very low
How long does it take to implement?	Quick
Does the solution pose a people risk?	Very low
Does the solution pose a negative environmental risk?	Very low
Does the solution pose a customer risk?	Very low

Timing of when to select solution from the ones generated is a function of solution reversibility, certainty, complexity, cost, time to implement, risk to people, risk to environment, and risk to customers.

That solution selection timing function plays out different ways. For example, we may jump to implementing a solution prematurely before a root cause is found if the right conditions are met. Table 9.1 shows some timing thought process questions and corresponding answers driving premature decisions.

To understand this case, examples help: A fundamental problem-solving team was launched on a quality problem that was becoming a crisis. Initial mass balance work uncovered information that the early troubleshooting efforts missed. This happens a lot with mass and energy balances. A hypothesis around the critical nature of getting the right amount of water to make the chemistry do what is supposed to do was on the table. The simple offset in the water loading that occurred after a meter was changed out explained the current facts and data. Green on the hypothesis onion, but the onion was just starting out and it was VERY early days. The team did not wait. Adjusting for the water meter offset was very reversible. The team could at any time in one batch adjust for the offset and in the next one not. This solution was almost free. Water adjustment was just a few minutes of the operator's time and a small procedural change, so could be done NOW. Customer, environmental, and people risks were all low. The team implemented the offset and it was a partial success. Later work on the science validated that the change theoretically and empirically would help. The premature decision was worth the risk because at any time, the system switchback to baseline could be made.

The bottom line is savvy project leaders do not have a fixed milestone of exactly in a sequence when to pick a solution. They flex with the need. Flexing accelerates a project and decreases its cycle time. On the scale of premature, early, normal, and late solution selection sequence, think on these considerations. Premature for this purpose is solution selection before root causes are found. Early is selecting solutions prior to fully exploring solution options. Normal solution selection timing is after solution options have been generated and are well explored. Late solution selection is the solution options have been fully explored but will continue development parallel to the implementation phases. Timing may seem more art than science, but just think logically. What is my purpose? Why am I selecting now? Why would I wait to select a solution? What are the risks of selecting earlier or later? Understand your situation and it will make sense.

Here are some considerations:

Is the solution easily reversible?

If the solution is essentially a switch that can be turned on or off, then it becomes easy to try. Solutions that answer this question "yes" tend to lean to the early or premature side of selection timing.

Solutions that are hard to switch off or flip back to baseline will need to work hard at making sure they will work and minimize risks that impact to people, environment, and customers. These tend to occur at a normal timing.

How certain are we that the solution will work?

The more certainty of a solution working, the earlier we can make decisions. Piloting can often accelerate solution selection because that is exactly what happens. We pilot a solution in the concept phase because we like it and it works. Solution selection will accelerate.

Uncertainty around a solution will postpone selection and tend to drive it toward the late timing. It can push it into the solution selection into the implementation phase. The team may be doing detailed designs of two solutions just to cover the uncertainty risk. In fact, we have seen multiple solutions built and tried because of this uncertainty. It was cheaper to build two different solutions that could be tried instead of taking the risk of building one that if it failed would dramatically delay the fix. That is an extreme of late, but where there is uncertainty, there is work that needs to run parallel to design and building of the solution that removes that uncertainty. It can be lab scale, pilot scale, or modeling work that ultimately lowers the uncertainty concerns.

How complex is the solution?

More complex solutions tend toward normal or late timing of solution selection. Complexity drives longer solution implementation timeline and often can drive higher costs. Complex solutions therefore cannot be taken lightly. They often will have knock-on effects that need exploration. Often, complex solutions are a multistep selection because they not only have concept selection but will have multiple options of subsystem selection.

How costly is the solution?
For projects where solution cost is important, these solutions will drive toward normal solution timing. They will also focus heavily on piloting and validation that the solution will work. Because of this, it is difficult to push it earlier in the sequence.

In some projects, solution costs are completely dwarfed by the burn rate costs. That can mean that other factors select solution selection sequence.

How long does it take to implement?
For projects where solution timing is important, these solutions will drive toward normal solution sequence. Because solution timing in these situations can dwarf the time it takes to select a solution, quick decisions are not the project focus. Bridging solutions will need exploration for these.

Risk factors affecting solution selections:

- Does the solution pose a people risk?
- Does the solution pose a negative environmental risk?
- Does the solution pose a customer risk?

These risks need to be mitigated or understood prior to final solution selection. The rate at which the project team can understand these risks will often drive the sequence of solution selection. For example, you may have a reversible, cheap, quick to implement solution that sits while the team develops understanding around the people, environmental, and customer risks. Because this work can be done in parallel to solution generation, a possible premature or early solution can push back to more of a normal timing. Watch these risks to keep them out of the critical path of your project. Consider sub-teams to drive risk mitigation.

One size does not fit all in terms of when to pick your solution. It really is a race where many factors drive the ultimate timing. The most common sequence is after solution generation with some validation to ensure that the highest priority solutions are likely to work; then, the solution to move forward with is selected. But nearly as often, be prepared for pushing for early selection or late selection.

9.5 How Do We Select Solutions?

Selecting solutions is all about understanding "What is success?" We at this stage have multiple answers that will likely work. If the team has done a spectacular solution-generation process, then theoretically the stakeholders could flip a coin to pick and the team would be successful. Having said that, we have not met very many key stakeholders in the coin flipping mood. We need to pitch options to them that help understand the implications of their choices. That pitch becomes the basis for their solution selection. To correctly do that pitch, we need to understand our stakeholder's decision criteria. We will then need to have answers for those basic questions. Then, our team and our stakeholders will be armed with what is needed to do a good job picking a solution.

9.5.1 Solution Selection: What Is Success?

A successful solution to a Fundamental Problem will contain three main elements.

1) **Quality** – How well is the solution going to address the fundamental problem? Are the knock-on effects minimized (unintended effects)? How will we know the solution is successful? Is it safe to people, customers, and to the environment?
2) **Time** – How long is it going to take to implement the solution?
3) **Cost** – How much is it going to cost? Ongoing costs and initial costs both need consideration. Is it worth the benefit?

9.5.2 How Do You Determine a Solution's Quality?

In many ways, quality is the only criteria. Saving or spending time and money on a solution that does not work is essentially waste. Solutions have to solve your problem, or they are waste. Unfortunately, there is not a device that you can wave over a selected solution and it will tell you if it works or not. Instead, we need to get creative. We need to figure out ways to test solutions before we implement them. Specifically, we need this for solutions that are hard to reverse and cost significant time and money. Testing solutions can be done in two major ways. Theoretical-Paper (or computer modeling) or Empirical-Piloting (lab-scale piloting, intermediate step piloting, or full-scale piloting). At least one of these needs to be done, to test the quality of a solution. Most often both are done.

Theoretical-Paper Exercises (Computer Modeling)

This is the most common test of if a solution will work. It is also the step that most every solution selection starts with in solution evaluation. Do the math. Get down to the first principles and calculate out the outcome. For example, if more heat is needed to complete a reaction to address the root cause of an incomplete reaction that is poisoning downstream catalyst, adding heat through increasing the size of a heat exchanger is going to take some heat transfer math. The team will do the math on paper to check initial sizing and understand the problem. The team will then likely go to the computer modeling their company uses to confirm the sizing and fine-tune the parameters to improve the design. If the problem is urgent, it may occur backward. We may have available spare heat exchangers that we are trying to understand if they might be something that would work. Checking to see if we have something that will work will mean the sizing and layout is fixed, we will just see if it mathematically will attain the heat transfer we need to occur. It would be impossible to put all of the different math that can and should be done on solutions to validate that on paper (or in a bunch of zeros and ones!) that the solution will work.

As a minimum, the mass and energy balances will occur in solution validation. Often, the process of killing hypotheses has resulted in a model getting built. For example, the same mass balance that proved trace contaminants concentrations in the feed were not large enough and not the right order of magnitude to be the source of your troubles could still be used during solution selection. That same model could be used and modified to see if the ultimate solution can fix the problem. Imagine that a solution is a carbon bed being used to absorb the contaminant that is causing the problem. That model mass balance can now be applied to the carbon bed solution.

There is a significant error rate in both paper calculations and computer models. This is the main risk with this type of validation. Mistyping a number can cost millions. Messing up units can cost millions. Errors are going to happen, count on it. To mitigate calculation errors, it is helpful to do independent parallel calculations. If one engineer completes a calculation alone, and that engineer's average error rate was 1 calculation error/every 10 calculations completed that residual risk is significant. If independently another engineer completes calculations on the same problem with a similar error rate, the odds of calculation errors are greatly reduced. To reach the same outcome from calculations, the theoretical error rate for 2 engineers working independently shifts from 1/10 to 1/100; and order of magnitude reduction ($0.1 \times 0.1 = 0.01$). Sometimes, there is pushback from team members on running independent calculations. It is useful to show them the math and then give some of the shining star examples of money wasted due to calculation errors. Sometimes, when building complex computer models is done, having two people independently build modes is not practical because there are not two experts. In those cases, and for a best practice in personal work, we recommend that the completed work be set aside. Then, after a cooling-off period, redo the model or the calculations from scratch. Although there is dependency between the calculations, often that dependency is offset by the complexity. If it is a hard model or calculation, it is difficult to remember how the first one was completed; so, the next set of calculations is almost independent. This serves to decrease error rates. Doing this for years has uncovered many calculation mistakes and provided opportunities to learn how to reduce them. Count on errors when building models or completing paper calculations.

Do the math on paper and in modeling programs; it will mitigate solution failures. Solutions based on first principles can be verified with those same principles. Think of our moon landing. The scientists that did those calculations had to do them before the task had ever been done. It is possible to calculate success or failure.

Empirical-Piloting

Piloting Success Criteria

At the risk of sounding like a broken record, piloting needs to have the same set up that we have used many times. We have to answer – What is success? What is necessary for success? And What is rate determining? Figure out your success criteria prior to the pilot. For example, if we have an oxygen species contaminate that averages 30 parts per million in our feeds and we have discovered that is BAD, then what is success? Success is to lower it. Success is likely not to eliminate that. The cost of taking a contamination down from 30 parts per million to 30 parts per trillion is likely VERY expensive and unnecessary. How low would solve the problem? If it was <5 ppm, then that is what success looks like.

There will be other success criteria. For example, if our solution was filtration, we might have a filter that does the filtration, but if we are changing it out every 20 minutes, that is likely not success. Success criteria will be several measures that tell us we are on the right track with the solution we are piloting. Other success criteria may include measurements aimed at:

- Operability
- Nonideal conditions (stress testing)
- Cost
- Safety
- Environmental
- Ease of solution adoption

Success criteria must be developed prior to starting a pilot. It is handy to put that into a scorecard that included your definition of the measurements and the levels that would be considered a success. Success criteria can be greater than, less than, or on target criteria. Greater than or less than criteria are self-explanatory. For example, less than 1 ppm formaldehyde for 24 hours of pilot operation as measured by product sampling every 2 hours would be an example of a "less than" success criteria. "Less than" and "greater than" success criteria can include a variation component such as Cpk and/or Ppk, and/or Z score. On target criteria will need a variation component such as Cpk and/or Ppk, and/or Z score. On target will also include a measure to determine if it is on target. For example, a viscosity target of 500 centiStokes might have success criteria such as having a t-test between actual data collected and target showing no conclusive differences. Or, having less than 5% deviation from a 500-centiStoke average. Deviation from target needs to consider expected measurement variation and the practical business needs. For example, if one of our success criteria was catalyst life, then obviously more would be better. But, setting a target of one million days would set the team up to fail or at least never see the end of the validation process. Instead, a cost calculation can help understand the time that the catalyst needs to last to be cost effective. That cost translation into a measurement value becomes the success criterion. So, more reasonable numbers such as >100 days to make a cost target are possible. Success criteria are practical values; not wishful thinking.

Setting success criteria takes time initially but will save time later. This helps stop the postmortem argument about was the pilot successful or not.

Lab-scale Piloting Go to the lab. Go to the lab. Go to the lab. These are three of the best pieces of advice that can be given in piloting solutions. There are challenges in shifting scale from lab to full commercial scale, but the primary science often can be proven in lab-scale experimentation. It takes creativity to think of how to validate a solution in a lab. It takes work to go through safety approval processes to do the experimentation. It takes time and resources to do validation. The payout is that solution failure is made with low costs. In general, each scale costs an order of magnitude more, so lab-scale solution failure can be 100 or 1000 times cheaper than going to full scale.

Other Piloting Scales Intermediate step piloting or full-scale piloting is just a bigger version of lab-scale piloting. The advantage of increasing the scale is that it decreases scale-up issues. It is rare to be able to directly scale from lab to commercial scale. Size matters because of wall effects, mass transfer issues, and many other first principle scale problems. Because of this, scaling-up generally will do the best job confirming that a solution will work. Cost will typically go up as a team scales up a solution, so confidence in outcome comes at a price. Build that cost into your stakeholder's expectations as much as possible. If they know that the main way to answer their question about "Will this work?", is to test and see if it will work, then they are motivated to support piloting. Also, cost is really a creativity question. Permanent pilot plants for a technology are helpful and businesses that maintain pilot plants for this purpose are wise. If your team has access to one, drive on. But most teams do not have that luxury. Fortunately, we do not need to build a parallel plant identical to the one we are piloting to be successful. We just must think critically about what we are piloting. For example, if the distillation column in our system is functioning fine, why would we add that to our pilot plan? What is the critical component change we are making? For example, if we have discovered a trace solids contaminant in our feeds is causing downstream corrosion, then **remove solids** is the critical to success function we are piloting. We may pilot with a bridging solution such as renting the equipment while the design for the permanent installation is in progress. So, can we scale our pilot by renting a filter and installing it on a bypass line or even on the full stream? This full-scale or fractional-scale pilot will cost us rental, set up, procedure writing, labor, etc.; but will not commit us to the likely hundreds of thousands or millions of dollars the full scale would cost us. We can test it and give us room to adjust the final design.

Most fundamental problem-solving projects will need to pilot at some scale. Prepare your stakeholders and your team by talking about it early on and make it happen when needed.

Measurement Plan Value in Piloting Developing a measurement plan is critical for success in understanding pilot results. We will be measuring the pilot success criteria for sure; but there is more to the measurement plan than just the success criteria.

Creating SIPOC (Suppliers, Inputs, Process, Outputs, and Customers) maps that map out the input, process, and output indicators are very helpful; so, read up on how to create those and do that. They are really measuring the Xs that drive the Ys. $F(x_1,x_2,x_3,\ldots) = Y$. The Y your team is interested in is project dependent on what is success. Every measurement plan also needs to have the 5Ws and an H for the measurements needed. **W**ho is going to measure, **W**hat are they measuring, **W**hy are they measuring it, **W**hen (and what frequency) is the measurement going to be done, Where is it going to be measured (sample or instrument location for example) and last but not least, **H**ow is it going to be measured.

Your measurement plan will also contain other Ys, other output indicators that tell you about the product. Mitigating knock-on effects is critical. Often, there are unintended consequences of solutions. At a pilot scale, developing a robust measurement plan looking for those consequences will be critical. Therefore, the other Ys, the ones you want to stay the same, need to be measured. For example, suppose your problem was that your product was cloudy when it was supposed to be clear. If your team figured out that trace salt was coming out of solution, then that turbidity or maybe a direct salt measurement would be your main Y measured in the pilot. But, maybe other product quality needs such as product viscosity may also need to be tracked. We want to fix turbidity without breaking viscosity. Finding out which of the solutions being tested have the fewest of these consequences may be the deciding factor on selecting a solution.

The measurement plan will also set you up for success with your ultimate control plan. It is a Plan Do Check Act cycle. Some of those measures will be critical to know if your solution is working or if there are adjustments needed to keep it working. Your team will: Plan the pilot. Do the pilot. Check the results against the success criteria. And then, it will Act on that information by correcting the less-successful components or deciding that full-scale implementation is the next step.

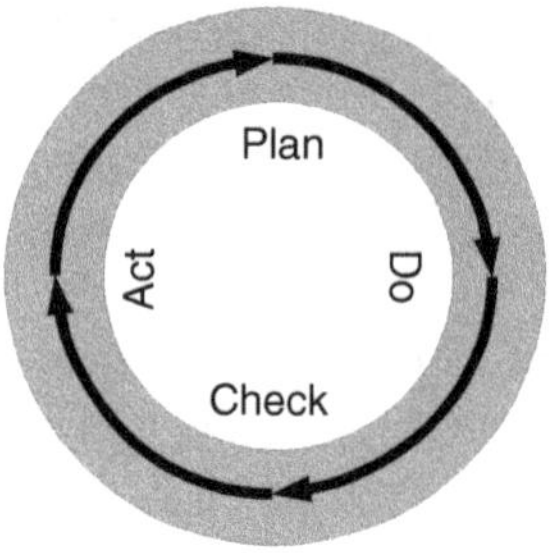

Figure 9.1 PDCA cycle used in piloting a solution.

The PDCA cycle (Figure 9.1) your team is following on the pilot is exactly the same cycle your process owner will follow when you hand over the keys on the solution.

When selecting success criteria for a pilot, it is a balancing act. Too aggressive of a level in success criteria and the pilot will be measured a failure, but potentially could be a commercial success. Too easy of a success criterion and a pilot can be deemed a success that should have undergone rework. Strangely, it is personality that pushes where teams lean. Personality-driven levels of success are the enemy of great results. To give an example, if a pilot result were to indicate that implementation would result in a $1 million/year savings and an aggressive team wanted it to be $1.5 million, both sides can be argued. $1 million is not as good as $1.5 million, so the team is right. Right? Go back to the original "What is success?" question. Was it actually to make the most money? Or, was it to get back on track to being able to supply a customer. Priorities matter. If this mess was all about getting the customer supply, then as long as the other success criteria are hitting the mark, maybe your pilot is a complete success. Maybe if it was a push on cost savings, your sponsor would still be ecstatic.

This brings us the second point on success criteria. Timing matters. Success criteria levels need to be set prior to doing the pilot work. It is a strange human thing that if we measure success and then set the goal, the measurement will often be the level that is considered successful. Or, again if the personality type is never satisfied, the measure is never enough. Put the line in the sand in advance of the pilot work. This will be the same when we set full-scale implementation success criteria.

9.5.3 How Do You Determine a Solution's Cost and Timing?

Cost and timing estimates are often handled within the organization's workflows. The company you work for is constantly estimating costs and time (schedule) for projects. This is the time to leverage that capability. Get help. At the solution selection phase, perfect estimates are costly and unnecessary. We need at this stage only enough understanding of costs and timing to communicate to stakeholders which of the solutions best meet the needs as well as the trade-offs involved. It is more important at this stage to get the estimate for all of the prioritized solutions relatively right (on the same basis, same method) than actually right. We are selecting between a group of solutions. If we keep them on the same relative basis, then we will not bias a solution selection by shifting the method around.

Often, cost and schedule are estimated with historical measurements applied to the current project. For example, TIC factors (Total Installed Cost) are sometimes used to estimate total cost. The cost of the equipment × the TIC factor = the cost of everything to get the equipment installed. Sometimes, standard programming or coding complexity factored estimates are used. The bottom line is that your company likely has methodology for estimating cost and timing. Use them. Perfect estimates are the enemy. Invest only in as much estimation as needed to move forward on the problem and improve the quality of those estimates as you get closer and closer to the final solution implementation. Using current cost and time estimate methods used by your company are the big take away. Keep solutions on the same basis and bias will be driven out of that part of solution selection.

9.5.4 How Do You Do Final Solution Selection?

Once critical information has been gathered on the quality, cost, and timing of the solutions, a decision needs to be made. A solution or sometimes a parallel set of solutions needs to be selected. The selection process has been set up for success if good quality, cost, and timing data have been generated. Your list of solutions has a high probability of success because of the paper and real pilots that have been done. To make good decisions, it takes good information. It also takes stakeholder buy-in. Hopefully, throughout the project your team has been communicating relentlessly to the level or better yet, slightly more than is necessary to keep stakeholders understanding the situation and progress being made.

Decision Timing

Although we will be discussing solution selection, timing of when to make the decision is important as well. The solution selection timing can come earlier than you think. For example, maybe we have done the paper/computer modeling exercise and some lab work around our solutions but need to scale-up to a larger pilot. If there are four solutions in the running, it is unlikely that the stakeholders will want to pilot all four solutions. The decision timing needs to occur before the piloting expansion. The rule of thumb on when to do solution selection is whenever you and your team are entering a new chapter for spending or risk. As the stakes escalate, bring your sponsor and stakeholders along for the ride.

Here are some typical solution selection timings:

- Prior to major modeling exercise with an outside vendor doing the modeling.
- Prior to a pilot escalation to full scale.
- Prior to a lab-scale trial that is high risk with many safety reviews.
- Prior to full-scale implementation.
- Prior to implementing a bridging solution.

Keeping Options Open and Closing Them

One of the biggest lessons to learn early on is when to narrow options and how narrow do they need to be. For example, if your team has come up with 30 ways to solve this problem, please do not drag 30 ways to solve the problem to your sponsor and stakeholders! There will be an initial selection within the team. Stakeholders can generally stomach two to seven solutions. Also, the team can only stomach getting detailed cost, timing, and quality data on two to seven as well. So, your first step is to get to that reasonable number for stakeholders and for the team. Sometimes, you start with that right number. For example, if you have dead obvious solutions, you may start in the right number range to bring to stakeholders. If you had to dig deep into creative solutions, you may have your hands full of options. To narrow the solution field initially with the team, start first with data, facts, benchmarked best practices, and all of the most tangible measures. Many folks will vote on other means to limit solutions, but only go there in the absence of data. It is OK to use process expert opinions to initially reduce the number if there are no data; but the purpose of that selection is to just prioritize what data to get. When will the first round of cost, quality, and time data gathering be done? Use opinions to narrow the work of the team but be flexible and prepared to cycle back to options not initially chosen if the cost, quality, and time data point in that direction.

When talking to sponsors and stakeholders, it is critical to not "bet the farm" on just one of the solutions. Solutions fail for many reasons. Having a second or even a third candidate may be the difference between success and failure. Sometimes, on CRITICAL projects with high stakes, teams should be running parallel solution implementations. Two independent ways of solving the problem should be implemented. If one fails, the next is ready to go. High burn rate projects can flow that way. All projects need to have backups at some stage of development. Usually, just in concept or high-level design where we understand the scope of the work and enough detail to pass it on to those that could implement them.

There is also a time to close options. We cannot go on with many solutions forever. We cannot keep tweaking solutions forever. Projects are defined as work that starts and **ENDS**. So, think about when the solution needs to undergo a hard freeze . . . no changes allowed. Perfection is the best enemy of the good. Project leaders show leadership by galvanizing teams to action. When the flow of information and data turns into strong confidence of success, more work can often just be slowing things down. Do the 20% of work to get 80% of the gain. That last 1% improvement will take as long as the first 99%. Processes will be improved more than just once in a continuous improvement culture company. Perfecting an untried solution is waste. Project leaders need to drive closure of options as well as at times driving keeping them open.

9.6 Solution Implementation

In traditional project management, these three elements (quality, timing, and cost) are critical in a team's ability to understand any project. To illustrate this, think of a simple weekend project you may have done or at least seen done on home improvement television programs – Building a deck. There are six ways a deck project could roll.

1) **Money** is a fixed non-negotiable budget and quality is also important.
2) **Money** is a fixed non-negotiable budget and timing is also important.
3) **Timing** is an absolute deadline and cost needs to be under control.
4) **Timing** is an absolute deadline and quality is also important.
5) **Quality** is non-negotiable and cost need to be under control.
6) **Quality** is non-negotiable and timing needs to be under control.

As two examples of these situations:

a) Lottery winner
You just won the lottery, so you are going to buy the deck of your dreams. And you also would like to also minimize the time of the disruption to your home life while it was being built. Then you would pick **option 6**. In this option, the scope gets set by what your dream deck needs to be. Maybe you want it to have a built-in sound system. Heck you just won the lottery, dream big. That quality of deck is then set in stone. That is what is going to get delivered. Your contractor (why build it yourself if you have money burning a hole in your pocket!) would then negotiate with you on timing. Cost is just the output of the negotiation of timing and quality. If you negotiate for an earlier timing, cost likely must go up as your contractor needs to work longer hours or hire more people or must expedite building supplies.
b) House flipper
Assume you are attempting to get rich by buying a house, fixing it up, and reselling it as a profit. Cost is critical to your goal, but timing is also a consideration. You might want to pick the **option 2**. A rock-solid budget would be set and the timing is something that is negotiated. It would need to fit in with the schedule of the rest of the project. Quality is what then floats. For example, if the deck size could range between 20 ft. × 20 ft. and 15 ft. × 20 ft., the budget could drive the quality to selecting the smaller cheaper option.

All projects and project solutions flow like this. There is a primary driver. The secondary driver is then negotiated, and the last factor is a consequence of the other two drivers. Most sponsors want all three. They cannot have them.

Otherwise, everything would be free, in place now, and the highest quality possible. As a project manager, your job includes explaining this to your sponsor.

In FPS projects, most of the solutions fall into option 4 or option 5. Sometimes it tends to be a huge burn rate (crisis project). For example: The project is costing $1 million/day. At that burn rate, cost is rarely a driver (option 4 – **Timing** is an absolute deadline and quality is also important). Or, it may be the nagging never been solved problem, costing us a lot of money, but that cost is now accepted as the cost of doing business (chronic problem project; option 5 – **Quality** is non-negotiable and cost need to be under control). We will cover the solution strategies for both options. Obviously, there are many situations that fall beyond those situations, but these two cases cover most FPS problems, so we will cover these in detail.

9.6.1 Solution Implementation in Crisis Project Type: What Is Success? What Is Necessary for Success?

Crisis projects have high burn rates; we are losing millions of dollars a day. This is an option 4 situation – **Timing** is an absolute deadline and quality is also important; Cost is the outcome of the first two. Who cares what it cost to fix it? Who cares about managing costs? Burn rates are high. We need quick solutions that work. Having said that, on a crisis project, solid cost management strategy is necessary for success. This seems contrary to intuition. Saving a dime on a crisis project is the least of your worries. This is NOT about saving a dime. This is about effectively spending that dime.

Cost management on a crisis project is kind of the opposite of most projects, because most projects are set up with layer after layer of cost controls. It can be difficult to change a culture ingrained in constantly pinching every penny. Workflows that work for most projects are often completely dysfunctional on high burn rate projects. Budgets that managers are rewarded for controlling are going to get blown away and there can be budget casualties. As a project manager, these are real barriers to your ultimate success. Your cash spigot could be ultimately turned too low to be successful. Navigating these waters is important as well.

On a real project situation, a team needed to spend an order of magnitude of hundreds of thousands of dollars on installing a solution. The problem burn rate's order of magnitude was millions of dollars each day. North American sales for this product segment were rapidly driving to zero. Major customers were going to shut down. To get funding approval for the new equipment needed, the workflow in place commanded that the project be presented to the right people, that the team justify the spending, that the team calculate a net present value, etc. We all know the drill. The person in charge of this workflow was considered good at the job because of they were a relentless stickler for following the workflow. In most projects, that was the right answer. The project leader was focused on the north star of success in getting the plant back up and running. The workflow was in the way and everyone in that workflow were bugs to be stepped on and that is what happened. A crisis is really no excuse for disrespecting people just trying to do their jobs. There are better ways to handle this.

The two main successful strategies are:

1) **Work with your sponsor and accounting/finance people to get a special cost center for crisis project spending up front.**
   There are several reasons this is a smart option.
   a) You may have to track the costs for legal reasons.
      Often, in these crisis projects, there can be legal or insurance actions. For example, your catalyst supplier knowingly sent you out of specification catalyst that ultimately caused your problem. Legal could be involved. Companies have various kinds of insurance. For example, the fundamental problem could have been a reactive chemicals explosion that needs to be understood prior to returning to operations. The lost production and cost to get back up and running may be under an insurance policy.
   b) Normal cost/budget tracking is blown
      It is highly unlikely on a crisis project that zero cost is the outcome. Management will want to understand "What did this thing cost us?". Separating a cost center for tracking it efficiently answers the question.
   c) Normal spending approval workflows get in the way
      Setting up a separate cost tracking system can also enable a separate approval process. You can work with your sponsor and prevent crazy spending with some gate keeper approval and yet still not be bogged down by workflows not designed to handle a crisis. An example of a simple approval process is for the larger $ spending. If the sponsor, project manager, and somebody in the finance/accounting realm agree, the spending is approved.
2) **Work with your sponsor and key workflow people to set approvals on a fast track.** This option is really about greasing the wheels. Involve the folks within the system to pull out the stops. Having a meeting that explains the situation and puts everyone on high alert can go a long way.
   There are a number of reasons this is a smart option.
   a) Working with the existing system can be faster than setting up something outside the system.
      Most of the work involved in our systems is waste. The work grinds through bureaucracy and rarely touches value. That stinks in normal situations, but in a crisis, it can be crushing. To minimize the effect, often do a blitz. Get

everyone that needs to move the ball forward into the same room and run through the steps. You will be surprised how little work really goes on relative to the time a typical workflow will take. Fast tracking can be as simple as getting the right people in the room to just get the job done. With the burning platform of a crisis, mountains of bureaucracy can be moved.

b) Working with the existing system can help gain buy-in.
Managing a crisis project is exciting, potentially one of the best things to happen to a project manager's career, and ironically . . . potentially deadly to your career. As a project manager, you will be under the microscope of everyone affected by the crisis and everyone on edges of it. Water cooler quarterbacks will critique your every move. Frankly, nobody is very happy during a crisis. As a project manager, you can be the punching bag for all of that pent-up stress. One of the best ways to deflate negative energy is to give folks a job to do. Pulling in people and giving them a mission like, get this spending approved will pull them into the team. People focused on moving the ball forward are much better than the folks that feel like they are on the sidelines.

c) When the problem goes away, everyone will forget the situation existed.
Burn rates are VERY strange beasts. It would seem like simple math. We are not producing because of this problem. We make $1 millions of earnings every day we produce. Obviously, if we are producing ZERO, this problem is costing us a $1 million/day. Fixing the problem and getting back up and running fixes the burn rate. Simple. Unfortunately, as a project manager or project team, you will find that there is an amazing amnesia that will permeate the organization once the crisis is over. The burn rate is forgotten, and the new focus will be what it cost to fix it. The focus naturally goes there because it is unplanned. Think of it this way. If your car is totaled in an accident and you no longer have use of the car, that fact permeates your entire life. How will I get to work? How will I get to my doctor's appointment? The pain of finding a ride from a friend or having to hire a taxi weights heavy on the mind. Once the car has been repaired, that pain is quickly forgotten. The new pain and focus becomes the cost of the repair. It is just human nature to move on from past pain to current pain. With a crisis project, that natural tendency can cause pain and suffering for you and your team. Once a project is done, everyone that is a stakeholder is just staring at the cost to fix it wishing it would have been zero. Zero of course is impossible, but sometimes humans wish for the impossible. We rewrite history almost as fast as we make it.

The strategy of fast-tracking existing workflow can help when this moment comes. The very folks that could be problematic were almost certainly in the existing workflows and were signing off on the decisions as they were made.

Having stakeholders deeply involved in the spending decisions helps keep the dialog on the burn rate, not on the spend to fix.

For example: A team was working on a crisis problem, we had a color-causing contaminant in our product that was making all of the white product lines pink. Obviously, making our biggest seller white was no longer an option. The bridge strategy (strategy to treat the problem until the final solution is found) was to do submicron filtration. Temporary filters were brought in and money was flowing out the door until the fundamental problem solution was found. Through an extensive measurement plan, the problem was solved by simply closing some valves and ultimately longer term removal of some carbon steel line. The problem was complex, but the solution was simple. Because the solution was simple, the rewrite of history in the minds of some of the management and others outside of the project inner circle was that the costs should be practically free. That was not the case because the bridge strategy was absolutely anything but free. Because the team followed the existing spending approval workflows (fast tracked!) with standard approvals, they brought in the high-level decision makers into that bridge strategy decision. These decision makers quickly stopped the garbage and history rewrite going on after the project was done by rewarding the project with a high level and respected technical award. The high-level decision makers had the ultimate buy-in. Documented decisions that they made.

Cost Strategy Is Chosen. Now What?

Once you have selected a strategy for managing costs, it will take relentless communication about the burn rate with all your key stakeholders to keep that cash spigot open and resources flowing without barriers or blockages. For example, procurement needs to know that the things being bought that are on the critical path need to be expedited and bird dogged until delivery is confirmed. Make sure your communication plan for solution implementation includes the cost discussion. This cost discussion may feel like wasting time.

This cost discussion may feel like wasting cash. To be clear, it is not. With the fundamental problem-solving methodology, the money is very laser focused. The cash is aimed at fixing the true no fooling root cause of the problem. This is much different than many crisis situations. If solid science and systematic approaches are not being used, then the cash being thrown around is aimless; jumping from one problem symptom to another; getting tossed at this moment's guess at the problem caused. Cash flows FASTER out the door when knee-jerk problem fixes are driving actions.

> ***Wisdom Tidbit: A high spending profile need to be LASER focused on the solution to the problem.***

9.6.2 Solution Implementation in Chronic Problem Project Type: What Is Success? What Is Necessary for Success?

Chronic life of the plant-style problems have a tough uphill battle during the solution implementation phase due to credibility. Remember, these are problems people have hit and hit again. We have seen 70+ year-old problems successfully solved with FPS, but problems that old often are completely accepted as-is. They are like gravity. They are going to be that way forever. When a team goes in and understands the science of what is driving the problem, it is exciting. The old problem is discovered to be fixable. The new battle becomes convincing operators, mechanics, engineers, chemists, management, and the rest of the army of people that have come to accept that this is the way it will always be. Because of this quality is almost always the focus of this type, we need to make certain the solution will work.

Go slow to go fast. Focus on good pilots and measurement plans. Focus on independent validation of results. Quality of data and getting respected experts on board with the solution will win this war. The effort and time you take in making the validation work bullet proof will pay dividends. Your sponsors and stakeholders have to believe enough in the work to foot the costs and to be advocates to make the changes. A key tactic to use is peer reviews of the work. This is a different kind of stakeholder review with the technical experts instead of management. Get the rocks that are going to get thrown at your solution out in the open. What are the technical expert's major concerns? What work needs to happen to alleviate those concerns? Unlike normal stakeholder reviews, these will include much more time and details of the work. It will include some alone time for the experts to do the work to understand the data.

As a project leader, the work to convince stakeholders is good enough game. Hundred percent of the people will not buy into any solution if you have a healthy productive workplace that values independent thought. That means 100% support should never be your goal and that is not what success is for these projects. Success is that there is enough stakeholder support to move the solution with sound science behind it forward. Work to get the quality on quality-driven projects.

Cost is the negotiated factor on these projects. Work with your sponsor on spending available. Calculate the cost–benefit analysis backward. How much is this costing us and how much can we afford to spend to fix this? Then, calculate it forward. Calculating the cost–benefit analysis on the cost of the solution and the benefit that will result will tell your stakeholders how much it is worth (forward analysis). For example, if you calculate that saving $1 million/year for the next 10 years is worth a $4 million investment, you then also calculate the cost of your solution and say, for example, it is $0.5 million. Clearly your team could spend more on the solution implementation and still do well. Because solutions have already been screened on cost, the solutions that cost more than their benefit have been eliminated as non-starts. Solutions remaining have room. *There better be room!* That means there is a gap between what you plan on spending with a solution and what you *could spend* and still have happy sponsors. The negotiation part is how much are your sponsor and other stakeholders willing to "risk" on solving this life of the plant-style problem. The negotiation needs to take place because certainty is never 100%. Your team hopefully has made great strides toward the target of certainty; but pulling the trigger on a solution is the only endgame that will get there ultimately. Your sponsors and stakeholders need to be prepared and that preparation is a negotiation. The value that your team is negotiating is more than the cost you plan on spending on the solution and less than the most you could spend and still be happy with the outcome. The purpose of gaining this understanding is that on a quality-driven project, there are often solution iterations. Life of the plant problems have been that way for reasons. Although your team may have killed hypotheses and validated the solution during some scale of piloting, there is still implementation of risk. Knowing the tolerance your sponsors have for spending is very useful for project teams. Your team may need it. Communicate ongoing costs relentlessly with sponsors. They need to be involved. This is a negotiation, so there can be a trade-off to get the quality with cost.

9.6.3 Solution Implementation Planning

In general, at the solution implementation phase, your company's workflows take over. For example, if a better filtration system needed to be installed, you will be following your company's capital workflow. If a new catalyst activation procedure is needed to fix your problem, your company will have approval workflows for procedure changes. If your solution changes the control strategy to fix an overconstrained mass balance around a distillation column, there will be approvals and workflows for that. If your solution was to switch raw material surfactant supplier to one that did have the right distribution to do a better job keeping particles in solution, then there will be qualification workflows that your team will follow. So, for this part, we will discuss the implementation planning that is **not** in the standardized workflows your teams are likely follow. Those workflows will be specific to the company. We will instead cover the implementation planning specific to FPS projects.

FPS projects will include in addition to standard workflow solution planning several additional elements. These include:

1) Success criteria definition (What is success?).
2) Overall plan generation, monitoring, and adjustment.
3) Stress testing planning.
4) Change management plan.

5) Solution-phase stakeholder communication plan.
6) Control plan generation and validation.

Success Criteria Definition (What Is Success?)

Just like the pilot phase. Measures of success need to be put in place. Reread the success criteria part of piloting. This is the same, but the success is just at the larger scale. Pull out your original "What is success?" definition. Your success criteria need to have this front and center. This will be a PDCA cycle. Some of your success criteria will likely not be met and will need adjustment. Put that into your planning. Often, writing this adjustment cycle into the plan and communicating it to sponsors will defuse the pain that can come from everyone expecting perfection out of an imperfect world. Success is a matter of persistence.

Overall Plan Generation, Monitoring, and Adjustment

To generate the plan, there are a couple of concepts and techniques that help teams out. Balancing between the two errors of planning, natural planning, and lean/just in time planning.

Two Errors of Planning

There are two errors typically made by teams: overplanning and underplanning. Overplanning is thinking of everything and creating a project plan with order of magnitude hundreds or even thousands of lines of milestones and tasks. Overplanning is overall a better error to make but does fail. One reason overplanning fails is because future predictions always fail. There will be changes to the plan. We have seen teams spend increasingly high levels of activity just on the plan trying to fix it due to changes. Planning is not value-added work; it can enable efficient work if done well, but it is not value-added work. Another reason overplanning fails is that overcontrolling every task often misses the big picture. What outcome are the tasks trying to achieve? Get the outcomes right and the folks doing the work will find the most efficient path to the outcome. For example, if the solution needs another temperature measurement installed in the reactor, so that the initiation can be more tightly controlled to maintain high levels of conversion, then the outcome is an operational temperature probe in the right spot of the reactor. Focusing the plan on the outcome allows the folks that are tasked with the detailed work to do their personal planning on how to accomplish the outcome. Maybe the engineer in charge of this outcome discovers that there is a nearby thermocouple that is no longer being used. The wires are pulled and a slight re-engineering and the outcome of installing the temperature probe in the right spot is achieved. Overplanning could have focused on running the wires or designing the loop; the right thing to do was none of those things. Avoid overplanning by making outcomes clear and stay out of the weeds of how they are achieved. That does put the work into making certain the outcome is clear. Note that planning needs names to outcomes; people need to know they are responsible. Keep it simple. Plan to right level.

Underplanning has many of the same outcomes as overplanning, but for the opposite reason. People need to be able to understand what tasks need to happen to do them. Coordination of effort is difficult to achieve by guessing. Guessing is what everyone on a team does when there is insufficient planning.

Natural Planning

Natural planning is a combination of benchmarking and just using the human potential and talent of your team. In natural planning, the team gets yellow stickies and writes down the major steps/sub outcomes needed to achieve your overall outcome. Then, these are grouped on a wall to eliminate duplication. The team then lines up the timeline. The big outcomes may have been done before. For example, putting in a new procedure for shutting down to prevent plugging, may have been done before. Shutdown procedure writing and approvals are likely a workflow. In those known well-trodden paths, then brainstorming step is not needed. Just benchmark and use history to determine steps and how the plan should flow. Starting with a benchmarked baseline is quicker. It likely still needs to be adjusted to the current situation; but having a successful starting point is useful. Assigning owners to the individual high-level plan steps/outcomes is also essential. People need to know who is doing what. Natural planning is simple, but it works.

Lean/Just in Time Planning

Getting it right on planning is a balancing act. Some average rules of thumb that are helpful are plan execution check in and plan adjustment happens zero to one week. Set a cadence such as Monday morning or Friday afternoon the team checks in on the solution implementation plan to make the plan adjustments and communicate details to each other. This is a PDCA cycle. Detailed planning should be done on the work that is one to six weeks out. High-level planning is usually enough for further out in time. Capture the details folks think about; but getting that perfect months out is not useful. The exception is on planning parts that are critical/high risk. Those elements deserve early attention at a detail level. Capture the details of those and talk about those elements every planning meeting. Note this scales to the speed of the implementation. For example, if our solution implementation timeline was only two weeks, the cadence of the planning meeting would be faster; for example, daily.

Theory of Constraints Planning

This complements the other discussions in this section. It is well described in the book Thinking for a Change by Lisa Scheinkopf. It describes well the use of buffers in setting schedules and managing the variability in completing subprojects.

Stress Testing Planning

Include in your solution implementation the extremes. For example, we have seen project teams bird dog every step of the procedure. That is helpful in getting it right; but long term it is unlikely that teams will be surrounding an operator every time they do a procedure. Instead, during an implementation, often it is better to train and then watch minimizing intervention. Stress testing is pushing your solution toward the conditions it will see when your team is disbanded. Plan to do this work. That may mean artificially inducing situations. For example, maybe there is a regularly occurring event where the distillation column goes on total reflux. If your team excludes that event from your solution implementation, then your team will not have a handle on if your solution will work under those conditions. Stress test planning is incorporating the extremes of your normal operations into your solution testing and will help prevent crisis.

Change Management Plan

Solutions are going to change what we have been doing and will make us do something different. That is inherently counter to our creatures of habit nature. To make changes stick, it is critical to start early in the solution phase. Put yourself in the shoes of the folks having to make the change. Ask them about their concerns. Bring those folks into the planning phases. Address the concerns. Change management plans are all about thinking through the impact of the change.

People involved in the change will all be asking WIIFM – What Is In It For Me. Have answers and understanding ready. Have a sponsor for those changes. Build a case for why the changes need to take place. Making money for the company is not a case for change. Aim higher. For example, suppose your problem is that the reactor keeps crashing every week and we are having to reload it just to run. The case for change for your solution is NOT to make the millions of dollars your company may be losing with the problem. The case for change needs to be focused on the impact that problem has on those fighting it. That will get buy-in.

Cast a vision of what it will be like once we go through the pain of implementing the solution. For example, we will be able to run for months instead of weeks without having the reactor crash with us fighting to get it back online.

People also need the skills to do the change. Have a robust and effective training plan. Pull out the science developed in understanding the problem. Your team needs to be education ambassadors communicating the new science found. New procedures need to be practiced. Skills need to be in the hands of those affected by the changes prior to the change happening.

Folks need incentives as well. Solutions need to have a positive feedback loop. It is good to make things easier than they were before. Solutions that add to complexity and difficulty of jobs are uphill battles that take much more effort. One key question in the solution phase that needs to be considered . . . Is this going to be harder than before? Is the medicine worse than the disease? Other incentives in addition to the solutions just being better than before also need to be there. Did you reward the work in developing the new skills? Did you celebrate successes along the way during your solution implementation? Feed the excitement and you will get excitement.

Resources must be in place, or your changes will fail. Error on the side of extra help. Be ready to adjust plans. Implementations fail 100% of the time when the resources to get the job done are not there. This should be part of the regular cadence of planning. Lean on your sponsor to do their job.

The plan needs to be in place and communicated relentlessly to those impacted by the change. When people know what is coming next, there is less push back and there is more alignment of activities to help put in place the work it takes to get the implementation done. Have a communication route for folks to escalate problems. The plan will flow better with adjustments made from those escalations.

Ultimately, a control plan is needed to keep your solution in place and working. We will cover this in detail in the next chapter. The control plan is really casting the team's thoughts 5–10 years in the future to think about how entropy will creep into the solution. How can that be fought, and the gains sustained? Control plans are the ultimate critical to success part of all changes.

Change management planning will make the difference between great solutions that succeed and great solutions that fail. Failure to manage change is exactly losing control of your change. A successful change management plan needs a sponsor providing resources and support, a case for change, a vision of the future after the change has been implemented, new skills need to be in place, incentives should drive the change forward, resources to make the change happen have to be there, the change needs to follow a well-communicated plan, and ultimately a control plan has to be in place to sustain the gains. Plan your change or your change will have its own plans.

Solution-phase Stakeholder Communication Plan

Your overall communication plan will need to change as solutions move forward. More people need to know what is going on with the work. Expand it to include the folks impacted by the solution. Communication methods will likely also need to be updated. Communication with your business leaders sponsoring the work is different than communicating with the operators and mechanics implementing solutions. Ask the folks impacted by the solution, what do they absolutely need know? And what would they like to hear about. Prioritize your communication content by those needs. Effective communication is a critical to success factor in the solution phase.

Control Plan Generation and Validation

This element is what drives sustainability and needs to be in your overall plan. We will cover this in the next chapter.

9.6.4 Solution Validation

After the plan has been executed and your solution is running, your measurement plan being pushed up against the success criteria your team has defined will determine if the solution needs more work. Obviously, if everything is going as planned and working, this phase is done. As stated before, this is not always going to be the case. Communicating with your key stakeholders that there will likely be adjustments to the solution is prudent. The Plan-Do-Check-Act cycle is your friend on those adjustments. Solid workup to this point generally makes this simple and straightforward. A few things could have been done better, fix those and roll on. Sometimes, easy is not the road forward and it will become important that your team be prepared to make the decision on what next. Although these cases are rare, being prepared to quickly move forward is a matter of rigorously looking at your success criteria. This will help you answer the basic questions. Do you keep going trying to make your solution work? Or, do you recycle to the solutions you had put on the back burner?

Solution Adjustment vs. Going Back to Your List of Solutions

Preparing your success criteria before you hit this stage is very powerful. You have the line in the sand on which side success lies. Teams without those lines in the sand will often flounder. They can argue over good or bad. Remember your team has spent time and effort to get this far and would rather be done. That line says done or not.

To know if adjustments are what it will take or is the solution dead in the water is a question of fatal flaws. Are there reasons that the solution cannot ever get better? Is the distance to success too far? Learn from those flaws and look at the solutions you set aside for ones that will be more robust against those flaws.

Gather your results, pull the stakeholders together, and present what you know. They need to buy into more work on the solution or the abandon and recycle direction. Get a decision. These are tough meetings.

9.6.5 Solution Phase Summary

Solution phase is flat out hard work. Generating, selecting, and implementing solutions takes planning, discipline, and teamwork. Communicating and doing good change management will help those impacted adopt the change. The outcome of this work can be very rewarding because this is when the problems go away. This is when the project gets done. It is time to celebrate. Plan for that. Plan to pause and celebrate your team's success.

10

Sustaining the Gains and Control Plans

Why does entropy seem to win on problem fixes over the long haul? Talk to folks with time under their belt around your company and you will likely hear laments about having to repeat work to fix the same problem again and again. Solutions just do not feel final and permanent. As we bring about a close to our project, it becomes time to think about how to make our now proven solutions stick. They need to stick in ways that prevent future rework. We want them to keep solving our problem for many years to come. This is where control plans come in to play. Control plans that include clear documentation will be our foundation to sustaining gains.

To think about sustaining the gains, we also have to think about the complete opposite first. We have to examine failed unsustainable solutions to help move our projects away from that space and into the space of longevity. Once we understand the root causes of unsustained gains, then we can construct project control plans that do sustain the gains. In this chapter, we will work to understand why solutions fail to sustain gains and then we will examine the work it takes to construct a robust control plan to sustain the gains with proper documentation, multi-generational training plans, robust solution change management, sustaining metrics, while minimizing the sustaining energy needed, and also share other corporate memory loss prevention methods.

10.1 Why We Fail to Sustain the Gains

To understand solution longevity, we must look back at old projects and their solutions. We need to do that on a big enough data set to understand the drivers involved. In this section, we will cover the main root causes found during that

Fundamentals of Industrial Problem Solving: A Practitioner's Guide, First Edition.
Zdravko I. Stefanov, Eldad Herceg, Carla Schmidt, David M. Jacobson, Dana Livingston, J.P. Chauvel, Sunil Kumar Chaudhary, and Christopher Paul Christenson.

study that contribute to shortening the life of solutions. We will tackle understanding of **Why We Fail to Sustain the Gains** because it will help set the stage for building a robust working control plan.

This feeling that knowledge is eroding, and we are losing control of our processes is hard to shake. Like many things within company cultures, it has two parts: **Perception** and **Reality**. In gain sustaining, these two parts need to be attacked differently. **Perception** is a measurement, a communication, or even a marketing problem. It will have different root causes when compared to trying to fix **Reality**. **Reality** goes deeper and is a much more costly problem and has its own set of root causes.

10.1.1 The Perception Problem

The first perception problem with solutions has to do with that our improvements are often only one step on the journey. In an example, an engineer asked to be coached on a project to decrease catalyst usage. The catalyst was very expensive and reducing it would also have added quality benefits to our customers. It was a good project. As the project was discussed, a sinking feeling hit. He was working on a problem that was "fixed" a couple of decades ago. The second it was over; the coach was diving into his old files. What went wrong? How could things have fallen apart? Did everyone forget how to run the process? After finding the old reports, that sinking feeling quickly shifted to euphoria. The baseline "poor" catalyst usage that he was being sponsored to fix was BETTER than the best performance that was achieved with the earlier work. The coach realized that this was just another step on a longer journey. There was not backtracking going on; just more moving forward movement needed.

Thinking about the ***important*** work, we quickly realize that many of our most important works ***must*** keep going. For example, what is your car's gas mileage? And do you want better gas mileage? We suspect that some version of this question will be there many generations from now. It may be measured in kilowatt hours/mile; but it will still be there. We will want our transportation to continue to be more efficient. Chemical processes face similar long-term needs. We want everything better, faster, and cheaper. We want less waste, lower energy usage, higher quality, better conversion, higher production rates, better productivity, and so on. So, part of the perception problem faced in sustaining the gains is rooted on the need to continuously improve things that are important today and will likely be important tomorrow. Auditing old projects has shown that a significant part of poor sustainability is related to a ***perception*** issue because more work continues to be done to improve the same metric. This issue occurs even though the gains have been successfully sustained. To keep moving an important metric over many years in the direction of better will almost always require solution modification or even replacement. Future teams modifying your team's solution for the better is a good thing that can generate a bad image for old work. Keeping this perception problem away from dogging your FPS project requires that ***there needs to be a history of that critical metric, a history of the changes that have been made to improve it, and it needs to be well understood by those folks impacted by the metric***. The full story of that metric needs to be told over a long period of time. These components need to be part of your control planning, otherwise there can be a perception that the work failed when in fact what is really happening is that the next generation is just continuing to improve.

The second perception problem is a variation problem. Imagine you are working on a team that is put in place to increase the capacity of a plant and achieve higher production levels. Current average capacity clocks in at 100 metric tons/month and the plant is totally sold out. If your team can make more, the customers need it; so, there is more money to be made and the potential to make your customers happy. The current bottleneck is distillation, so the team focuses there. Your team figures out the fundamental issues that are constraining distillation and successfully implements solutions that increase overall plant capacity by 10% by debottlenecking distillation. Using gamma scans, the team found that due to an installation error a distributor tray has a significant slope, causing the system to have a skewed distribution. That distribution problem ultimately leads to capacity constraints because to meet specifications, the distillation feed rates needed to be limited. The team corrects the sloping tray problem. With solutions in place, the plant successfully demonstrates an average of 110 metric tons/month for over a year. More product is sold, the customers are happy, sales records are set, and the FPS team is rewarded. There is a nice celebration held that is an event that people will remember.

This success sounds great! This is why we got into this game! What happens to that system as time goes by? Does that plant continue to experience 110.00, 110.00, 110.00, 110.00, 110.00, . . . production month after month forever? Maybe on paper or in management's dreams! The reality is even with a system that is in a state of statistical control, there **will be** month-to-month variation. The months don't even have the same number of days in them! How could rate ever be perfectly the same? This and other factors will cause variation. Processes have supplier, process, and measurement variation that all will result in month-to-month variation. An average of 110 is more likely made up data that is a lot messier. . ..

Table 10.1 Actual production rate data.

Month	Jan.	Feb.	Mar.	Apr.	May	Jun.	Jul.	Aug.	Sep.	Oct.	Nov.	Dec.
Production	110.8	113.8	114.2	106.7	102.7	118.5	108.1	105.9	108.0	102.8	111.3	119.5

The data set shown in Table 10.1 is normal. This is the way processes run. With this data set showing the average at an improved level, we can have the expectation that this will roll forward and the next 12 months should be at the same improved level. But folks that are less clear on the role of common cause process variation will point to December and ask . . . "Why can't every month be December?" Or others may point to April and tell the team, "The fix your team is has put in place is no longer working!" Common cause variation can cause a control plan perception problem. To avoid a common cause variation perception problem, include control limits (the variation component) as part of the solution control plan. In the case of the example Table 10.1, the solution delivers an average of 110, but individual months will vary + or –20. Note: Xmr calculations were used to calculate upper control limit of 130 and lower control limit of 90. See Figure 10.1 for what that Xmr plot of these data look like. So, when January rolls around and production clocks in at 95 . . . there should be no panic. There should be a calm expectation that the average will hold unless something special is happening. If January rolls around and we produce 70, then it makes sense to ask questions because that is unexpected. Educating folks on common cause variation will help prevent this source of a perception problem.

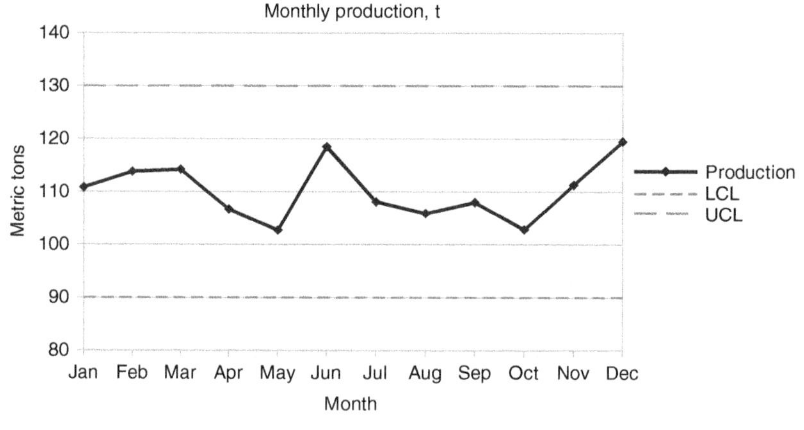

Figure 10.1 Xmr plot of production data.

The reality is there is also special cause variation in all parts of the system including parts that the team did ***not*** fix. That variation can and will happen even though the process has improved. For example, there is variation in customer demand. What if a few years down the road there is a downturn in the economy and your customers are no longer needing as much? They only need 50 tons/month. Running at full rates would just build inventory and costs money. The sensible thing is to decrease rates to match customer demand. The right answer of lowering production rates can create a perception problem for the solutions the team put in place. Running at reduced rates of 50 tons/month will just plain look bad even though the solutions are still working.

What if a pump in another part of the plant that feeds the distillation column becomes unreliable due to a string of bad replacement parts? The plant experiences a new bottleneck and a new source of production rate variation that has nothing to do with the distillation column. Your team's fix to the distribution tray did not make every piece of equipment in the plant perfect. Special cause variation outside of the fix can create a perception problem.

What if a new problem creeps into your distillation column system? For example, what if under insulation corrosion is found during a reliability inspection and the distillation column cannot be run until the corrosion issue is fixed. Fixing every possible future problem was not in your team's scope. Restricted rates due to an unrelated problem within the distillation system can create a perception problem.

A ROBUST reliability tracking system is needed to help prevent special cause variation outside of your teams fix or even within the process that your team fixed from causing a special cause perception problem with the control plan. Look not only at trying to sustain the gains of the fix, but also to help future generations or people outside the team to clearly understand when the fix is working and not working. This understanding is critical to chasing the right problem without messing with fixes that are still working. Your team will not fix everything that can possibly happen moving forward. They can and will only fix the root causes of what is happening now. There will be new work to your metric. Plan for it and plan to manage the perception problems that these future special causes will cause.

Perception problems caused by variation are often a culprit in the failing to sustain the gain. It is a problem area that seems to come as a surprise to teams. We are deep into understanding our problem and therefore fail to plan for it. To sustain gains, your team will need to manage not only the reality of sustaining the fix but you will also need to plan for perception problems. To include common and special cause variation sources to the perception problem dilemma, ***there needs to be a history of the critical metric that includes a variability component, a history of the changes that have been made to improve it, a robust***

understanding that separates when the fix is working or not vs. other parts of the system working or not, and it needs to be well understood by those folks impacted by the metric.

These key perception components need to be part of your control planning or your solution can suffer the fate of failing to sustain the gains.

Summary of the *Perception* problem components of a control plan:

- History of the critical metric to show progress over time.
- History of the normal variation to expect with the critical metric.
- History of changes made to improve the critical metric to show how improvements have been made.
- Robust plan to validate when the solution is working and not working.
- Robust plan to identify new problems and trigger a targeted response to the new problem.
- Communication plan to keep stakeholders involved and informed over the long haul.

10.1.2 The Reality Problem

The **Reality** problem is the worst of all worlds. Solutions do fail to sustain gains. History repeats itself. It goes beyond the painful, but not so costly **Perception** problems. **Perception** problems cost us mostly in credibility that can cause sustaining continuous improvement programs harder. But because perception problems are just that . . . perception, they do not initially cost us in bottom line dollars. **Reality** sustainability problems are costly. They are solutions that truly fail over time and problem costs carry forward despite the investments made to solve them. Our burning platform keeps burning. Problems haunt our processes repeatedly. Solutions that work get kicked to the curb. It is kind of depressing to think about all of the hard work throughout industry that ultimately is being wasted. How much of all the cost in making a product we use every day just goes into the ongoing problem costs of problems that have already been solved? Unfortunately, failure patterns repeat themselves. But also ***fortunately*** failure patterns repeat themselves making them easier to prevent. In this section, we will look at common control plan failure modes and how to prevent them. Common solution sustainability failure modes include:

- High-risk problem types
  - Low frequency
  - Counter intuitive
  - Reward conflicting

- High-risk solutions
  - High-energy solutions
  - Soft changes
  - Lone wolves
- Poor solution implementation
  - Poor change management and buy-in
  - Roles and responsibility breakdown
  - High-skill training breakdown
- Outside forces conspiring against sustainability
  - Rotten apples
  - Business cycle failures
  - Reorganization disease

Thinking about these modes prior to building your control plan will help your team sustain gains.

High-risk Problem Types

Some problem types are almost crying out that they want to be unsustainable. We will cover the three main types, but obviously there are other cases. As a project leader and as a team, just asking the question whether this problem and solution likely to struggle holding its gains is a very useful exercise for the team. It will get them thinking about the risks and how to prevent them. You can also use this chapter to prompt the team on these common failure modes.

Low Frequency

The first of the three high-risk problem types are driven by the frequency the problem occurs. Problems come in many flavors. One flavor is the problem is happening every minute of every day of every month of every year. Another flavor is the problem only happens on the order of magnitude scale of years or decades. For example, floods or hurricanes can totally disrupt your business; but these generally sit on the scale of years or decades or even centuries. For sustaining gains, the problems that are less frequent tend to be the hardest to sustain. By their nature, they are easy to drop from corporate memory. For example, a fundamental problem-solving team was called into a reactive chemicals event where a process experienced an explosion. That explosion was the first of its kind. The team discovered that contaminants had adsorbed into the vessel wall. The contaminants under normal ambient conditions did not react. But the shutdown when this explosion occurred was during an especially hot summer day and ambient temperatures were not normal. Clean out procedures and system monitoring were changed to address the issue. Obviously, this is a high impact problem. Nobody wants to repeat that problem. But think about what happens 30 years from solution implementation. The people involved have mostly moved on to other jobs, retired, or left the company. The ones that are left may or may not have been deeply involved in the event. Sustaining gains under these circumstances is going to be difficult. Low frequency of problems is clearly a high-risk factor and is a common failure mode.

The root cause of low frequency problems is the slow feedback loop involved. If there is near instantaneous feedback that a problem is happening, then it can be quickly corrected. For example, a fundamental problem-solving team was tasked with improving the yield out of a reactor system. The team changed the heating profile to initially be cooler and then throughout the course of the run to increase temperature. The team was taking advantage of the kinetics of the main reaction and the side reactions. Improved yield made more money. Yield was tracked every run. If that temperature profile deviates, the low-yield situation provides immediate feedback. Compare that to the feedback loop in the explosion case. Doing the right monitoring and the right cleanout procedures results in nothing happening. Doing the wrong monitoring and the wrong cleanout procedures also ***almost always*** results in nothing happening. Ouch. That feedback loop is going to lead to problems.

We will cover project documentation as a control plan element later, but ***how and what*** you document when faced with a low frequency problem is important. We want to make it as visual as possible. Early in his career an engineer was working in an area that many years previously had a major incident. In those days many years ago, the plant Nitrogen supply was a trucked in supply. During an offloading period, a driver hauling Oxygen modified the loading arms to be able to off load the Oxygen instead of the Nitrogen set up. A plant that was dependent on Nitrogen inerting was instead fed Oxygen. The entire tower ultimately burned down. As part of the insurance claims, they had to photograph the damages. The engineer shares "Because it was much earlier than my start date, I had nothing to do with the event, but I still remember the event decades later. Those pictures left an impression on me that I cannot forget." That kind of impression is what needs to be put into a corporate memory to help sustain the gains on low-frequency problems. There are many types of problems that can benefit from that visual picture of actual consequences. For example, a significant number of fundamental problem-solving projects have been color-related quality problems. If the product is supposed to be white and it is hot pink, a picture of what bad looks like helps future teams remember why they are putting in the ongoing effort to prevent the problem reoccurrence.

For some low-frequency problems, your team will need to be creative to show the consequence in a visual way. Once a team was trying to communicate the impact of the increased waste that occurred for a low-frequency problem. Their answer was to show how many miles of waste removal trucks parked end to end the problem caused. A graphic with little trucks parked end to end and a mile measurement did the trick to have the impact desired of never ever wanting to have that problem again. Maybe the toughest one to make visual is when the impact is only money. In most cases such as the waste and burned down tower examples, it is better to send the message with something other than money. This is a human systems strategy. Employees do care about high-level concepts like the profitability of the company; but what they care most about is the impact to their daily life. It is not that people are bad or do not care about the company's success. It is that soft impacts like money and profit are more remote than hard impacts like "For the next six weeks at work I will need to be loading waste disposal trucks to get rid of this mistake." Soft impacts are not kickable. You cannot touch them easily. They only impact our daily lives in thinkable ways. Put yourself in the shoes of an employee maintaining a solution that fixes a low-frequency problem. Which impact motivates you more? Note: Each of these are the $ impact or the personal impact of the same problem.

1) $150,000 of a waste disposable cost or that you will have to suit up in full hazard gear risking your life to do a cleanout.
2) $1 million in lost sales due to a quality problem or that you will have to drop everything you were doing for the next four months to answer customer complaints and retest stacks of returned samples.
3) $50 million in lost sales and capital spending to rebuild a process that blew up or that you died in an industrial accident.
4) $1 million in lost sales and maintenance spending due to a fouled heat exchanger or that you will have to work on your days off to prepare the plant to fix this unexpected problem.

Effective control plans are about motivation. With low-frequency high-impact problems, motivation is the biggest hammer you have in your toolkit.

> ***Wisdom Tidbit - To mitigate the low-frequency problem risks, it becomes critical to document consequences as visually and easy to access as possible.***

Counter Intuitive

The second type of problem that can be highly susceptible to having the right solution tossed aside has to do with a population of problems that have root causes going against the grain of basic principles of science, common sense, or conventional wisdom. An extreme example of how what we know is not always right are phenomenon like Bose-Einstein condensates. If we put a solid block of any substance on top of another solid block, the two blocks stack. Most of us as children or parents of children have stacked blocks. Our buildings are mostly solids sitting on solids. It is clearly the way the world works. Nope. Not across all temperature ranges. As we approach absolute zero, the atoms slow down enough that the two solid blocks just fall together. This is an extreme example, but it illustrates the point that everything we think we know is not going to be always true across all situations.

In the context of having control plans fail, these counterintuitive issues cause a fight. For example, we are taught Arrhenius plots in kinetics showing reaction rates increasing with temperature. We raise our chemists and engineers on this and beat it into them with homework, testing, and lectures. It is critical to do this because it is an important principle for sizing reactors, understanding how to run reactors, and many other critical to technology success factors. But what happens when increasing temperature is slowing down the reaction rate. For example, a fundamental problem-solving team determined that reaction rates were slowing down as temperature increase due to a thermally degrading catalyst. Good science that makes sense. Note that there are also many other ways this can happen.

The issue becomes that increasing the temperature to try and speed up the reaction is baked into the DNA of all the folks who will be involved in the process after the fundamental problem-solving team long ago rode off into the sunset. The failure of this control plan goes like this. A new manufacturing engineer starts their first day on the job. The boss immediately tasks the new engineer with increasing capacity as job one. Customer orders are backlogged. Asking around, the engineer discovers that the bottleneck is the reactor. They know to work on the bottleneck. Looking at the situation the engineer tweaks the temperature up to try and push the system harder. The next sample result looks good. The reactor is moving in the right direction. A great day all around! A quick stop by the boss's office to let them know about the solid progress and everyone is happy. Longer term the science established by the fundamental problem-solving team wins. The initial sample result looked better only because of normal process variation and sample timing. Now what? It is possible that the temperature change sticks. People are busy. The engineer may not have given it a second thought. It is possible that the new sample results are noticed, and the engineer figures out something is wrong and worth digging deeper. It is also possible that the new results are noticed, and it puts the engineer in an awkward situation. He told management things are better. Does everyone have the intestinal fortitude to admit that the initial stab at the situation was wrong? In this situation, the people are trying to do the right thing. They are not bad people. Sometimes doing the right thing just makes the problem worse. Our best efforts can go wrong.

The problems that need fundamental problem solving are often still there because the process is behaving in ways that are against the grain of basic principles of science, common sense, or conventional wisdom. Examples include:

- Heat exchangers getting worse heat transfer as the viscosity of the fluid decreases. The exact opposite should be true.

- Distillation columns getting better separation as reflux ratio is decreased. What? Increasing reflux ratio should make separation better.
- Putting finer filters in results in more solids getting into the product. Huh? Better filters should equal cleaner product.
- Fresh solid catalyst bed works worse than the stuff that is five years old. Does that sound right? No! Fresh clean catalyst surface should have great activity.
- Increasing the pump speed results in lower flow rates. Faster is slower. No way!

Unraveling these "Gordian knots" are challenges. Sustaining the gains for counterintuitive problems are often an equal challenge.

The way to think about, explain, and communicate these situations is **INCLUDE** conventional thinking in the control plan documentation such as the troubleshooting guides. All of our processes from accounting, to chemical plant operations, to machines, to computers follow one simple concept. The output of any process – Y is a function of several Xs. $F(X_1, X_2, X_3, \ldots) = Y$. The functions change, but the conceptual model holds. A simple example is one used earlier. Gas mileage. Physics for example tells us that the heavier a pickup truck, the worse the gas mileage. Our friends in aerospace engineering would also say that the more aerodynamic the body shape, the better the gas mileage. Thermodynamics would tell us the hotter our engine, the better the gas mileage. And many more other Xs play a role. Gas mileage Y is a function of these Xs: Engine temperature, vehicle weight, aerodynamics, and the others. These X functions can be stronger in some situations vs. others. Some are more obvious and likely to be conventional thinking. Suppose one asked which would get the best gas mileage; a pickup truck loaded with a 1-ton load or one that was empty. Most folks would quickly conclude that the empty one would get better gas mileage. Temperature would be less obvious. Aerodynamic might sit in between the two on how well it would be understood. Now, let's put it in the context of counterintuitive or contradictory Xs. The gorilla vote is that higher weight is bad for gas mileage. If putting a 200-pound cover on the bed of the truck is proposed to help improve gas mileage, there are going to be arguments about if that is better or worse. Remember that arguments sit outside of factual science. Debate is just that. A solid well-done scientific study would tell what is happening. If my control plan training and other documents just explain that the bed cover improves aerodynamics, then the control plan is going to be at risk to fail when the first person who believes weight is the one true factor comes along. That person would be susceptible to the intuitive answer of lighter is better and might go with the gut response and undo a control plan. If instead, we communicate that we have proven that the gas mileage gains due to improved aerodynamics clearly have offset the penalty of the increased weight, then this inclusion of conventional thinking lets the future people involved in the process know that this is a special case that deviates. Referencing supporting data will also help. Leaving out the conventional thinking just leaves folks the option of concluding that the original team did not think things through. The bottom line is that complete documentation and training that includes all of the factors considered, especially the most likely ones has the best chance of sustainability success.

> ***Wisdom Tidbit - Good solid communication that includes conventional thinking and robust training plans are overall the best strategy to sustain gains in counterintuitive problem situations.***

Reward Conflicting

In people systems, we always need to think about sustainability in the context of normal human responses. **If sustaining the gains is punishing to those people involved in the system, ultimately the gains will fail.** As an example, an improvement team determined that the age of one of the raw materials was adversely effecting product performance due to a slow chemical reaction making contaminates. As part of the solution, a first in-first out storage system with minimal onsite storage was implemented. This minimized the average age and decreased the variability of aging. Unfortunately, because of the storage location and the way raw material was purchased a system was set up where most of the time the first in (oldest material) was hidden behind newer drums of raw material. In order to get the first in out of storage, the other newer drums needed to be moved out of the way. Obviously, the system can and did work for some time. People can have operational discipline and move the drums to always use the oldest material. But that system is prone to failure due to punishing the humans doing the right thing. There are ways to make the system work, but why swim against the tide? How about we swim ***with*** the tide? It is much easier and sustainable. For example, changing the storage location system to sloped roller shelving loaded on one side and unloaded on the other sequentially feeds first in, first out without having to shuffle drums. Equal access shelving where old and new are exactly the same work to obtain, could also swim with the tide.

Understanding how systems can punish the people involved can take some detective work in the motivation of people. What drives the folks accomplishing the work? An improvement team was working on a product that was initially out of specification 40% of time. A first-time reject rate of 40% was costing production rate and the expense of additional testing. The process was inconstant in shipping to customers due to the rework step's huge impact on batch time variability. Rework of the out of specification material required an adjustment be made to the product by adding a component and then the product was mixed for eight hours. Rework was a major process capacity loss. At first blush, improving first time reject rate seems like something everyone could get behind. Motivations should be aligned because getting it right the first time is less work.

A deeper dive showed more complexity. The packaging part of the operation was highly labor intensive and easily the most manual work of the job. It was a highly hands on operation. Everyone hated doing the packaging. Because there was no packaging being done when the product was out of spec, rework days were a refreshing break from the grueling packaging work. What motivation is there for that team to reduce rework? Zero rework would likely mean everyday was packaging day. To be successful in sustaining gains, the improvement team not only had to solve the problem causing the rework; they also had to streamline, improve ergonomics, and increase automation on the packaging system. These changes were small potatoes costs when compared to the benefits of decreasing reject rates. The team did both and the gains have been sustained for decades. It is unlikely the gains would have held without the non-problem specific changes to the work of the folks within the process.

> ***Wisdom Tidbit - To sustain gains in people systems, learn the impact of sustaining compared to not sustaining on the people involved and work to make gain sustaining the easier of the two.***

High-risk Solutions

Just like some problems will tend to struggle in sustaining the gains, some solutions will also struggle. There are strategies to help handle these solutions.

High-energy Solutions

Some solutions by their nature take more energy, more effort to sustain. They are not all created equal. When your team is selecting solutions; the robustness of the solution should have been part of the equation. But despite best efforts, not all problems can be solved with low energy easy to maintain system. The difficult science behind the problem and the complexity of the problem often drive complex solutions. FPS problems are often multifactor and have interaction effects. With multifactor problems and interactions, multiple factors will need to be controlled to achieve success. In general, the more factors involved, the more complex the control plan. Compare these two cases. Case 1: To maintain the desired activity of the catalyst, we need to limit the water fed to the catalyst. Case 2: To maintain the desired activity of the catalyst, we need to balance water, acid concentration, and air intake. Any one of those factors out of balance with the others kills the catalyst. Obviously, case 2 is the harder control plan. But what drove the two cases? Nothing the project team has control over. It was the underlining science of the problem. That is why this particular risk needs to be thought through when developing your control plan.

Pilots and astronauts are faced with some of the most difficult control problems with some of the highest consequences. A pilot error can kill an entire plane full of people. An astronaut error can kill a crew. These systems are all multifactor and difficult to control. The control plan on these systems are exactly the kind of control plan needed for complex problems. Pilots standardize and train using detailed step-by-step checklists. Well-designed instruments and systems with redundant backups keep them knowing where they are and how things are running. Work is standardized and responses to problems are standardized. Standardized work means that winging it is not the norm. Operating outside of key parameters is frowned upon and can get the pilot fired. Procedures are followed.

For complex control plans, we will need to be like the pilots and astronauts. We will need well-designed standardized work. Good standardized work is visual. At any moment in time, the person running the process knows the step they are doing and status of the system. Good standard work is clear and easy to understand. When to take corrective action and under what conditions are highly prescribed. This is discipline. Most systems do not need this level of discipline. For example, driving to work. Your average person does not calculate the fuel it takes to go to work compensating for number of passengers and route planned. Instead a glance down at the fuel gauge and a quick decision of if gas is needed sooner or later is made. The decision outcome will vary based on urgency of the drive, how the driver is feeling, and other factors such as if coffee is needed or not needed. Not very disciplined. It does not need to be disciplined to be successful. Most people every day can prevent running out of gas with a pretty undisciplined process. Contrast that to the fuel calculations a pilot does before take-off. How much is needed with a prescribed safety factor is calculated. More passengers and luggage added at the last minute will delay take-off until everything is recalculated and fuel may be added. Very disciplined. There are many how-to manuals on creating standardized work. Use this concept for complex problem controls.

Wisdom Tidbit - To control problem complexity, use the simplicity of standard work.

Soft Changes

Soft changes are changes that are not kickable. The solution cannot be touched. These solutions are procedural or cultural or reside in documents. Hard changes can be touched. They are equipment, objects, piping, things. Contrast a soft change to a hard change. Case 1: A solution requires an operator to procedurally close and open the right valves in the right sequence. Case 2: Another solution accomplishes the same thing as Case 1 by hard piping a new line with an automated valve that automatically opens in the right sequence. Both solutions for Case 1 and Case 2 can accomplish exactly the same outcome. One is a hard-kickable solution. The other is procedural. Hard changes stick because of the work to move them. They stick because objects are difficult to forget or leave out. Hard changes remind you every day that they are there. Soft changes can and do struggle. Case 2 by its nature will tend to last longer. For example, an improvement team stopped large swings in a distillation column composition by implementing a direct contact heat exchanger that closed the energy balance and prevented swinging. Existing piping routed with existing valves in a new way accomplished the outcome. Within a year, the swings were back. The old procedure was accidentally or on purpose reinstated. Luckily, the project team was still nearby. The following shutdown, the solution was hard piped the correct way and the old piping allowing the old procedures to be done was removed. The solution has now stuck for decades.

Hard changes stick. Simple but true. Often the initial solutions are procedural or soft. Once proven, then that should be a different story. Procedure changes can be hard coded into software to make them more sustainable than purely procedural changes. Visual indicators can be put into place to make certain control is kept. For example, it was discovered that to control contamination a raw material filter needed to be changed every batch. Putting error coding and production stops in place when a new filter was not entered a stock withdrawal made producing the batch impossible without first consuming a new filter. A soft change was made hard. Be creative. Soft changes can be converted to hard changes in most situations. If it is critical, making it happen 100 times out of 100 is the right answer.

> ***Wisdom Tidbit - To make soft changes stick, wherever possible make them into hard changes.***

Lone Wolves

Some solutions are so unique that there are not systems in place to control them. These are lone wolf solutions. Companies have systems in place to do many things and control changes. When changes are made to the product composition, there are systems in place to notify customers, test that the changes are working, and to lock down the new qualities. When procedural changes are made to documents such as standard operating procedures that tell the operators how to run the plant, there are systems in place to approve the changes and to control the documents, so that only recent copies are available. When new raw material specifications are made, there are controls to ensure that the incoming raw materials meet specification. These are typical systems in place in nearly every company. But unfortunately, solutions do not always play into existing systems. Sometimes the right solution lives in a part of the company that is not so neatly controlled with existing long-standing systems. These one-off controls called lone wolves are at high risk of failure.

Lone wolf solutions often occur in the cutting edges of new software or other information technology solutions. For example, today it is common to have cloud-based systems to share data between groups. At one time, the same functionality would have been a custom program. This changes quickly. Those custom programs become obsolete; so, unless there is a concentrated effort to duplicate that functionality in new systems control failures can occur.

Other areas that lone wolf solutions occur are in the parts of a company that are non-core: the areas that are not critical to production. For example, a project team implemented a solution in a specialized sub-department of the tax folks. Total employees working in the area was one. The control plan was robust and held for years. Pretty much as long as the employee was there, the control plan was going to work. The control plan fell apart with a rapid reorganization in which poor job hand over times were defined. The work was handed over, but the new employee was not in position long enough to realize the importance of the work. Fortunately, the documentation was there, and the control plan and process were restarted. Two main factors were the saving graces of this lone wolf solution. First, the worth of the project from a tax savings standpoint was quite high. Money talks. Solutions on small problems dying do not shake the world. Solutions that fail on the big stuff are a different story. There is a drive to get those trains back on their tracks. The second factor was good documentation. Because the documentation was in the right place and showed exactly how to continue, the fix was simple. Just do what used to happen.

> ***Wisdom Tidbit - Lone wolves' solutions need to error on the side of <u>over</u> documentation.***

Poor Solution Implementation

The poor solution implementation category of high-risk control plans appears to be self-evident as to why it would be a problem and seems to have a simple solution. Always do great at solution implementation. Who would do poor solution implementation? However, your teams do exactly that sometimes. Obviously, for the most part we all want to do our best and strive to do great things. But at the end of the day, everything happening on your project will be subject to trade-offs. Great solution implementation can be a trade-off decision that your team will need to make. If your problem is costing your company \$1 million/week, you may be taking some shortcuts that trade-off solution implementation from being great for OK or even poor in the interest of speed. That trade-off decision can have long-term implications to your control plan.

Poor Change Management and Buy-in

Good solution implementation should include good change management for the folks impacted. You are changing their work, so to make things stick there is often some convincing through a communication of the case for change, some work to get everyone involved in the guiding vision of the change, input from the folks impacted is gathered, and in short, the change is planned carefully. This work takes time. Instead all of the work could quickly take a short cut by using edicts by management. Thou shalt follow this solution or else. This and other common short cuts get made in the interest of time. Quality is traded off for speed. If your team makes these trade-offs, your solution can still be very successful. The issue will come later. If your project team has had to step on everyone's toes and caused hard feelings, what happens when the team is done and leaves? The answer to that question may or may not mean a sustainable solution. Additional work will need to happen to help sustain gains.

The first step in preventing this failure mode is finding out if it is happening. How can you tell if the solution implementation was a change management mess? The simple answer is to ask the folks impacted by the solution. Honest feedback is needed. For example, a project team implemented a new heat exchanger installation procedure to prevent failure. The procedure tripled the installation time due to the careful measurements and alignment needed to get the installation right and prevent the earlier stress issues. The people doing the installation had to do things differently for this heat exchanger compared to other installations. The procedure changes were implemented quickly without much change management. Effectively, the changes were forced through. The fix worked, but the mechanics involved were not happy with the changes. Talking to the people doing the work, the team realized the problem. Meetings were held acknowledging the problems and collecting the feedback on how to improve communication and the procedure itself. Feedback after the fact lead to changes that incorporated the needs of the mechanics. The revised procedures had more buy-in and ownership and have lasted for many years.

The steps for a post solution implementation change management course correction are straightforward.

1) Find out if the problem of poor buy-in exists with post solution interviews.
2) Create a case for change based on solution results and outcome.
3) Communicate and acknowledge the shortcomings of the solution to key stakeholders.
4) Gather feedback on any improvements needed.
5) Modify the solution and/or solution communications to help establish buy-in.

Wisdom Tidbit - To succeed in sustaining gains for cases of poor change management, do the change management effort after solution implementation.

Roles and Responsibility Breakdown

Knowing who is going to do what is critical to complete any job. Knowing who is responsible is also critical. Solution implementation involves a shifting of roles and responsibilities from the project team to the long-term folks running the process. That shift is not always smooth. Project teams often fail to find the right owner for pieces of the solution in their rush to get something done. For example, it is common for engineers to "hand hold" the process during solution implementation and often this goes on for several weeks or months after the implementation. That is a transition role, not the actual role of the engineers involved. Ultimately, operators will run the process. If the role and responsibility of running the process is left ***in-between*** those people (operators and engineers), something is going to fall between the cracks. Two people responsible for something is the same as nobody responsible. The transition will ultimately need to be made and made sharply to prevent this mode of breakdown.

One of the most common role and responsibility breakdowns sits squarely on the project leader's shoulders. Project leaders fall in love with the limelight of reporting progress on the goals at the end of the project. During the early stages, it is generally a different story. Being the messenger reporting that the problem is still there, and the team doesn't know why is a tough day at work. At the end, however, when a problem gets solved, it is great to be the messenger. Stakeholders love the message and the messenger. Project leaders often get the perception of being the most knowledgeable about the problem due to often being the "voice" of the team. By the end of the project, the reporting of metrics and current activity is easy for the project leader to do. The measurement system is at their fingertips. These and other factors will drive project leaders to be the defacto owner of a process. One of the key definitions of a project is that it must have a start and end point. What the project leader is doing during the project that does need to continue after the project needs to transfer to a new owner.

This issue can be fixed by gaining agreement and having the key stakeholder make decisions on who the ultimate owner is of the control activities. To accomplish this, the following needs to happen:

1) The project team and sponsor need to clearly define what reoccurring activities or tasks need to continue once the project is done in order to sustain the gains. Note: This activity is covered in the next section on control plan development.

2) Those reoccurring activities and tasks need to find homes within the roles and responsibilities of the folks running the process or within the work of leadership. This means there is a decision that needs to be made for each one.
3) A transition plan needs to be made to transfer ownership from the project team to the permanent owners.
4) A cut-over date needs to be selected in which there is a formal acknowledgment that the work is transferred. Ideally, there is some memorable event such as signing a piece of paper or even things like handing the project files over. Be creative. People need to remember that the transfer of ownership has happened. Think of events like buying a car and having the title transfer occur.

Preventing the breakdown of roles (who does what) and responsibilities (who makes sure what got done) is an important activity in sustaining gains. It is possible to fail to do this within a project and return to put the control plan back together later, but obviously it is better to do this earlier.

> ***Wisdom Tidbit - To fix or prevent a roles and responsibility breakdown, stakeholders will need to participate in transition planning and complete the needed role and responsibility decisions.***

High-Skill Training Breakdown

By their nature, fundamental problems that have stumped and perplexed earlier efforts can be highly technical. This inherent property of a problem can lead to difficulty in sustaining the gains due to the need to pass that information forward through time. For example, in the chemical industry, we sometimes find reactions and side reactions that grow so complex in their nature that only a handful of experts within the company really understand them. Typically, those experts are not running the plants. This is the high-skill training challenge.

The high-skill training challenge is at its core taking super complex knowledge and codifying it into an algorithm to make it usable by folks who are not going to get that PhD on the topic. The algorithm can take many forms. It may be a well-written document. It may be a troubleshooting guide. It may be a flow chart of troubleshooting. It could be computer code with IFTTTs (IF This Then That) that produces the right answer under the right conditions. It could be a control system design. It could be a detailed training program. It could even be a simulator that trains people; most pilots could not design an aircraft, but a simulator can sure help train them to fly them.

To codify the complex, the following steps need to occur.

1) Define what knowledge transfer needs to occur. Be specific.
2) Develop the flowchart that explains the cause and effect relationships that the experts understand and that the users need to know.

3) Pick a media leveraging your company's capabilities to communicate this complex knowledge. Start simple. If you can't put it on paper first, you can never build something more complex like a simulator. Go no more complex than it takes to be successful; 10% of the work will get you 90% of the gain.
4) Pilot the results on people that have not been involved and use the PDCA cycle: plan-do-check-act to fix issues. The desired outcome is that the complex knowledge is successfully used.
5) Institutionalize the results. Roll the work out to all the folks that will be using this.

FPS teams are often stuck with this problem, so being successful at taking the complex and making it useful can be the make or break moment for sustaining your gains.

Wisdom Tidbit - Fixing the high-skill training challenge takes creativity and work to codify the complex into effective learning or control measures.

Outside Forces Conspiring Against Sustainability

Even with the best and most effective control plans possible, there are outside forces that can flat out dismantle them. Outside forces are the forces beyond the FPS team's control. For example, during a buy-out situation, the large company that took over a smaller company destroyed hundreds of control plans effectively destroying years of quality advancements that ultimately wiped out customer satisfaction. Customers went from delighted to complete frustration. Market share declined sharply, and the gains made by this previous work were not sustained. It is a sad statement on the world, but these kinds of outside forces are here to stay. There is still hope for control plans despite these winds of change.

Wisdom Tidbit – It is impossible to foresee the outside forces that can and often will wipe out the gains but doing the best documentation possible can save the day and sustain the gains.

Here are some of the most common winds of ***negative*** change.

- Rotten apples
- Business cycle failures
- Reorganization disease

It is interesting that each one of these negative change winds can have an opposite wind that is positive. For example, the buyout example above could have exactly the ***opposite*** outcome. If for example, a great company with a tremendous culture for continuous improvement buys a company that is suffering due to a poor culture, better control plans could have been introduced, quality could have improved, and market share could have increased. So, although this section will talk about negative outcomes, positive outcomes are equally likely.

Rotten Apples

Leaders come in many flavors. Some leader flavors are vanilla ice cream. Vanilla is popular world over because it delivers on what it was made to do. Some leaders are like that. They are shining stars that folks just want to follow. Other leaders have a different flavor . . . asphalt. We will try and not use the term leaders for these folks. Let us call them simply managers. These types of managers are rotten apples. Apples used to be stored in barrels. During the storage process, it was critical to inspect individual apples because if one was bruised and damaged, it could rot and cause a chain reaction rotting other apples around it. An entire barrel could easily spoil. This rotten apple process happens in companies and cycles through waves. It can happen just to a department, but it can grow to divisions or even the entire company or if you look globally, entire countries will see similar patterns in history. The reverse can happen and one leader in the right position can sort out the bad apples changing the game. Bad barrels can go back to being delicious apples. This pattern appears to be a human condition pattern.

There are many casualties of poor managers. Stress goes up with the employees causing some of today's most common killers like heart disease, customers suffer, costs increase, and many other negative outcomes. Among the outcomes, control plans can also be a casualty. For example, a brand-new manager took over an area and immediately started making changes to make a name for themselves. Not a servant and listening leader at that moment. One of the first changes made was a control plan change. The manager had the troubleshooting sections removed from every standard operating procedure and replaced them with the simple instruction to inform management when there were problems. Years of wisdom on how to make great product (control plans) was set back to zero in the blink of an eye. Being in control (in power) of as many decisions as possible was deemed as critical to that manager. Power can be a drug. Ultimately, the manager moved on to other pastures and the old troubleshooting sections were eventually reinstated. Good document control was in place and it preserved previous versions; so, this control failure got set right. This theme of good documentation is the most important lesson for good control plans. Years later, that same manager developed and grew becoming a great role model and truly a ***leader*** people wanted to follow. Rotten apples do not always remain that way; but from a control plan and sustainability standpoint we have to be prepared to weather this storm. It is happening now in your company, because there are good, great, mediocre, and bad leaders. Steps need to be taken to handle all kinds. Document, document, document.

Business Cycle Failures

Business is not always bad. Business is not always good. It cycles and changes. It evolves as technology and other driving forces make change. These cycles can directly impact control plans. For example, a team needed to increase the capacity of a system because it just plain could not keep up with the incredible customer demand and growth. The problem was solved by decreasing the longest cycle time step by accelerating the chemistry involved. This speed was accomplished through a higher purity, but more expensive catalyst. The gain was sustained for many years. Then 2008 happened and the industry that the product fed, automobiles, was on the rocks and demand plummeted. Capacity was no longer a problem. Cost minimization became the priority and cost savings flipped back to the lower purity lower cost catalyst. The gain in capacity was not sustained because business drivers made it obsolete. As the economy ship righted itself, demand returned. Good documentation was the only thing that enabled the older gains to reestablish themselves. Document, document, document.

Reorganization Disease

There is an age-old story that has a strong truth within it that helps understand the reorganization disease problem. It goes something like this. A new manager starts her first day at the office by meeting with the old manager. As the old manager leaves, they hand the new manager three envelopes and says "If you get into trouble, open the first envelope and it will get you out of trouble. If it happens again, open the second envelope, and if it happens again after that open the last envelope."

Things went well for the new manager. No trouble. But then, the worm turned, and things went wrong. In desperation, the manager opened the first envelope. A note inside said "Blame the previous manager." The instructions were simple and worked like a charm. A person to blame diffused the situation and things got back on track. After a while, more problems popped up and things looked dire again. It was time for the second envelope. The note inside the second envelope said, "Initiate a reorganization." Reorganization worked like a charm as well. In the chaos of everyone changing jobs, the problems that were so dire now seemed insignificant. Again, things went great. But managing is never without problems, so eventually trouble came again. This time it was more desperate than ever. Her hands were shaking as she opened the third envelope. The note inside said "Prepare three envelopes for the next manager."

After observing and participating in dozens of reorganizations, we believe they can be characterized as always claiming to be better than the previous organization, always making a claim of being a new concept, and never having a measure that validates the claims. The patterns cycle around and eventually the old organization structure usually returns and becomes the next new and improved organizational structure to be implemented. That makes the claim of being better to be theoretically true in half of the organizational structures. In our experience, about half of the organizational structures are better and represent good change. The other half are worse and make things worse. But better or worse, the implementation execution must be done well, or else your team's control plan can be put at risk. The pace of these reorganizations in industry feels to be increasing in frequency; so, it is likely that the problem you are solving will need to weather this storm at some point and more than likely will have to weather many of these storms.

During reorganizational storms, roles and responsibilities change, skill mismatches occur, workflows change, and who is going to be doing what is not always clear. In short, everyone's daily work can become chaotic and confusing. This puts control plans at risk. For example, during a reorganization part of a person job was left out of the new organizational structure. The person was the person who answered the phone for billing errors. It took weeks to figure out that omission and the answering machine had filled. An entire job was missed. It took months to fix all the problems and the new organizational structure had to be revised. That is the kind of problem this storm causes. Control plans are all about doing the right things. That at its heart relies on the right people who are properly trained being assigned to do these right things. Good documentation is the only thing that will enable older gains to reestablish themselves. Document, document, document.

> ***Wisdom Tidbit – Repeat! It is impossible to foresee the outside forces that can and often will wipe out the gains but doing the best documentation possible can save the day and sustain the gains.***

Summary of the *Reality* Problem Components of a Control Plan

In addition to perceived issues on sustaining the gains, there are also real problems. Here is the reminder of the most common and the advice to prevent them.

- **High-risk problem types**
  - Low frequency
    - *To mitigate the low frequency problem risks, it becomes critical to document consequences as visually and easy to access as possible.*
  - Counter intuitive
    - *Good solid communication that includes conventional thinking and robust training plans are overall the best strategy to sustain gains in counter intuitive problem situations.*
  - Reward conflicting

- **High-risk solutions**
  - High-energy solutions
    - *The complexity of the problem can dictate a complex control problem, maintaining gains in these situations will take turning the complexity to standardized work.*
  - Soft changes
    - *To make soft changes stick, wherever possible make them into hard changes.*
  - Lone wolves
    - *Lone wolves' solutions need to error on the side of over documentation.*
- **Poor Solution Implementation**
  - Poor change management and buy-in
    - *To succeed in sustaining gains for cases of poor change management, do the change management effort after solution implementation.*
  - Roles and responsibility breakdown
    - *To fix or prevent a roles and responsibility breakdown, stakeholders will need to participate in transition planning and complete the needed role and responsibility decisions.*
  - High-skill training breakdown
    - *Fixing the high-skill training challenge takes creativity and work to codify the complex into effective learning or control measures.*
- **Outside forces conspiring against sustainability**
  - Rotten apples, Business cycle failures, and Reorganization disease
    - *It is impossible to foresee the outside forces that can and often will wipe out the gains but doing the best documentation possible can save the day and sustain the gains.*

10.2 Building Robust Control Plans for Solutions to Fundamental Problems

With a clear understanding of the mechanisms that cause control plan failures, it is time to turn our attention to putting those lessons into action and building robust control plans for solutions to our Fundamental Problems. The first concept is that Fundamental problem control plans have near perfect overlap with typical control plans. If all the team did was put in a standard control plan, it will mostly work and mostly be right. Not the end of the world, so starting there is good advice. Let's start there.

There are volumes written on quality control systems. The purpose of this section is to help frame and understand the slice of the pie that is slightly different in FPS project than that in the typical control problems. Figure 10.2 helps to illustrate the concept.

Typical Control Plans

To illustrate the concept of a control plan, let's look at a simple example that most of us have seen. If you were running a coffee shop, you would need to have some basic controls in place just to have a chance to stay in business. To illustrate this, here is a SIPOC (Suppliers, Inputs, Process, Outputs and Customer, sometimes a COPIS) shown in Figure 10.3. All control plans are based on customers, outputs from the process, steps of the process, the inputs to the process, and the suppliers of the process. The control plans then use measurements within each of these pieces. By controlling the key input and process measures and steps, the outputs follow and deliver the results that we want.

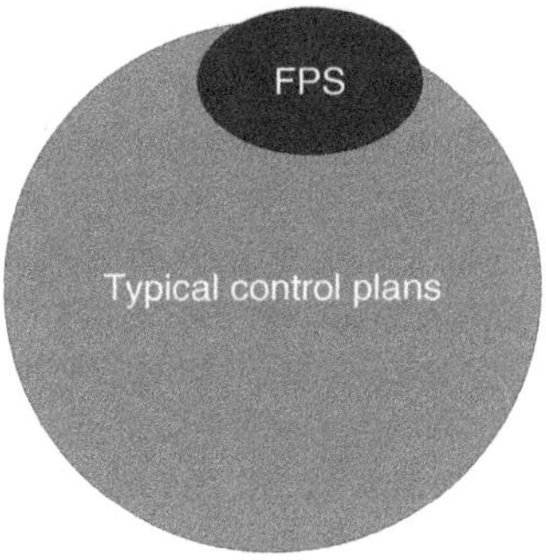

Figure 10.2 FPS control plans vs. typical.

Sheet 1

Start boundary- coffee machine =>

<=End boundary- poured cup of coffee

Suppliers	Input	Process	Output	Customer
		Coffee process		
Electric Co.	Electricity	Turn on power	Grinds and spent filter	Garbage collector
Coffee vendor	Coffee	Add filter	Cup of coffee	Coffee drinker
Filter vendor	Filters	Add Coffee		Coffee cup and pot cleaning process
	Clean cups and pots	Pour water		
	Water	Start brew		
		Pour coffee		
		Dump grinds		

Input indicators	Process indicators	Output indicators	Critical customer requirements
$ Electrical	Water Brew Temperature	Coffee temperature	Customer Satisfaction rating
Taste rating of coffee	Water Level	$ to make/cup	Coffee taste rating
Micron rating of filter	Water or coffeee spilled	Coffee usage scoops	Coffee temperature rating
Coffee shelf age	Coffee process cycle time		Purchase $/cup
$/bag coffee	Time of coffee in pot		
$/box of filters	Pot temperature		
Cup cleanliness inspection rating			
Water microbial count			

Figure 10.3 Coffee process example SIPOC.

To stay in business, we need to make good tasting coffee or else our customer satisfaction will drop as will our business. There are undoubtedly many things to making good coffee. To illustrate the process let's put controls on just one of them. In this system, we want our output to be a hot cup of coffee that tastes great. But, the first question is what is hot? So, the process starts with the customer. What do they need? That is the starting point for all good control plans. In this case, a little research into the customers show that most customers want to be drinking coffee between 120 and 140 °F to maximize flavor and to enjoy the warmth without the burn. This becomes the target of controlling this part of our process. We then need to understand what Inputs or process step lead to this output. These are driven often by fundamental principles. For example, if our coffee is stored in a pot that retains heat with insulation instead of using a heating coil to retain heat, then we need to understand heat loss to understand temperature. Air flow around the pot, time in the pot, thickness of the insulation, evaporation, initial temperature of the coffee entering the pot, and so on are what control the temperature the coffee will be when it leaves the pot. Good control plans will focus on the vital few and where possible put hard fixes on the rest. For example, selecting the right thermos storage pot and replicating that selection when new ones are needed will make controlling that part of the equation seamless and easy. Of the remaining factors, initial temperature and time in the pot could be calculated as being the main factors that control our output indicator of temperature in the cup. A control plan to maintain this would need to have controls to hold the right brewing temperature and the right time. Brewing temperature would be a range to optimize taste. Say 195–205 °F. But that same temperature would set the time on when to sell the coffee. Too soon and it will be too hot for customers. Too long and it will be too cold. Once the controls are planned, then more details follow. Somebody needs to be responsible for making sure that these controls happen. Somebody needs training on controlling the process. There needs to be a troubleshooting plan. What do you do when brew temperature drops to 190 °F? Maybe a clean out procedure is needed to remove scale and a test pot needs to be made to validate that the troubleshooting fix works. Somebody needs to validate that the folks who are responsible for doing these steps are in fact doing them. These are the elements of a typical control plan: a fundamental linkage between the inputs and processing steps with the output that drives quality, procedures to control quality, training on the critical to quality procedures, and auditing that the execution is happening.

Companies will have these pieces of a control plan put in department roles, job roles, standard operating procedures, ISO auditing, supplier specification, customer specifications, quality assurance procedures, and many locations dependent on what is being controlled. Typical control plans will use these systems to ensure that the process is controlled. All of this is the same when setting up the control plans for an FPS problem. Go from the Xs that control to the Ys and make certain that the people or machines doing that job . . . do their job. Simple.

FPS Control Plan Emphasis

As stated earlier, fundamental problem solving is diving deep into the science to find what is going on in the system. The relationships are often quite complex.

10.3 Document, Document, Document

We learned earlier the critical nature of documentation in sustaining the gains. Knowing that it is important is not the same as getting it done. Documentation is a lot like painting a room. When it is done, you are glad you did it; but getting started and the process itself is about as fun as watching paint dry. Every surface really needs to get covered or else it feels like a sloppy job. Imagine if it was painting a room that was only used every three or four years. Why do the work? Who would even remember who was responsible for getting the job done? This motivational struggle is something your project team will need to overcome to sustain your projects gains.

The problem is there is an asymmetry to the payout of failing to document. The team that does a lousy job or zero work documenting can disappear before there are consequences. They have a positive feedback of less work with near zero consequences. Even if team members are still in the process area, when lack of documentation-based problems show up, often the old team members can rescue the system from failure by pulling up files saved locally or within their memories of the issue. Eventually, these rescues fail, but again low cost for the folks that WERE involved (high cost to those currently involved!). Think about the do nothing pay out . . . the cost/effort is near zero and for the foreseeable future, life rolls on. Project teams need to be motivated to move forward in the documentation. They need to understand why they are doing the documentation. They need clear scope of what needs to be documented, so that there is an end in sight. They must have a plan on where the documents will be stored and how they will get maintained. They also need a time driver on getting that work done in a timely fashion. Project leaders must rise to this challenge. The purpose of this section is to help the team and project leader through this motivational maze and deliver world-class control.

How to Motivate Teams to Document

The documentation motivation game is a **WIIFM** game . . . **W**hat **I**s **I**n **I**t **F**or **M**e? Humans need to be motivated to act. We prefer to do the minimum work needed. Motivation is the force that drives the action and each person is going to have their own drivers. For example, some people are motivated by the positive feedback loop of teaching. If somebody can pick up the documentation and learn something new, then that can provide positive feedback. If I have written something down to teach a concept and somebody reads it, uses it, and reaches out to let me know that it helped then it is a rush. It feels good. Other people are motivated by advancing their career or technical status. Publishing can be a way to gain the respect of others or management. Some folks just like telling a story. Maybe the work was really an interesting or almost unbelievable set of events that happened. Telling the story gains an audience reaction that can be a positive feedback. Some folks get a rush from finding the best way to get something done. Following a very exact procedure correctly will get exactly the right results and the quest to have everything done right is motivating. There are many reasons people get motivated by documenting things. Your job as a project leader is to find that fire within your team.

Chances are that as you read the list of the different types of motivation to document, there were one or two that truly spoke to you. The answer of what motivates is unique to everyone. As a project team leader, to understand that motivation just ask the team.

Ask each team member during a team meeting: What is your motivation for documenting this project? What excites you and drives your desire to get this information out?

Take the time and record their answers. This will help in several ways. First, as a project leader whenever a team member falls behind and struggles to do their part in the documentation, you can read this back to them. Remind them of why they thought this was important. Often little nudges are helpful. Second, knowing what motivates a person can help align the work to the person who is going to be the most motivated and passionate in getting the work done. For example: A team member really wants to get the story out, because they love the story telling aspect of documentation. Maybe they write things like the introduction, summary, and conclusions. Third, it can give you clues on how to help team members grow and develop. Often project leaders will fall in the trap of giving documentation work only to the person who has the job role that ***should*** do that work. Certainly, that can be why somebody does the documentation. But during a project giving team members opportunities to grow can be more important. For example, maybe one team member is in research and development and has never written a standard operating procedure. Obviously, we do not want non-experts to have the final say on the documentation; but that is an accountability question. If your research person would like to try their hand a writing a procedure, they can do the first draft and work with the expert in production who ultimately has the accountability to get the job done correctly. That expert in production will grow because they can see a new perspective on how to write a procedure and can get the chance to help somebody that might ultimately work in a production environment. The bottom line is finding out what motivates your team members to document is going to make this phase flow better.

Speed is also critical in motivation. Two of the mistakes teams can make on documentation is to go slow when they need to go fast or to go fast when they should go at their own pace. Figuring out how fast the documentation should happen will help with motivation because a mismatch is a root cause of **DE**motivation. For example: A project team needs to document the new procedures prior to initiating the changes. On a problem that has a VERY high burn rate, getting that done quickly is critical to getting the solution to run. The project team decided to do a blitz activity that involved key folks sitting down together in a conference room for a day straight with the sole purpose of getting those procedures done. The team did "dry" walk-throughs of the procedures with operators that would do the procedures. They also arranged for the procedure approvers to be there at the end of the day to complete the approval process. In this case, the speed of the need matched the speed of the work. If this same approach was used for a technical, detailed, and thoughtful, problem research report, the urgency approach would likely demotivate a team. Pick the speed plan that matches with the need. Sometimes documentation is fill in work that fits within other higher priority work. Sometimes it is the most critical work. Know which it is.

One useful technique is to put on the calendar periodic progress check-in points. Dates and deadlines can be very motivating, as well as peer pressure. When the team periodically meets, those that are on track completing their documentation will help with the motivation.

Perfection is the enemy of progress. Minimum effective documentation is a critical concept to hit your teams with during the progress. People often imagine the worst when a task or work is put in front of them. Ask the team members what is the minimum documentation that they can do and still be effective. Get them to think hard about doing exactly what is needed, but no more. Rough drafts should be exactly that . . . rough. So, help guide the team to eliminate rework waste by doing less in the beginning. Get a start of a document out as soon as possible and refine from there.

> **Understanding what motivates your team will help you complete this phase of control.**

5Ws and an H

When you get to the documentation portion of a project, we are going to ask the 5W&H question. Start here with the team during a team meeting or e-mail chain.

Why are we going to document this?
This question not only helps motivate the team but defines the purpose of the document. A technical report has a completely different purpose than a startup procedure. Focus on the purpose of the documents will help drive the right behaviors in the team. The startup procedure won't become a dissertation on the chemistry at play when the team understands the **Why** question on each piece of documentation.

What are we going to document?
Scope is critical to documentation. There are several ways to figure out **What** needs to be documented. Often a great tool to understanding the **What** question is to use a mind map. Capture the team's input and let them build off each other. Another technique to answer the **What** question is to have the team write down on sticky notes everything they think needs to be documented. Then after they finish, read them off and group the like items together (affinitize). This process will result in the **What**. Often companies will also have documentation checklists that share each system available and the team can just go through the checklist and see what applies to their situation. This can also be extremely effective at getting through the **What** question.

Here is a checklist that might be helpful in your FPS project:

- Data Quality Objectives (or charter)
- Hypothesis onion
- Good hypothesis documentation
- Hypothesis details both right and wrong
- Testing done to kill hypothesis.
- Clear explanation of science
- Supporting data
- Test methods used
- $ impact
- Clear statements of causation to customer impact
- References used on the science being tested

Where are we going to store this documentation?
For most documentation needs, this is defined within existing corporate processes. Having said that, do not let the existing corporate processes hold the team back from doing the right thing. Visual factory concept is that at a glance the person doing the work should know if the system is doing what it is supposed to do or not. Think of your gas gauge on your car. At a glance, you know if you need to start the process of adding gas to your car. Document in the most effective place. If that is on the machine, do it there. Troubleshooting guides that pop up when trouble is happening are far more effective than the ones on some dusty shelf. Engage your team and make the right outcome happen.

When are we going to get this done?
The team needs to understand the documentation process is going to start AND finish. Establishing a plan is essential. Because the value of documentation is often an investment, some slippage will not typically make or break a project and your team members will know that. By creating a sense of some urgency, you will help the team move on to other projects. The **When** question will help the team focus on finishing.

Who is going to do each part of the documentation?
Tasks are not tasks if nobody is assigned to them. Select first on passion and second on role responsibility whenever possible. Try and share the load. Within teams there will be over committers and under committers. Estimating the time for each of the documentation tasks and then tallying up the total time commitment next to the person's names help folks do a better job of load leveling. Answer the **Who** question and revisit it often.

How are we going to go about the documentation process?
Horses for courses is a saying that means some horses are made to jump and some are made to sprint. The **How** question is about figuring the logistics and practical matters. Where will the working documents be stored while the team works on them? It is likely that storage spot will not be the document's final resting place. Will they be on a network drive? Or are we going to do this with an email chain? Are we going to blitz this document and get it done in a day or is this fill-in work we will be doing in between other work in the coming weeks? The **How** question helps teams focus on the process of doing the documentation.

There will be many parts to each of these questions. Your team will likely need to be doing some small system integrated documentation. For example, odds are that your company has procedures and procedure documentation controls. Part of the documentation will go into those systems. Your team may have changed a startup procedure. There will be a place predefined to put that information. Maybe your team introduced a new DCS (distributed control system) sequence. There will be a place for that already in place with controls around it.

Because FPS is often developing or validating science, odds are that there will be a technical report that needs to document that science. Use good technical report writing formats and follow your company's process. Carefully consider your search terms, this will help people find the report you are writing. Get that technical report published!

Summary
Documentation is often thankless work, but clearly important work. With the right attitude, a clear understanding of your team's motivation, and some simple techniques this can be a project step that is enjoyable to complete.

11

Project Wrap-up

The Final Stakeholder Update communicates with the business, operations and site leadership, the root cause, the status of the implemented solutions, the control plan, etc. It may occur a year or more after the control plan is in place. It also communicates the delivered project value. This builds credibility and is done preferably with the help of the Finance department. The update ensures the continuity of the implemented solution and the continuity of the FPS capability. Finally, people recognition is discussed, as being a major factor in team management and assuring cooperation of various SMEs (Subject Matter Experts) in future projects.

Proper project closing or wrap-up is a very important step in the problem-solving process. This step should be completed without failure. Many projects done well but without proper closing leave open questions and uncertainty. So in a proper project closing one can write a Data Quality Objective (DQO) for the closing and ensure all pending items are assigned and action items are completed and documented. Following are some of the recommended best practices for project closing:

Follow the principles of DQO to outline what is expected in wrapping up the project. Keep it to the point and any new ideas should become new projects not tagged on to old one. Specifically to the current project show what are the steps associated with closing the project and who is assigned those responsibilities. Sometime proposed solutions are longer term and hard to validate immediately. In that case the project should be closed with clear recommendations covering all potential scenarios and follow-up validation actions to be taken. The project closing is also negotiated, documented, and communicated to all levels, and agreement is obtained with the stakeholders that the project is being wrapped up. Usually the DQO captures the background, the current and the projected path for the completion of the project.

Fundamentals of Industrial Problem Solving: A Practitioner's Guide, First Edition.
Zdravko I. Stefanov, Eldad Herceg, Carla Schmidt, David M. Jacobson, Dana Livingston, J.P. Chauvel, Sunil Kumar Chaudhary, and Christopher Paul Christenson.

The final steps of the project must include the final stakeholder update, which should review the agreed-upon success statement. This also prevents scope creep and project drag-on. Once the project closing is completed, the teams move in consultation mode and do not work on the project actively. Expectations are that key team members will be available if needed but the project is considered complete. The final stakeholder review is a good resource and becomes part of the documentation, which should include key findings, key validated hypothesis, root cause of the problem, validation status, and any uncertainty in validation or barriers encountered. This is kind of a roadmap of where we started, what we found, and what is in store in the future. A good idea is to include details of the control plan, so the stakeholders are convinced that the fix is permanent and the same problem will not appear again. As said earlier, the team may not be able to achieve some validations immediately, but this should be documented and communicated to the stakeholders.

When claiming accomplishments, a validation by the business adds to the credibility, and this includes any claims regarding immediate or potential financial gains. At times, even after good effort and success of the project, financial gains can't be realized due to change in business conditions; if this scenario exists, it must be communicated. This does not mean the project was not successful. Usually, the business leadership fully understands these constraints. There are times when in the big scheme of things, financials are less important and stakeholders are not interested in detailed analysis as long as the problem is solved; in that case, the team does not spend lots of time and effort in tracking. It all depends what the stakeholders desire. One rule always apply "Never surprise stakeholders," be factual and, even if success is met partially, communicate to the stakeholders and proceed according to their wishes.

Sometimes the project has to be terminated in the middle or before completion, due to affordability issues or due to changing business priorities; in that case don't drag on and have a communication with the stakeholders and get a new success statement or document, report the reasons the project was terminated and the work done to date, so in the future if the project is activated again, work is not lost and can be picked up quickly.

Project closing includes the completion of the control plan, other planned reports or documentation, and expected timeline to complete any final pending items, any intellectual property where patents have to be filed or intellectual property has to be preserved. The team acts according to the stakeholder agreement.

This is good time to include any other potential improvement opportunities identified during the investigation. A brief list with enough details so future teams can understand should be communicated to the stakeholders during the final project update. Depending upon what is the agreement with the team and the stakeholders, it may be necessary for this team to charter future projects, identify potential team leaders and members for smooth transition. If this step is ignored and not well documented, then confusion can occur in the future.

Project wrap-up also includes team recognition, which depends upon a case-by-case basis but it is highly recommended to do some sort of celebration at the end of the project. This helps in creating motivational and happy environment, and everyone wants to create an environment where folks want to work again for the team or the business. The project manager in consultation with other leadership determines the level of recognition, and it should be done in timely manner. If recognition is not done in timely manner, its impact is minimized so doing it right after the project completion is the best time. There is no hard-and-fast rule of how, when, and what to do. Plan what makes sense. Some prefer to do recognitions in smaller steps throughout the project and some do it at the end. Lots depend upon the culture of the organization. Be very inclusive in recognition. Erroring on the side of over recognition is lot better than forgetting someone.

The final update should also consider sustainability of the solution and, if any capital or extra resources are required, stakeholders must be aware of it and it must be clearly documented in key recommendations. When thinking about key recommendations, they can be divided into must do, nice to do, or very optional for future improvement. Well documented recommendations are key in project closing. Do not leave open ended or vague recommendations open to interpretation – make it as clear as possible.

Section III

Examples

12

The Big World of Problems

The problems presented to the industrial problem solver could be grouped in many ways. This discussion will divide the opportunities into those of existing processes and those of new processes. The various classes of problems will be outlined, and the skills and tools typically used to resolve them will be discussed. While this cannot ever be an exhaustive list, it will present many of the key issues and serve as a template for others that arise.

12.1 Existing Products and Processes

Existing products and processes typically run well for extended periods of time. This sets an expectation on how they perform and sets budgets to manage them effectively.

12.1.1 Common Problems

Plant Product Is Out of Specification

The most common problem by far is a plant that is running "normally" but is not producing "in spec" product or a product that while "in spec" doesn't work as needed by the customer.

In this situation, the major chemistry and unit operations of the process are well known and understood. The side reactions that impact yield and previous specification issues are known, but the mechanistic details may be sketchy. The trace side reactions (typically, ppm to ppb level impurities) are not well identified or measured. The mechanistic chemistry that produces them is often a pathway

Fundamentals of Industrial Problem Solving: A Practitioner's Guide, First Edition.
Zdravko I. Stefanov, Eldad Herceg, Carla Schmidt, David M. Jacobson, Dana Livingston, J.P. Chauvel, Sunil Kumar Chaudhary, and Christopher Paul Christenson.

that would have been marked incorrect on your organic chemistry exam in college. It is often these minor reactions that prevent the utilization of your product by your customer.

Color

Color problems arise because many products when pure are colorless and/or the customer will carefully add color to make attractive consumer products. Color chromophores can easily show up because the customer views the product in a jar with a long path length using the Mark I human eyeball that is very sensitive to slight changes in color or particles scattering light. The absorbance of many common chromophores is often very large. These three factors make color bodies important at the ppm to ppb level. When the chromophores are attached to a polymer molecule (as is typical in urethanes and other polymer systems), the conventional separation methods to isolate the impurity fail. Further, the UV-visible spectroscopy has limited specificity compared to NMR (Nuclear Magnetic Resonance), IR (InfraRed), and MS (Mass Spectrometry) tools.

The favored approaches for small-molecule system include LC (Liquid Chromatography)-MS using a full spectrum UV-visible detector before the MS. When the MS has extended mass capability and high-resolution time of flight detection, the problem may become more tractable. The color can also come from inorganic impurities with iron being the most popular. A good trace-level inorganic analysis is strongly recommended.

Today, with the development of rapid computational chemistry, the problem solver can calculate the UV-visible spectrum of a long list of hypothetical impurities. This list is developed by skilled chemists with a well-developed imagination. This makes the problem much more tractable. The experimental UV-visible spectrum can be compared to candidate molecules and reaction pathways. The failed candidates are excluded and experiments are designed to test the remaining hypotheses.

The team for this class of problems needs to include engineers to model the process, especially the separations and recycles, spectroscopists, and separations scientists to do the measurements and a computational chemist to calculate spectra and estimate reaction kinetics.

Odor

Odor problems are another common complaint by customers. Again, the capability of the human nose, while not in the same class as many good dogs, is easily capable of detecting ppb levels of many organic impurities. These range from the notorious sulfur compounds to esters, aldehydes, acids, unsaturated hydrocarbons, and amines. Further, subtle changes in the stereochemistry of the impurity can change the odor profile greatly. The odor profile itself is complex, and a function of the concentration of the impurity. Molecules that are pleasant to acceptable at one level can be a major problem at a different level.

Fortunately, odor molecules need to get to the nose and thus be in the gas phase. This enables the use of GC (Gas Chromatography)-MS as a major problem-solving tool when combined with an odor port before the mass spectrometer. Modern gas chromatography has exceptional separation capability and can be further enhanced using multiple columns (2D-GC). When the mass spectrometer is capable of MS-MS operation or high-resolution time of flight measurement, the modern GC-MS is an exceptional tool. The use of retention indexes in the extensive NIST GC data bases complements the extensive NIST mass spectroscopy databases. When these are combined with the compilations of molecules with odor descriptions (http://www.leffingwell.com/bacis2.htm), it is often possible to identify the odor molecule quickly. A good example of the complexity of odor systems and how to study them is given in a study of the odors and molecules from the vanilla bean (https://rucore.libraries.rutgers.edu/rutgers-lib/46341/PDF/1/play/). The remaining challenge is to accumulate the odor molecules so that they can be introduced into the GC-MS. In the last few years, substantial progress has been made in developing sorbent systems that can concentrate impurities from air, water, or other systems and conveniently introduce them into the GCMS (http://www.gerstel.com/en/flavor-fragrance.htm).

Trace Components – Evils of Recycles, Accumulation, and Trace Reactions

One of the classic ways to have a process get in trouble is for something to accumulate in a recycle loop. All plants have recycle loops to minimize waste streams and conserve products. The problem starts when an impurity builds up in a recycle loop. For example, a product may go overhead in the first distillation column with the bottoms going to a simple flash whose overhead recycles to the first column and the bottoms going to an incineration device. If there is an impurity that goes down in the first column and overhead in the flash column, it will build up till it goes out the top of the first column and sends the product out of spec. There are many variations on this and it may be missed in pilot/market development plants that don't have the full recycle turned on. The recycle may come from phase separation, precipitation, or distillation. The key is not how the recycle is done but that a small component builds up and sends the system out of specification. This can be particularly challenging if there is no specification for an odor molecule (ppb level) and no one smells the product because it is toxic. When the reactive product is made into a polymer, the odor molecule is released from the customer's product to the consumer. The delay in discovering the problem can make a major economic challenge.

These key impurities are not from the main reaction that was designed for the plant. They may not even be one of the important impurities that the separations train was designed to remove. The yield of the trace reaction may be negligible from a process yield standpoint. The issue is if the impurity goes out with the product and negatively impacts the customer.

Understanding the formation of these impurities may be very interesting and be unexpected by the scientists who designed the process and made an A in organic chemistry. These are improbable reactions and reaction products. They are only happening at trace levels (ppm or even much less). Developing these chemical pathways is often very interesting. Still the core organic reaction principles apply. But the constraints are relaxed and your imagination is stretched.

The key tool for much of this work is the mass spectrometer coupled with GC or LC. Currently, the mass spectrometer is often the high-resolution time of flight MS where the high-resolution spectra greatly facilitate identifications. Computational chemistry is also very useful. Particularly, in the estimation of kinetics that suggests the expected yield to these reaction products.

The component mass balance across key unit operations is the basic problem-solving tool. Remember if something is made, it goes to the product or the waste stream. It may accumulate for a while before seeking its exit. Sadly, there is no way to avoid this, while the impurity slowly builds up. This delay greatly confuses the usual cause and effect logic we regularly use. We think the problem happened because of a recent change in the process when, in fact, the plant was running steadily and steadily building up the impurity. The impurity can also build up and foul the process. This can challenge closing the mass balance. It may plug the system and lead to a process shutdown.

12.2 New Products and Processes

12.2.1 Products Are Often Developed Over Time Without Fundamental Understanding

During acquisitions or when building a new plant for increased capacity of a heritage product, there is a painful opportunity for unexpected results. This can happen in specialty businesses where complex product performance developed over time. The customer expects a product with good shelf life, complex rheological properties, and a cure to produce a strong product. This product development happens empirically overtime without a strong fundamental understanding. When the "new and improved process" is built promising higher throughput, better yield and good performance goes on line, unexpected results are likely.

When called to respond to this situation, the problem solver may be faced with no quantitative process models, limited or insufficient analytical capability and limited cross discipline understanding of the detailed chemistry and the materials.

Clearly, this is a challenge. Time pressure is high and information is lacking. The best approach is to build two or more teams. One team is to develop and implement the required analytical tools - both chemical and materials science tools. The other team is to develop the predictive models that link the chemistry to the materials science. These are some of the most challenging and rewarding opportunities. The ideal result is the documentation of the fundamentals, setting the process to produce in control on spec product and developing the trouble-shooting guide to facilitate a rapid return to good production from missteps.

Statistically designed experiments are essential to doing this quickly and effectively. Having a statistician and an excellent materials scientist in addition to the team is of great importance. Sometimes it is essential that the experimental design both estimate the parameters of a proposed fundamental model and is suitable for an empirical interpolative model to use while further work is done on the fundamentals.

12.2.2 Product Doesn't React Properly

Today, many products of the chemical companies are the raw materials of the part manufacturing process. The speed of the part production process may be controlled by the kinetics of the chemical products. These kinetics are frequently complex. They depend on the stereo chemistry of the product, the catalyst concentration, the age of the product... The list goes on and on. One of the "analyses" that is required is the measurement of the kinetics of the product. This may be the activity of the catalyst that was sold to make a raw material, or the kinetics of a reactive polymerization. Today, analyses include much more than determining the concentration of components. They include measuring kinetics of products or in-process kinetics to monitor plant performance and determine catalyst change outs. This can also involve the changes in rheology that develop over time (both long, shelf life, or short, cure kinetics).

12.2.3 Shelf Life

Our products have to be good when they go out of the door. They also have to be good when the customer uses them, for the duration of the use. This can be relatively short for a raw material or very long when the product is a durable part that lasts for years or decades.

The challenge is that our typical measuring tools are developed for moderately fast system (minutes, to hours) and the time scale has moved to months or years. The first tool often chosen is to accelerate the aging by doing things hotter. This is frequently successful. There are restrictions that must be followed. First, the reaction mechanism must stay constant and new reactions need to be excluded. Alternately, more sensitive tools can be used while maintaining the typical storage/use conditions.

Calorimetry can regularly be very useful here. This is especially true if the condition to be followed is the slow reaction with moisture or oxygen during storage or use and the heat of reaction is sufficient (it often is). Ultra-sensitive calorimeters are available that can measure these changes quantitatively (https://www.setaram.com/product_categories/calorimetry/). When this approach is taken, the predictive model is more secure since extrapolation was not required.

If the goal is to follow the degradation directly, then trace level detection of the key products is required. The usual first step is to go for GC or LC-MS because of the sensitivity and selectivity of the tools.

When the shelf-life problem is color, the problem is more complicated since the "color body" may be at ultralow concentrations or bound to a high–molecular weight component and not "separable." In this case, the solution may be to build model systems that capture the structure of the polymer in a segment of suitable size for GC or LC-MS coupled with a sensitive UV-visible detector.

12.2.4 Polymers

Polymeric products have a variety of key challenges. The molecular weight or molecular weight distribution may be wrong. This leads to processing problems particularly in rheologically sensitive areas such as film blowing where melt strength is critical. Polymeric strength is also a sensitive function of the molecular weight particularly when combined with a need to minimize molecular weight for ease of processing. These problems require a combined team of rheologists, chromatographers to assess the molecular weight distribution, and materials scientists to build the structure-property relationships.

Some polymeric products are designed to have sensitive rheological properties such as flowing reasonably, having a good yield stress when the flow stops to support the parts that are gluing together and a moisture cure time. This kind of multiple dimensional specification brings substantial problem-solving challenges. The rheology is challenging to measure. The molecular weight/distribution is challenging to measure. The rheology is a combination of the polymer properties and the interaction of the polymer with the filler. Finally, the system changes with time, and there is a shelf-life specification to be measured and controlled. Clearly, a well-integrated team is required. Don't forget to include your favorite statistician to ensure good statistical design of the experiments and testing of the predictive models.

12.2.5 Process and Plant Capacity

Production plant throughput is critical to profitability. Fouling is one of the common and challenging problems to be solved. Typically, it shows up in heat exchangers and distillation columns. In monomer production plants, the product that is being made is designed to polymerize (somewhere else), and frequently low levels of polymer build up and clog critical systems. Some of the most common are radical polymerizations of styrene, acrylic acid (and derivatives). Trace levels of oxygen, lack of inhibitor distribution, or loss of inhibition due to distillation all contribute. These can be very difficult to sample and analyze. Mass balancing and kinetic measurements are key.

Other processes produce molecules such as isocyanates that react with themselves and with impurities to make polymers that plug the process and require a shutdown to clean. Understanding the side reactions is facilitated by very good LC and GC-MS. The side reactions are complex and require considerable study.

12.2.6 Catalyst

Catalysts are essential to most processes. Modern catalysts are often complex compositions and degrade slowly over time in use. The analysis of catalysts composition is frequently done with X-ray powder diffraction combined with an electron microprobe. For some catalyst systems, the size of the active metal is critical and grows with use (reducing the surface area) and the reaction slows. For these systems, a transmission electron microscope is often needed.

Catalyst characterization often includes surface area and pore size distribution analysis. This is regularly combined with test systems where the composition of the product is studied as a function of temperature, composition, and time on line.

The catalyst is often supported on a carrier that has a well-defined surface area and pore size distribution. This may change with time and the catalyst support may lose strength and collapse causing an unacceptable restriction in the flow over the catalyst. Catalyst challenges that reduce the time between shutdowns can create severe economic difficulties.

Homogeneous catalysts are also an opportunity for challenging problems. Many of these catalysts are extremely active and are present in very small quantities. This is economically very useful. However, since they are only present in parts per million or less, impurities at those same levels are sufficient to deactivate the catalyst and shut down the process. Raw materials regularly have uncharacterized impurities at the ppm to ppb levels. GC-MS, LC-MS, and computational methods are particularly useful.

12.2.7 Reactive Chemicals

Chemical production plants are regularly energy-intense systems. The reactions used to make the products often are exothermic and the heat released must be managed carefully. Further, the products of many processes are themselves molecules with tens of kilocalories per mole. Thus, managing the energy balance is as critical as managing the mass balance of the system. Understanding the energetics of the components of the system is central to the design. Today, the energetics of a reaction and its kinetics are regularly measured with calorimetry (Accelerating Rate Calorimeters and others), as well as rapidly estimating the thermodynamic properties and kinetics with computational chemistry. When doing computations, it is essential to test the computations with similar molecules where the experimental results are available.

When the fundamental kinetics and heat flows are known, the process model is built (ASPEN or other process modeling tools) and tested. This establishes how things should work. The challenge is thinking about how the systems can fail. For example, what if the catalyst isn't added or is not active. How does the system ensure that the feeds are stopped before the potential energy accumulated in the system is too much for the cooling system when it starts to react?

Imagine distilling a reactive monomer under vacuum. The temperature in the column may be hot enough that small amounts of the monomer are reacting slowly. The heat of this reaction is small relative to the heat being lost by distillation and the process works smoothly. What happens if power is lost? The vacuum is also lost and the liquid in the column collapses into the bottoms where there is limited heat loss. The systems has changed abruptly. It is now an adiabatic reactor with a time to run away.

A similar situation happened in the Houston area after Hurricane Harvey. Peroxides were being stored in refrigerated storage. The heat generated at low temperatures was small and managed by the refrigeration. The plant had multiple sources of power that were cut off by the flooding. The storage then became an adiabatic reactor and the result was a thermal runaway.

The planning for reactive chemical events is complex. It requires a complete understanding of the heat flows and kinetics. But further, it requires the ability to imagine how things can go wrong. Single things going wrong may not be too difficult to understand. Understanding the multiple causes coming together, how to understand the chances and the consequences is difficult and important.

Chemical Reactivity Hazard is a special focus area for problem solving. While outside the scope of this work, it is important to understand. An excellent reference to begin this understanding can be found at the American Institute of Chemical Engineers site (https://www.aiche.org/ccps/resources/glossary/process-safety-glossary/chemical-reactivity-hazard). This class of problem solving needs to begin with the initial research work on the project and be part of the design, construction, and operation planning and operating discipline.

If a reactive chemical event occurs, the problem-solving team will need a diverse team of engineers, analytical chemists, and thermochemists working together to understand the kinetics and heat evolution and pressure build and relief for the scenario. Today, with the enhanced speed of computation, the partnership of an exceptional computational chemist complements this team greatly. This is typically a group of very experienced subject matter experts.

12.2.8 Materials Failure

Materials Failure is another complex area where multiple disciplines come together. In typical chemical plants, the main material expert work takes place during the design and construction phases. Knowledge of the major and minor components of the process and the temperature limits are critical to effective design and construction. Thermal treatment of welds must be carefully planned and executed to avoid failure. This shows up as a serious problem from time to time. The close coordination of the work by metallurgists, ceramicists, and those with a detailed knowledge of the process chemistry and thermal conditions is essential. In problem solving, the use of the electron microprobe, X-ray powder diffraction, and fracture mechanics experiments are common and challenging.

12.2.9 Waste Water Treatment

The wastewater treatment plant (WWTP) is one of the most difficult processes to trouble shoot. This plant combines the challenges of multiple feeds with process variability and the complexity of managing the biological process. Understanding the oxygen requirements, delivering the required oxygen, recycling the bacteria, and disposing safely the excess bacteria produced is a special exercise requiring a team skilled in bacterial kinetics, trace and bulk chemistry analysis (often dilute) in an aqueous matrix. The outfall requirements for the wastewater plant are often very strict and are becoming stricter. The key measure such as chemical and biological oxygen demand (BOD) are frequently slow tests (days) in a continuous process where the results of the data lag the discharge by several days. When the BOD limits are monthly, this can require careful predictive models to ensure compliance. In addition to the usual chromatography tools, NMR in water with suppression of the water resonance has become a particularly useful fast analysis for understanding the state of the WWTP. This can be complemented by solid-state extraction and desorption into a GC-MS.

In addition to the basic requirements of the WWTP, there are frequently special requirements for ultra-trace molecules including polychlorinated biphenyls (PCB) and chlorinated and brominated dibenzodioxins (Dioxins). This work requires special analytical procedures and equipment extending down to the parts per quadrillion levels. Mass balancing and trouble shooting at these levels is possible but challenging. It regularly requires special sampling procedures to get the reliable starting samples for the troubleshooting.

Another challenge of the WWTP is the odors that may be produced. This is particularly difficult, when the operation is close to residential housing. The emissions are irregular, the detection limits for the human nose go down to the parts per billion and parts per trillion level. The wind is variable and people come and go. It is among the most challenging problem-solving scenarios. The keys to success are excellent coordination with the people in the area to record what is smelled, where on a map, and when (including the wind conditions and production plant operating parameters). The working environment can be challenging. The neighbors are upset and the plant may not be smelling anything because the exhaust is from a high stack. Dispersion modeling is not trivial. Ultra-trace analysis begins with monitors that can accumulate these trace-level chemicals. The use of passive sampling similar to that developed by the EPA for fence line monitoring has proved very helpful (https://www.epa.gov/sciencematters/fenceline-monitoring). By the time the neighbors notify the consultant and they arrive at the scene, the odor may have vanished. Passive sampling integrates for an extended period (week) and provides the initial probe into the molecules of interest. The next step is often desorption of the passive sample into a GC-MS with an odor port. This way a skilled analyst can observe the odors, comparing their odors with what was reported by the neighbors and identifying the odor with the GC-MS retention index and mass spectra. Once the key odors are identified, the sources of the odors can be probed up wind and the rest of the problem solving can be done with mass balancing in the production plant. Then operating conditions and mitigation measures such as carbon adsorption usually completes the work.

12.3 Analytical Challenges to Problem Solving

12.3.1 Key Measurements Are Missing

When the problem is assessed and the Definition of Success is initially complete, one of the critical Necessary for Success elements is the list of measurements that are being done and the database of the results of these measurements. Measurements are done for many reasons - process control, safety, product specification. . . The measurements required for successful problem solving may be currently being developed, or they may be missing. If they are being developed, then an assessment of the quality of the measurements is required and run charts of the measurements with appropriate error bars. Frequently, the validation documents for the measurements are missing or the measurement method may have drifted from the procedure in the validation document. A critical first step is the assessment of the current state of the measurement system. If the key measurements are not available, it is necessary to set up these measurements. Figure out what is needed and the required precision and accuracy. Do the validation document. It will seem like a lot of work and you may feel pressure to skip this critical work. It is important. Do it before trying to do the correlations to solve a challenging problem. If the measurement system was adequate, you may not have been called to do the problem solving.

Getting the key samples properly can also be very challenging. The key is to figure out what samples are important and which are extra. The important ones must be collected even if difficult or if process analytical ports have to be added. You cannot control what you don't measure well. Take the time to make the measurements, do the control charts, and analyze the distribution functions. The distribution functions may not be Gaussian.

The timing of the analytical results can also be very important. Sample shipping is important and scheduling of the analytical resources so that the results come back in time to make decision will enable progress. When the key measurements are determined, developed, and tracked, you must remember to transfer the ones the plant will do long term back to the plant and to train the local analysts. This builds teamwork and ensures that the success you create will be sustained.

When you have good analytical data and process data, you are ready to bring in the data analysts. These team members may have already been involved in determining the state of the analytical information. They will have the skill to look at the complex data sets produced by the process data system and the analytical data to determine the characteristics of the "good" in-specification periods and the out of control periods. This is regularly very valuable in determining the direction for problem solving. They are also valuable partners in developing the predictive models for product performance and in-specification operation. The "modelers" and the statistical control partners accelerate the problem solving by seeing through the noise and enabling the team to focus on the crucial elements that deliver success. Without the in-control analytical, the system simply doesn't work.

12.4 Rare Problems

These "rare problems" may not be so rare to the advanced problem-solving team in a large organization. The "usual problems" are often solved by the plants and their dedicated support functions. The "rare problems" are typically the ones where more than one variable have combined to produce a very difficult situation. For example, a trace impurity may have begun to accumulate and precipitate plugging up lines that limit production. This may happen because of subtle changes in the feed or the process conditions or because a recycle loop was closed tighter than previously. Or a minor feed to a big wastewater system in an integrated site may show up as a major feed in the WWTP of a new site with only a few plants. So a minor byproduct, which was lost in the noise can put the small dedicated WWTP out of permit. This list goes on and on.

When dealing with the "rare problem," it is important to recognize that it is an outlier to the usual problem-solving experience and to "open" the hypothesis windows to include the fact that several low probability events may be combining to create this "rare problem." Don't open up the windows to the unusual until you have gone through the "usual suspects." But when the usual suspects are eliminated, start looking for the combinations.

Rare problems often do not have a planned response since two or more low probability issues happening at the same time is discounted in the planning.

The multiple cause/effect problems are very challenging and require a depth of problem solving skill and experience to recognize and solve. These are often the domain of the most senior members of the team. It is particularly important to involve "younger" members of the team in these challenges so that they gain this critical experience. In large companies, while seen as a whole they are rare, they happen frequently enough to be part of the problem response planning.

For example: Cosorb (a process to produce carbon monoxide from partial oxidation of methane) presented as a problem from cleaning heat exchanger tubes with water that caused even more heat exchanger fouling. The true root cause was a trace propylene impurity in the feed that reacted with the solvent to change the solubility of the absorbent. This could not be observed till an intermediate solution to cleaning the heat exchanger without water was developed. The problem of fouling was observed to continue, but at a much lower frequency. In the end, the root cause was a ppm level of ethylene and propylene that were reacting with the solvent to change the solubility of the copper(I) chloride. This illustrates the necessity of solving the immediate problem (cleaning that limited production) and then being able to observe the root problem when it wasn't swamped by the immediate problem.

12.5 Improving the State of Existing Processes That Are Running as Expected

12.5.1 The Reference "Good State"

The most valuable resource when problem solving is a complete round of plant samples taken when the process was stable and running correctly. Typically, these samples are not collected, analyzed, and retained. Having this resource is a substantial cost in planning and execution. Never the less, it is predictable that problems happen in almost all plants. When the problem happens, the first question asked is: What is different? Engineers go to the process logs and try to see what has changed. A central issue is how has the trace chemistry changed. Without the key samples, this is not observable. Problem solvers go to similar plants at other locations to search for differences. Plants are not identical and many "differences" are not instructive. Just saving the "standard analytical results" is helpful and must be studied. However, they are regularly insufficient since these results frequently do not probe the problem space. Only a good set of samples across the entire process carefully preserved (often sealed in very cold storage) is the optimal starting point. Then in-depth measurements are taken with these precious samples to observe the key differences and point to root causes.

12.5.2 New Product/Process Development (All Plant Modifications Are a Subset of This Class) Where the Fundamentals Are Not Well Understood

Today, production plants are carefully designed with the aid of complex computer programs. The advantage of this approach is the creation of a "model" for the plant that can be used in troubleshooting, as well as design. The challenge is that this model does not include many of the small side reactions that produce trace components that may interfere with the use of the product, polymers that foul the equipment, and buildup in recycle streams since their presence was not included in the model.

Pilot plants are a useful way to discover some of these challenges. Pilots are expensive, time consuming, and there is more resistance to building and operating them today. Their use as market development tools does enhance their value.

Pilot plants are useful to test the process models and the specifications required for successful production. Pilot plants can also be useful in determining the yield of on-spec product in closer to a production environment. This is especially true if the laboratory development work was done in a batch mode.

The fundamentals of the product/performance relationship are often complex. The conversion of a mature process to make a product to an "improved and lower cost or higher capacity process" has proved difficult. This is especially true in the case of complex specialty polymers where the performance requirements of rheology, shelf-life, and mechanical properties are challenging to measure.

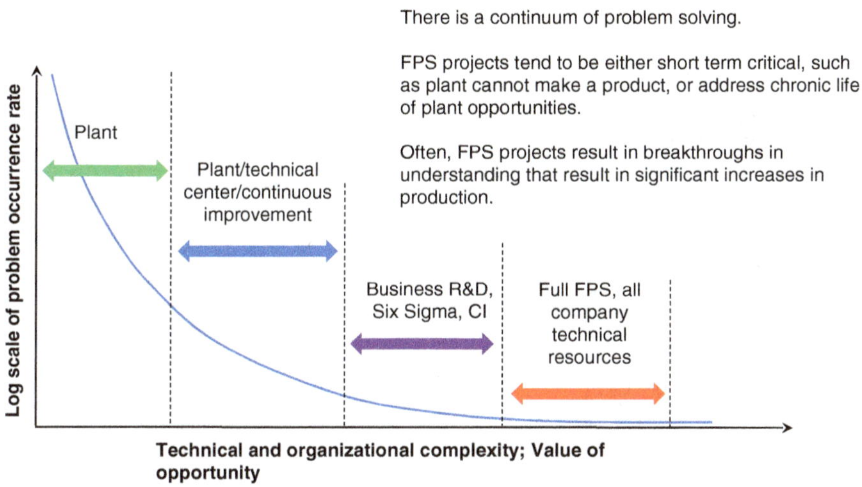

Figure 1.1 The problem solving continuum concept.

Fundamentals of Industrial Problem Solving: A Practitioner's Guide, First Edition.
Zdravko I. Stefanov, Eldad Herceg, Carla Schmidt, David M. Jacobson, Dana Livingston, J.P. Chauvel, Sunil Kumar Chaudhary, and Christopher Paul Christenson.

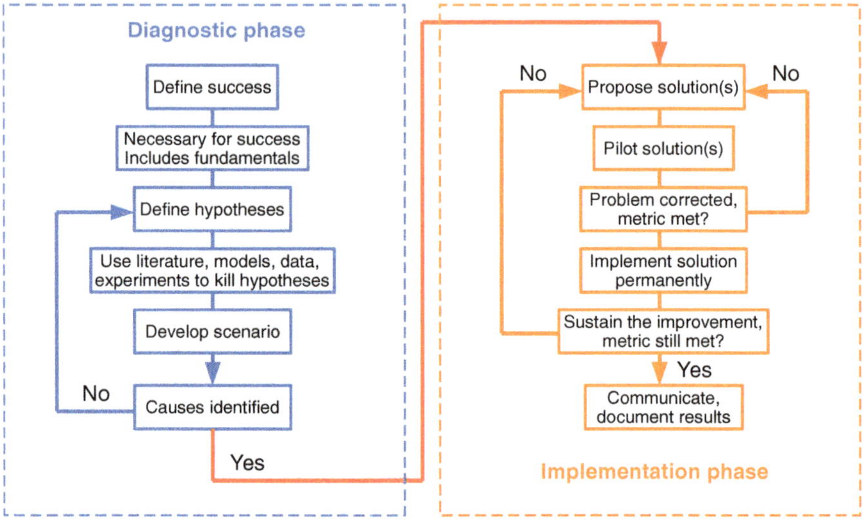

Figure 1.2 Phases of an FPS project.

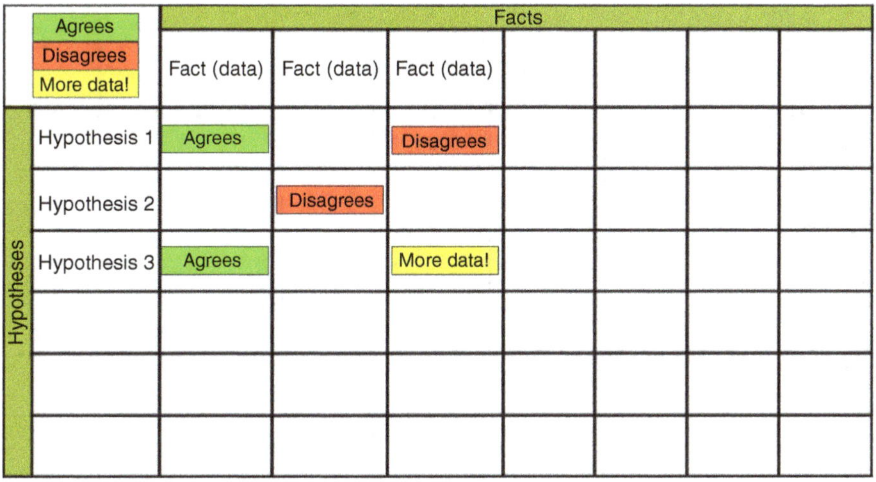

Figure 4.1 Example template for a hypothesis onion.

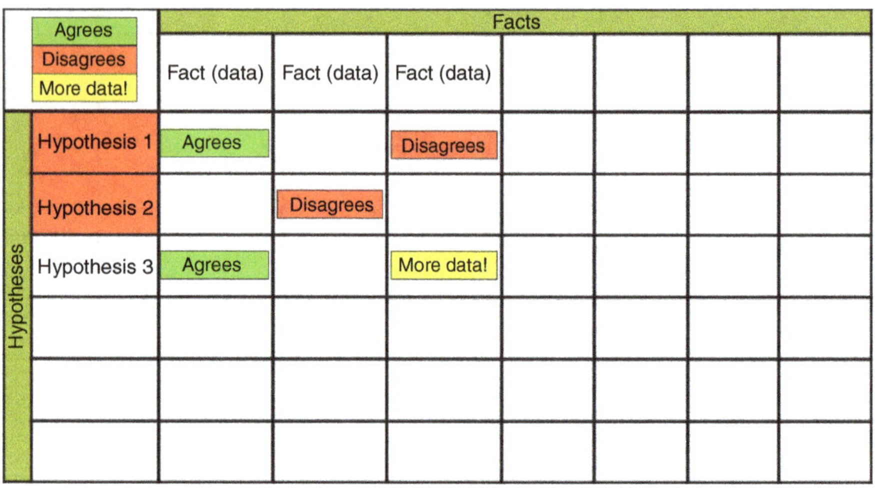

Figure 4.2 Hypothesis onion with refuted hypotheses.

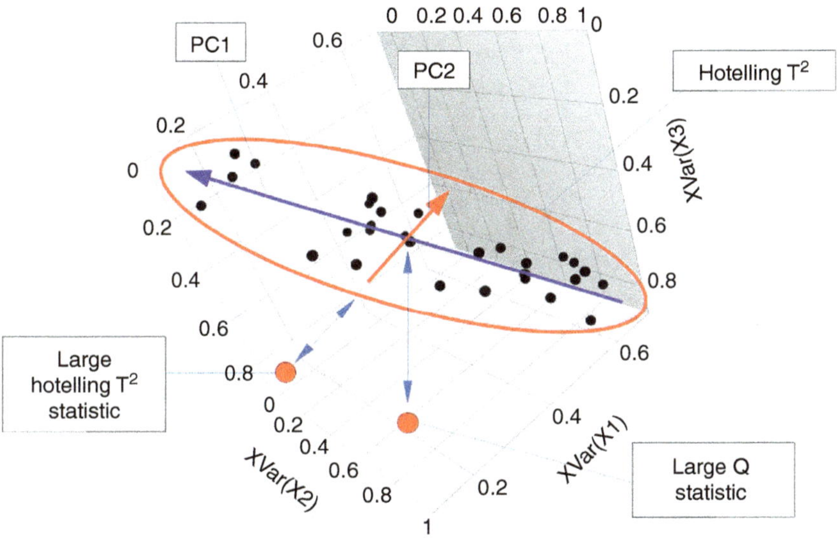

Figure 19.1 Principal component analysis.

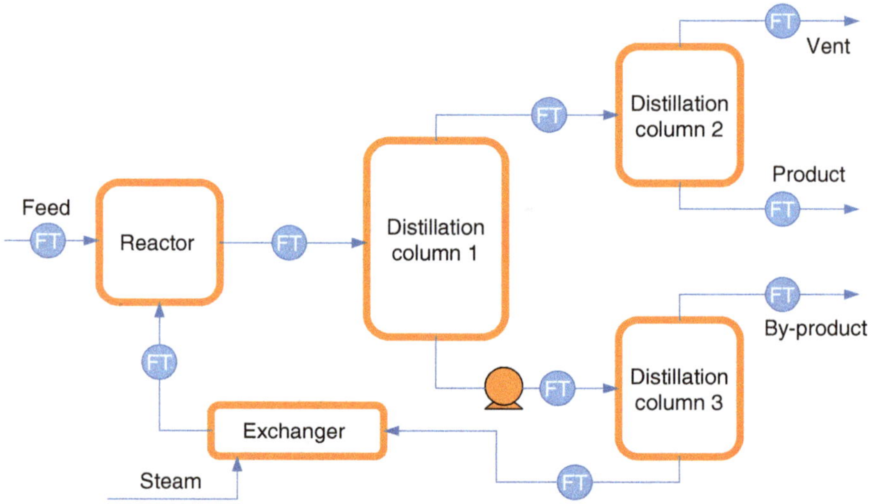

Figure 25.2 Nominal flowsheet.

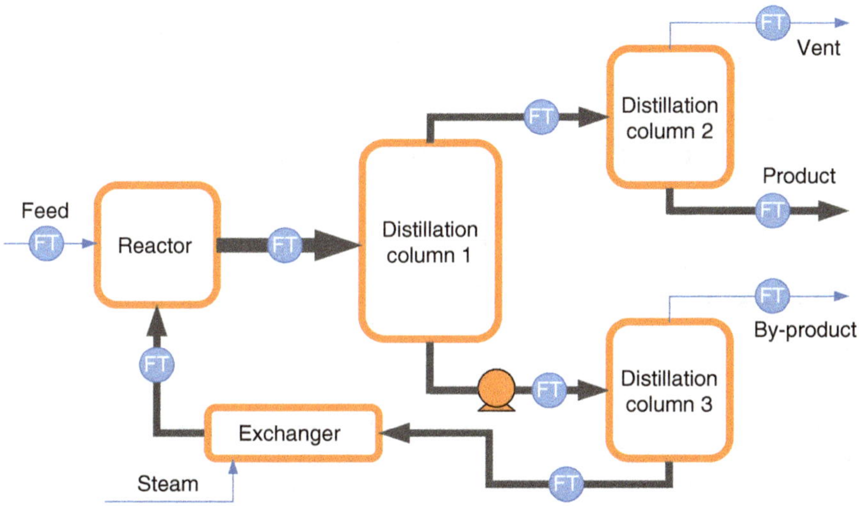

Figure 25.3 "Napoleon" chart flowsheet.

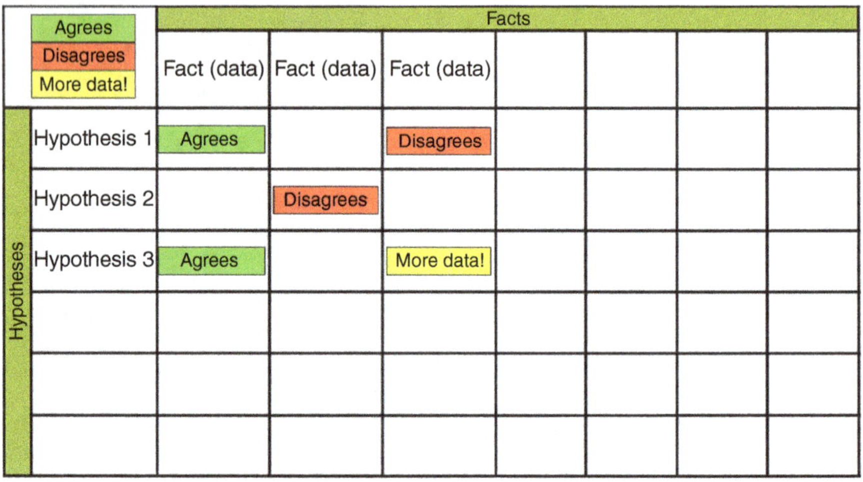

Figure 25.4 The hypothesis matrix is also a graphical tool.

13

Corrosion Problems

As already mentioned in previous chapters, the application of the FPS approach is not limited by the type of the solved problem. Therefore, FPS has been successfully applied to various corrosion problems.

Corrosion, unfortunately, is a never-ending problem in chemical plants (and in many other areas of industry as well). It is everywhere and most of the time there are engineering solutions available that can slow it down enough to not be a significant problem. Sometimes, however, there are corrosion problems that are quite sinister to require an FPS approach to solve. There are two major classes of such problem from the authors practice. The first class is the case where a corrosion that has been managed successfully for years is suddenly completely out of control. The second class is where the corrosion problems in one end of the plant create a catalytic compound that gets carried off and it creates an entirely different problem in another area of the plant that has never seen that kind of problem before.

In this chapter, several interesting examples are provided that cover both classes of corrosion problems.

13.1 The Case of the Corroded Drum

In this example, a certain pressurized vessel was determined to have suffered severe wall loss during a routine turnaround check. The corrosion was extremely severe and unexpected, as the equipment inspection history has shown no sign of such intensive corrosion in the past. The vessel in question was in such position in the plant that catastrophic failure would have led to both severe safety and financial consequences. This was enough to justify an FPS/SWAT (very fast

Fundamentals of Industrial Problem Solving: A Practitioner's Guide, First Edition.
Zdravko I. Stefanov, Eldad Herceg, Carla Schmidt, David M. Jacobson, Dana Livingston, J.P. Chauvel, Sunil Kumar Chaudhary, and Christopher Paul Christenson.

all-hands-on-deck FPS) approach to the problem. The problem was attacked by creating a detailed hypothesis matrix and killing the hypotheses one by one. One of the hypotheses was that the attack was due to the presence of a certain corrosive chemical element, known to be particularly detrimental to the metallurgy of the walls. It was a hard-to-believe hypothesis, as there was no obvious way how this particular element would make its way into the vessel of question. However, the facts (analytical data from the analysis of the corroded material) pointed beyond any doubt that this element was present. After analyzing multiple streams and chasing the corrosive element back to the source, it was determined that it was a reactor in a different plant on site. When the reactor product was separated, the side stream from this reactor was sent through series of other unit operations, carrying the corrosive agent along the way. No corrosion was observed in any of these unit operations, because of the apolar nature of the environment (pure hydrocarbons) and the corrosion agent required polar environment to work (water would do just fine). Finally, in the drum in question, the corrosive agent did meet with water and promptly proceeded to eat away the walls of the drum. The ultimate source of the corrosion agent in the reactor turned out to be a new and improved catalyst that happened to have a new moderator added. The new moderator was a compound that had the corrosive element in its molecule, and it was slowly degrading and releasing said corrosive element in the product stream.

13.2 The Case of the Corroded Reactor Tubes

In this case, a reactor that was operating for many years without serious corrosion issues suddenly lost about half of its tube thickness in about a year. The problem was on the inside of the tubes and it was much unexpected, as even if there is some water and oxygen in the stream, the operating temperature is significantly above the dew point of water for this system. There were various hypotheses involving impurities, changes in chemistry, etc. A detailed investigation of the failed tubes revealed that the corrosion product was what is usually found when carbon steel is reacting with water and oxygen, i.e. rust. All other hypotheses were killed using different methods, so the only living hypothesis left was that indeed the corrosion was due to water/oxygen attack. Deeper literature search was conducted, and this revealed that this kind of corrosion is possible even if there is no liquid water present. A layer of hydration water is enough to provide enough conductivity for the corrosion to occur. This fact is not common knowledge. To finally confirm the root cause, lab experiments with carbon steel coupons were conducted repeating the plant environment, and this proved beyond any doubt that the corrosion was indeed possible even with temperatures above the dew point of water.

13.3 The Case of the Corroded Rotating Equipment Seals

In this case, a rotating equipment seals were failing at a high rate. The failures were more or less irregular, but after several root cause investigations (RCIs), it became clear that all failed seals experience similar failure patterns. Investigation of the failed surfaces revealed that the reason of the failure was fatigue due to repeated stress. What was startling was that similar seals in a similar equipment in sister plants don't suffer the high failure rate, even if exposed to similar repeated stresses. A literature search revealed that the seal material can experience corrosion when exposed to certain chemicals. More literature search revealed that corrosion fatigue micrographs look exactly as what was observed in this plant. All clues essentially led to corrosion being the main reason for the seal failures. The process was one source of potential corrosive agents, but this hypothesis was killed, as sister plants running similar grades did not have the high failure rate. The only other source was the lubrication oil. After further literature search, it was discovered that a component in the lubrication oil was corrosive toward the seals material. The variability in mean time between failures (MTBFs) was coming from other sources (startup/shutdown/grades, run/seal material inconsistencies, etc.), but the high baseline was due to lubrication oil corroding the seals.

13.4 The Case of the Corrosion Catalyzed Fouling

In this final example, a rotating equipment failed catastrophically during plant startup. When the equipment was open, it was revealed that it was plugged by a semi-solid polymer of cryptic composition that prevented the equipment from rotating. An analysis of the foulant revealed that it is a kind of polymer that can form if the material normally processed is exposed to high temperatures. The required temperatures were much higher than the normal operating temperatures and analysis of plant data did not indicate any violation of the temperature limits. Further investigation revealed that the material contained substantial amounts of iron. Following that lead, it was determined that iron is a catalyst and can indeed cause the polymerization reaction to occur at significant rates at much lower temperatures. However, it was not clear where the iron was coming from, as the incoming stream contained very little. An examination of the startup procedure indicated that the plant starts by using recycle material until the main feed becomes available. But this is a standard procedure, and this plant and sister plants used this procedure for many years and did not have a problem. So the mystery persisted. Therefore, deeper investigation of the recycle tank material was performed. The recycle tank was emptied in ISO containers and their content analyzed. Some containers revealed very high iron content, some not. This fact indicated that the tank has the tendency to layer out different fractions and some happen to contain very high iron levels. Then, it was discovered that during that particular startup, the plant made a minor change in the startup procedure and effectively "mined" the recycle tank and "struck gold" (i.e. iron) exactly at the right moment. The slug of high iron material polymerized in the process and plugged the rotating equipment in question. More investigation determined that the iron in the recycle tank comes from turnaround activities exposing carbon steel to the atmosphere for short amounts of time. As during startup, the first several hours of production don't meet the specs, they are sent to the recycle tank, and that material just happens to wash out all of the rust created during the turnaround. Further R&D effort (lab experiments) showed that all plants using this startup method were operating at the edge of the razor and were in essence just lucky they did not experience the same failure at startup as in this particular case.

13.5 Summary

To summarize, corrosion problems are ubiquitous in chemical plants. There is always push to use the cheaper carbon steel. This leads to rust formation when improperly treated or exposed to ambient conditions. Another issue is of course the presence of various corrosive chemistries that cannot be avoided. In any case, corrosion problems can be quite insidious, as shown in the examples above. They are sometimes quite far from obvious, and it requires significant effort to weed them out, but the seasoned chemical plant problem troubleshooter should always consider corrosion as the root cause of the problem in one way or another.

14

Mechanical Issues

This book has been written by folks in the chemical industry. Many of the examples the authors are sharing have that slant. We tell the stories we know. This could lead to some confusion on FPS and lead the reader to think this is a chemical-industry-only methodology. FPS is independent to scientific discipline. We can attack biological, mechanical, chemical, electrical, nuclear, or physics-based problems with the same methodology. The methodology is based on the scientific method. This method is not steeped in only one discipline. It crosses the lines. If the authors were in the automotive industry, the stories would have more mechanical basis; but the pattern for solving problems would be the same.

The purpose of this chapter is to cross the line from mostly chemical-based examples to more mechanical-based examples. Hopefully, by crossing this line, the readers will see that although the science is different, the METHOD for solving the problems is the same. We could continue to cross the lines and cover electrical or biological or nuclear examples, but hopefully these mechanical examples will open the reader's eyes to the fact that the method is **EXACTLY the same** no matter the inherent nature of the fundamental problem. The only real difference is ***who*** the team members need to be to solve different types of fundamental problems. With FPS, it would not matter, one can help solve problems in that industry ***only*** with the additional help of experts in that scientific discipline just by following the FPS methodology. The experts need to be there to propose likely hypotheses and to deliver the real scientific basis for the problem solving. The experts will be critical in figuring out how to kill hypotheses and ultimately how to solve the problem. With the right skill set ***and*** the power of Fundamental Problem-Solving methodology, you can slay the big problems anywhere. This chapter will give some mechanical examples to help the reader understand that the methodology works beyond chemical-based problems.

Fundamentals of Industrial Problem Solving: A Practitioner's Guide, First Edition.
Zdravko I. Stefanov, Eldad Herceg, Carla Schmidt, David M. Jacobson, Dana Livingston, J.P. Chauvel, Sunil Kumar Chaudhary, and Christopher Paul Christenson.

14.1 Rotating Equipment Examples

Rotating equipment dominates the work of many of our mechanical engineers. It is at the heart of the discipline. Many mechanical engineers fell into their chosen profession starting with a love of cars. That first car felt like life's biggest choice. Then it was about fixing that car up to be something new, something better. To do that work, you must understand how brakes work, how the engine works, how the transmission work, and all the "how it works" questions. Behind those questions are fundamental physics. Rotational forces, frictional forces, and all the other forces that drive the inner workings of a machine. That same passion and pursuit of knowledge is exactly what it takes to run mechanical FPS projects.

Rotating equipment in the form of blowers, fans, pumps, motors, turbines, compressors, and agitators cause downtime when they fail. Long lead times and long repair times result in expensive maintenance costs and downtime losses. Shaft fatigue, bearing failures, seal failures, and many other types of failures occur. These failure modes can be identified and fixed applying physics/first principle techniques. In this section, we will share examples of applying first principles to rotating equipment mechanical failures.

Rotating equipment failures and problems will use hypotheses driven by process experts and standardized troubleshooting guides. As you usually pay a large amount of money for any piece of rotating equipment, you will receive a bonus. The bonus is a manual that shows details about the operation of the machinery and will also generally have a troubleshooting guide. Note that newer manuals are online instead of paper; but the format is not important, it is the content that is important. Your manufacturer will have seen many common failure modes. Because their customer base is often one or two orders of magnitude larger than your company's rotating equipment needs, their experiences can be extremely valuable. Pulling out that information should be a starting point literature search for the hypotheses generation. You can then take the known facts and observations to see which of those hypotheses are worth pursuing. Because FPS uses the hypothesis onion matrix, many hypotheses can be checked against the facts and observations simultaneously. In addition to hypotheses that can be pulled from the manual, there will be trade groups, books, blogs, and other sources of literature that will have common failure mechanisms. Process experts can then round out your hypotheses with their experience and understanding. The parallel processing aspect of FPS leads to a very complete pursuit of understanding. A surprising number of problems come from not doing a thorough and complete problem investigations. Teams take shortcuts and can miss important details. FPS is robust because it gets the facts and observations pushed up against many hypotheses at the same time.

For example, an improvement team was tasked with reducing bearing failure in an 8-pump system. Four pumps and their spares were failing at an alarming rate in a combination of mostly seal failures with the occasional bearing failure. Mapping the failures showed that most of the failures were occurring to only two of the pumps and their spares. The team formed several hypotheses but focused on high vibration per the vendor manual on pump troubleshooting. Fatigue, poor lubrication, seal fluid contamination, and installation issues were among the front runner hypotheses to be pursued first. Physical inspection of the pumps showed that the two sets of pumps with the highest failure rates were different from the remaining pumps. They were set on floating bases. Additional data were gathered to kill the high vibration family of hypotheses by measuring pump vibration with an accelerometer. To kill the hypothesis, the team needed to find statistically similar vibration levels between the high failure rate pumps and the low failure rate pumps. Often with mechanical issues, killing of hypotheses is done with a direct measure of the hypotheses. In this case, kill the high vibration family of hypotheses with a vibration measurement.

The team discovered high vibration on the floating-base-type pumps and failed to kill the high vibration hypothesis family. Further investigation showed thermal expansion design issues drove the original design decision to install two of the pumps on the floating bases. The clear scenario was that these pump bases were failing to provide the support needed to minimize vibration. The FPS team decided that decreasing the vibration by moving to a fixed base was the right answer for the credible scenario the team had found. The team brought in computer modeling of thermal expansion to see if there were alternate solutions to both reduce vibration and yet not stress the pumps and pipes when process underwent thermal expansion cycles. A significant redesign of the suction piping was shown to work on the computer models and the team moved to implement the solution of a redesigned suction pipe and a fixed base. The installation dramatically decreased the vibration levels and ultimately the pump seal failure rate was dramatically reduced. In this case, the main hypothesis and validation methods were in the vendor literature. This team did exactly what FPS teams do all over. They formed good hypothesis (literature search!) and tested the hypothesis as directly and simply as possible. Using modeling and good solid engineering principles, the team solved the problem. FPS works in mechanical/civil discipline problems.

Another rotating equipment example FPS problem occurred when the bearings of a major compressor were failing at an unacceptable rate. Each failure brought production down and was costing hundreds of thousands of dollars every week. A team of rotating equipment specialized mechanical engineers had worked for a few months on the problem and had not come to a resolution. The decision was made to utilize the Fundamental Problem Solving methodology to try and understand the problem deeper and come to a solution. The team was nearly the same. One difference was injected into the equation. A project leader with FPS experience was asked to lead the improvement work. The flip of trying to kill hypotheses vs. trying to prove them was instrumental in fixing the problem. This case study furthers the premise that FPS is independent of discipline, but inherently more effective than other problem-solving techniques.

The first move the reformed team did was to put the troubleshooting done so far into the form of specific hypotheses. This was extremely helpful in understanding the thought processes the experts were using. It was also very helpful in coming up with some dramatic and sweeping ways to kill hypotheses. For example, there were many hypotheses that were essentially around mistakes or wrong repairs that would result in the machines failing quickly. The mechanics involved were open to systematic repair mistakes but insisted that they were following manufacturer recommended procedures and had been doing the same techniques prior to the failures. The team prior to the FPS formation had taken measures such as watching all the steps of the repairs to see if the mechanics were following the procedures. This only lead to further questions on the best ways to do the steps and more questions about mechanic to mechanic differences. It was interesting but drove only hypothesis supporting behaviors. The FPS challenge was how to KILL the hypothesis. The radical kill technique the team decided on was to take the same model compressor that was working in a nearby service and completely reinstall it in the failing location. A repaired machine was then installed in the nearby service location. The logic was if a known to be working compressor failed quickly in the service change, it would kill the entire branch of all the different hypotheses centered around repair procedures. Many hypotheses could be killed with one well-designed test. That previously working compressor failed quickly and the repaired compressor installed in the nearby service kept running. The repair technique hypotheses were dead, and the team could then focus on more productive areas. The mechanics and engineers now could stop the arguments about the repairs the mechanics were doing and think harder on the real problem.

Another example of how this team focused on the mechanical problems of a compressor used killing hypotheses as a strategy to move forward came in the lubrication for the compressor. Poor lubrication is known to destroy bearings. The lubricating fluid was also known to degrade in service. Previous efforts prior to the FPS team forming had shown degradation was occurring. Unfortunately, that happens in these systems. The team did not know what amount of degradation could ultimately cause the bearing failures. Proving that degradation was occurring was just supporting the degradation hypothesis and did not kill the hypothesis. To kill this one for good, the FPS team replaced the lubrication oil completely with new batch. If the failures continued to occur before degradation had happened to the fluid, then this hypothesis could be killed. It died. Once again effectively killing a hypothesis was extremely helpful in moving the teams to focus on to other causes.

The lubrication degradation hypothesis showed the value of watching the measurement process in hypothesis work. The operator and the engineer that took the new oil sample during the investigation of the lubrication degradation hypothesis while the compressor was running noticed off-gassing was occurring. The sample was hot and whatever was in the lubricant was not showing up in the lubricant testing because it had left the liquid prior to that measurement. This observation led to a new hypothesis about a light boiler contaminating the oil and then flashing in the bearing ultimately causing bearing failures. The team worked to get a sample of the off-gassing material through a simple dry ice cooling system. The sample analysis identified the light boiler. Physical property research and simple modeling showed that flashing across the bearing was possible. Dewpoint calculations helped the team try and understand why the light boiler was getting in the compressor system. The calculations showed that it was possible to condense the material in the compressor's high-pressure side. That would allow the contaminant into the lubricant. The suction side could condense the light boiler only if the temperature was colder. Tracing the suction sideline resulted in the team seeing some brand-new insulation installed well downstream of the compressor. This physical observation caused team members to think hard about the temperatures in the system. One technique that is effective is imagining you inside the equipment and asking what would you see? The team did this and asked what you would see if you were inside the uninsulated line. They believed they would see fine droplets forming along the wall as the air cooled the gas. Those droplets would then flow in the direction of the gas, but along the wall. They would slowly gravitate to the bottom of the pipe due to gravity. Following the line, they realized that the droplets would stop following the gas as they hit the bypass line and they would travel down that line by gravity and accumulate in that line above a bypass valve. The liquid would then periodically get emptied out every time the bypass valve opened. Because it was a liquid, it would fall downward against the gas bypass flow and then eventually rejoin the gas stream, but beyond the compressor. The liquid effectively bypasses the compressor.

To kill this hypothesis, the team removed the recently installed insulation on the line. The team pursued this kill strategy with the highest priority because checking the records showed that the insulation was installed in the right part of the timeline where the problems began to occur. A seemingly unrelated event may have been related. Ultimately, the team failed to kill this hypothesis because the compressor continued to run and the bearings did not fail at an accelerated rate. The problem was solved. This deeper understanding led to another benefit. The team realized that the bypass valve did not cycle on a regular basis, so even with the normal compressor life, there could have been historical failures occurring by this same mechanism. The team designed and installed a small condenser and drain system to do what was done unintentionally in a planned manner.

The compressor performed better than ever. The failure mechanism was disrupted and no longer caused problems.

This example shows how when the cash is flowing fast, it becomes time to make bold moves. Swapping compressors and replacing the lubricant entirely are costly, but cost–benefit analysis and understanding the total costs help drive bold moves. It also shows that it doesn't have to be Chemistry to find FPS helpful.

14.2 Materials of Construction

Engineers select materials of construction with the intent of that material to provide durability to make the part last. It is a trade-off decision because often the physical properties that make the part durable in an application will also impede function or impact overall cost. Many mechanical FPS examples are rooted in failure modes unique or part of the science of the materials of construction. This is one such example.

A pipe blew apart causing a significant chemical release. Once the leak was contained, the source of the failure was found. The bolts holding the pipe flanges together had sheared. With an entire plant held together by bolts, the problem quickly escalated, and a team was formed to figure out what happened. The first step was to define success. The team decided to keep the scope small by just understanding why these bolts failed. Keeping implementation of fixes out of the initial scope helped the team stick to the problem at hand and not worry about every bolt in the plant. Once the root cause was understood, then the scope could be adjusted to the size needed to prevent future occurrences.

The team brought in expertise specific to materials of construction and bolt design. The team followed the FPS methodology of forming several likely hypotheses that included failure mechanisms in the original manufacturing of the bolt, over torquing, thermal expansion problems, and many others. One of the first steps was to validate the material composition of the bolt. The team did not make any assumptions about the bolt make up. This helped kill hypotheses that were dependent on certain materials of construction.

> ***Wisdom Tidbit – In any failure situation, saving all failed parts is critical to success of understanding the problem. Make that an initial priority.***

One hypothesis that was formed during the initial exploration was chloride-ion stress corrosion cracking. The team attempted to kill this by seeing if the failed bolts had cracks inconsistent with stress corrosion cracking via microscopy. This failed to kill the hypothesis because cracking as viewed by experts was deemed to be consistent with stress corrosion. The initial testing of bolt material was also a way to attempt to kill this hypothesis. This also failed to kill it. The material was consistent with material susceptible to chloride-ion stress corrosion cracking. Chloride ions were also found in testing. In the end, this hypothesis could not be killed, and so the preponderance of the evidence supported the hypothesis of chloride-ion stress corrosion cracking. The team could then move forward and plan a fix with the right scope and the right problem to attack.

Materials of construction matter and do cause problems. The FPS approach works on these forensic failure investigations just like it works on other types of problems.

14.3 Heat Exchangers

Heat exchanger mechanical failures are expensive and almost always result in expensive downtime, and sometimes can cause significant chemical releases to the environment. In addition to corrosion failure modes, with heat exchangers there are many mechanical failure modes. Metal fatigue, vibration, metal erosion, erosion/corrosion, thermal shock, thermal expansion, freezing-induced expansion, sealing failures, water hammer, steam hammer, over pressurization, and many other failure modes occur. These modes can be identified and fixed applying physics/first principle techniques. In this section, we will share an example of applying first principles to heat exchanger mechanical failures.

An exotic material heat exchanger was failing every month. Each time it failed, the entire production facility had to slow down or shut down. Failed exchanger costs were mounting and production losses were taking their toll on customers. The burn rate was high and a team was formed to understand the fundamental reasons for the failures and put countermeasures in to extend the life of the heat exchanger.

Thermal expansion mismatches between the exotic material leading to excessive forces was formed as one of the leading hypotheses. Other hypotheses were formed and killed, but the thermal expansion hypothesis was not dying. The system was complex because plant trips and restarts could lead to complicated thermal profiles across the heat exchanger. Fundamentals were needed to understand and try and kill this hypothesis.

Three models were created. The first was a force displacement calculation model that used first principles to calculate the forces involved with different temperature profiles across the heat exchanger generated with the different thermal expansion rates of the materials of construction. The team needed the actual temperature profiles to complete the model. The team then turned to modeling the heat exchanger heat transfer. This was a complex system with a cold liquid that was brought to boiling temperature and then completely vaporized. Multiple components in the mixture and different heat transfer regimes made this a challenging model. The model was completed and checked against actual field data to validate the model. Because the model was complex and consumed a lot of resources, the team faced a challenge to understand all the different run conditions. These run conditions could number in hundreds of combinations. To simplify and speed up the process, this complex model was estimated using design of experiment-based equations. This enabled the team to take a theoretical model and simplify it to accurate, but empirical equations. Armed with a force displacement model and a model of the temperature profile for any run conditions, the team could then put them to work on actual run conditions. The team downloaded months of run conditions, fed those conditions through the model, and looked at the forces calculated. This enabled the team to see when and how high forces were generated from this mechanism. This helped in two ways. First it helped the team set controls in place to try and prevent these high forces. The second was to help with redesign of the heat exchanger. Alternate designs could then be checked on the model to select the best performing designs.

A combination of controls set to prevent high thermal expansion mismatches and a heat exchanger redesign ultimately lead to normal heat exchanger failure rates and brought the plant back to normal production. Once again fundamentals and doing the work it takes to truly understand the problem came through. Fundamental problem solving takes these steps. It works.

14.4 Summary

Fundamental problem solving is independent of the engineering or science discipline. It is a methodology rooted in the scientific method and as such is highly specialized in its approach, but highly generalized in its application. This chapter shows multiple mechanical base problems that were solved using the disciplined methodology of FPS. There are many more examples, but hopefully these will help the reader keep an open mind. The authors encourage the readers to be bold and apply FPS to any scientifically challenging problem. It works by killing hypotheses rooted in whatever science applies to the problem.

15

Color

Color problems are some of the most frequent and challenging issues in many industrial processes since they are usually due to some trace side chemical reactions. They are frequently intermittent, they are present for a few days or weeks, and then they disappear, only to reappear again. It is quite often that products have specifications for color or yellowness index, often measured by the Pt/Co number and expressed as APHA color (for the American Public Health Association and defined by ASTM D1209). High color results in off specification product that either is sold at a lower price or cannot be sold at all. This causes supply chain issues as well. Low APHA color products are expected by many customers for many applications (textile manufacturing, electronics industry, agricultural chemicals, etc.).

A common approach, when facing a color issue, is to send good and bad samples to the analytical lab and try to figure out what the color is. This can work if the color bodies can be isolated to a level where they can be detected. This implies preparative chromatography (thin layer paper chromatography, liquid chromatography, etc.). Often the results of such an approach can be inconclusive. The color bodies frequently are a family of molecules that are structurally related but individually are <1 ppm in concentration, often in a matrix of similar molecules, making their isolation and detection difficult.

In this section of the book, we will describe a new methodology that has been used to successfully resolve many color problems for different chemistries. This more useful approach uses, among others, the process chemistry and side reactions, a UV-Vis of the colored product and quantum mechanical modeling of hypothetical color bodies. A real work example of a color in ethylene glycol (EG) plant is presented.

The EG process is a complex process with multiple towers, reactors, evaporators, and recycle streams (Figure 15.1).

Fundamentals of Industrial Problem Solving: A Practitioner's Guide, First Edition.
Zdravko I. Stefanov, Eldad Herceg, Carla Schmidt, David M. Jacobson, Dana Livingston, J.P. Chauvel, Sunil Kumar Chaudhary, and Christopher Paul Christenson.

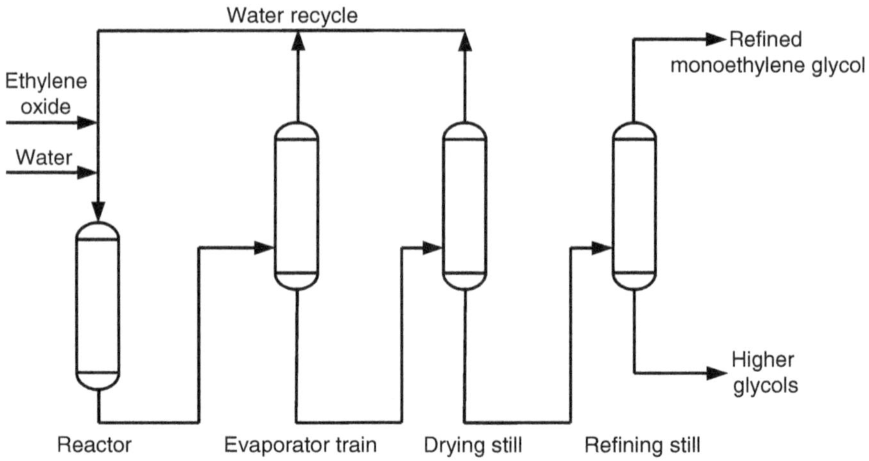

Figure 15.1 Simplified diagram of an ethylene glycol process.

The plant was experiencing a common color issue – an intermittent yellow color that is very faint. The normal Pt/Co readings were about two units. The product is considered to be off specification if it exceeds a reading of 5. In this case, the color was appearing in the final product tank of an EG plant. The plant has only two very large storage tanks. When a sample from the tank is off specification, the entire tank is not suitable for polyester grade EG. After a common approach of isolating and identifying color bodies has failed in solving the issues, a new methodology for color problem solving was developed. The methodology involved using multiple techniques, including chemometrics plant data analysis, computational chemistry, laboratory and plant experiments, liquid chromatography/mass spectrometry (LC/MS), and UV/VIS spectral analysis.

Success for this particular problem was defined as to have APHA color of <2 units. Necessary for success was the following:

- Process data and an analysis (model) of the color formation
- A fundamental understanding of the potential chemistry of color in the system
- Quantum mechanical modeling of hypothetical color bodies (or at least Woodward-Fieser rules)
- An UV-Vis spectrum of the color (preferably taken on a double-beam instrument)
- Proof by thermodynamics (reaction is allowed) and/or proof by synthesis
- Plant test

Rate limiting was the identification of color-causing molecules. As a part of the methodology, numerous hypotheses were postulated about the origin of the color, including the usual metals hypotheses (Table 15.1).

Table 15.1 Hypothesis list.

Major hypotheses
The material in the MEG storage tank is contaminated with something which causes the color.
Corrosion is the cause of the color.
The measurement system is not capable of detecting color changes – gives false color readings.
Nitrogen compounds (nitro, amine, etc.) are the cause of the color.
Recycles are the cause of the color.
The color is due to particulate matter, not molecules.
pH swings, especially high pH, is the cause of the color.
O_2 leaks into process is the cause of the color.

The metals hypotheses were quickly killed by ICP (Ion Coupled Plasma)-MS, which showed little if any detectable metals and no differences between good and off-spec EG.

Next step is to use process chemometrics (specifically cluster analysis and multivariate analysis) to examine the plant data in detail. In the cluster analysis, all process variables, typically data collected at one minute intervals, are examined during times of low and off-spec color. The clusters are typically in different regions of a high-dimensional vector space. The causes of the differences are identified as key variables for further examination. Multivariate analysis complements cluster analysis in that a mathematical model of the color is developed that uses process variables as inputs. The most important process variables are typically related to color formation chemistry (pH, temperature, residence time, etc.).

The process data contained over 250 process variables. A preliminary partial least squares (PLS) model was built to gain a general data set understanding and to remove outliers in the data. The top most important variables found by chemometrics analysis are listed in Table 15.2. They are sorted by their Variable Importance for Prediction (VIP) metric.

The data set of the final model was reduced to only 24 process variables. To validate the model, the latest process data was left out of the calibration set. After the model was built, the color of this set was predicted using the model with the result shown in Figure 15.2.

The prediction of the validation set is of good quality. A major color event (Figure 15.2, left part in the graph) was predicted very well. This confirms the quality of the model. The model quality is satisfactory, keeping in mind that the color data has low signal-to-noise ratio. This is because the standard deviation of the color analyzer was 0.3 color units. What is important here is that the predicted color follows the trend of the raw color, even if the high-frequency variation is not captured by the model.

Table 15.2 VIP metric values for the most important variables.

Variable	VIP
Ethylene Glycol Reactor Caustic Feed	1.81
Ethylene Oxide Reactor Inlet Dichloroethane	1.52
Downstream location 1 pH	1.36
Ethylene Oxide Reactor Inlet Ethylene Oxide	1.24
Distillation Column Overhead Flow	1.12
Second Column Overhead Flow	1.09
Ethylene Oxide Scrubber Overhead Ethylene	1.09
Downstream location 2 pH	1.07
Ethylene Oxide Reactor Pump Ditch Flow	1.00

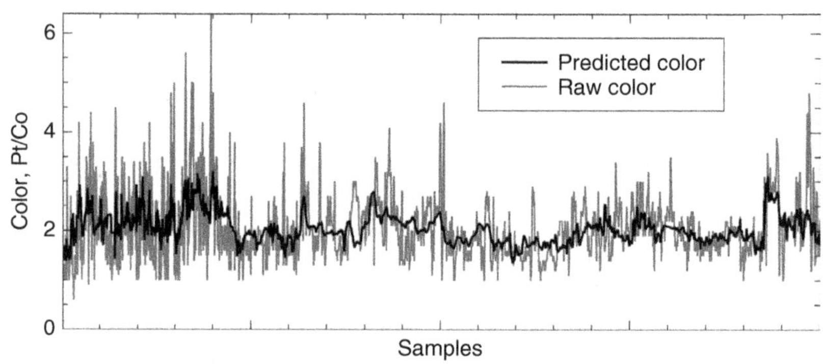

Figure 15.2 Observed vs. predicted color.

The data analysis shows that the main process variables statistically related to the color were caustic related. The top candidate is the caustic flow to the EG reactor. Another variable that seems important is the EG reactor discharge pump flow, which points to the residence time in the glycol system. This could also point to a higher concentration of aldehydes being fed to the system due to a higher pure EO production. Overall, process chemometrics was able to quickly identify pH as a key variable. When certain pH values were above about 8.5, the color was likely to occur.

Once it is known what process variables correlate with the observed color, next step is to postulate a most probable chemistry that can lead to the formation of color bodies in the process. In our example, the possible source of color is a chemical reaction that is catalyzed by caustic, i.e. high pH. The process conditions observed in the glycol system include basic pH, high temperatures (135–180 °C), low water concentrations, relatively high residence time, and acetaldehyde impurities. These conditions are favorable for aldol condensation reaction to occur. The aldol condensation products are long-chain unsaturated compounds with conjugated double bonds that absorb visible light and therefore can create color in the product.

To prove the aldol reaction hypothesis, first a literature search was performed. The literature search provided results that shed light on the conditions necessary for the aldol condensation reaction to occur and the necessary precursors. It was known from historical plant data that at the operating conditions of the plant, it is possible to have acetaldehyde formation. An important observation from the literature is the pKa of the beta proton in acetaldehyde, which is shown to be ~16.7. This means that alkoxide is necessary for the reaction to occur, in other words the reaction is catalyzed by base. The activation energies for the addition and for the elimination of water in base are about 13.9 and 14.1 kcal/mol. This means at 135 °C, this reaction is going very well. It indicates that concentration of acetaldehyde in regions of the plant that contain alkoxide could be of importance. A change of 2x in the reactant for a bimolecular reaction will change the make rate of product 4x. The series aldol condensation mechanism is summarized below:

$$CH_3CH{=}O + OH^{(-)} \rightarrow {}^{(-)}CH_2CH{=}O + H_2O$$

$$^{(-)}CH_2CH{=}O + CH_3CH{=}O \rightarrow CH_3CH(OH)CH_2CH{=}O$$

$$CH_3CH(OH)CH_2CH{=}O + \text{base / heat} \rightarrow CH_3CH{=}CH{-}CH{=}O(\text{crotonaldehyde})$$

Repeat to build series $C_xH_{x+2}O$(Aldol series).

A version of this chemistry is when nitrogen compounds are also present. NO_2 in the cycle gas produces about 1 $HOCH_2CH_2NO_2$ for every 6 $CH_3CH{=}O$.

$$HOCH_2CH_2NO_2 + OH^{(-)} \rightarrow HOCH_2CH^{(-)}NO_2 + H_2O$$

$$HOCH_2CH^{(-)}NO_2 + CH_3CH{=}O \rightarrow CH_3CH(OH)CH(NO_2)CH_2OH$$

$$CH_3CH(OH)CH(NO_2)CH_2OH + \text{base / heat} \rightarrow CH_3CH{=}C(NO_2)CH_2OH$$

Therefore, it is possible to incorporate an NO_2 into an Aldol chain.

Next step is to perform quantum mechanical modeling of hypothetical color bodies to investigate the effect of the chain length and the presence of nitrogen in the chain of the aldol condensation product. To do this, a computational chemistry model for the absorption of different compounds was built. The results for two different cases are summarized below. The first set of molecules investigated was crotonaldehyde and its higher homologues. They can be expressed with the formula $C_xH_{(x+2)}O$, where x = 4–18 (x increased in increments of 2). Results from theoretical modeling can be found in Figure 15.3, which visualizes calculated UV/Vis spectra. From Figure 15.3 can be seen that the molecules having 6 conjugated double bonds (5 C=C and 1 C=O) or more exhibit appreciable absorption in the visible region ($\lambda > 400$ nm). The first candidate with significant absorption in the visible spectrum is hence the C_{12} product.

Another class of molecules studied that is going to be shown is similar to the previous one with the difference of having nitro functionality inserted on a second C atom from the methyl side ($NO_2C_xH_{(x+1)}O$). Outcome from this modeling can be found in Figure 15.4. It can be seen that this class of molecules leads to a significantly different set of UV/VIS traces when compared to the $C_xH_{(x+2)}O$ compounds. Each molecule leads to appearance of at least two intense absorption peaks. The most noticeable difference is the length of the chain required to reach molecules who exhibit a color. As mention above, for $C_xH_{(x+2)}O$ aldehydes, this happens when $x \geq 12$. However, for the molecules having NO_2 functionality x reduces to only 6.

The results from the computational chemistry modeling confirm that the aldol condensation products with chain lengths from C6 to C12 can be contributors to color in the product.

Thermodynamic calculations should be used subsequently to confirm that the formation of color bodies is thermodynamically allowed. For the reaction of two

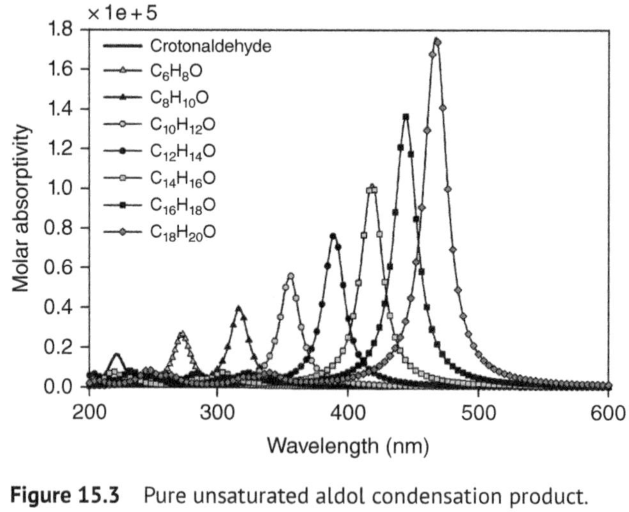

Figure 15.3 Pure unsaturated aldol condensation product.

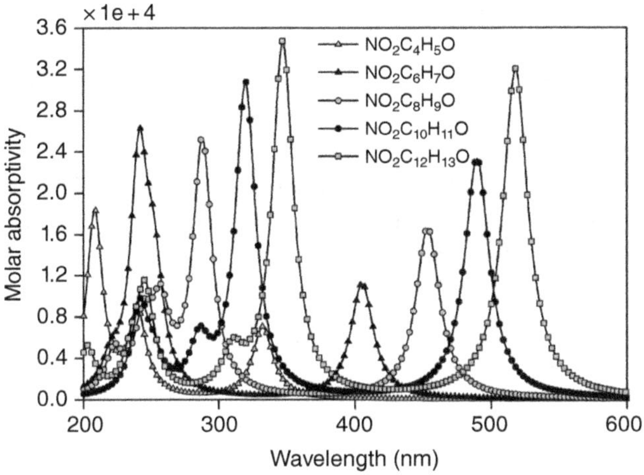

Figure 15.4 Nitro group added to double bond.

acetaldehyde molecules to make crotonaldehyde and water, ΔH is about −5.1 kcal/mol, ΔS is about −4 cal/mol*K. This means ΔG is −3.5 kcal/mol at 408 K (135 °C), indicating reaction is spontaneous at this temperature. The subsequent addition steps to make the C6, C8, etc., adducts have similar parameters. Dehydration of the product in a column reboiler will drive this reaction even further to the right (to the product). The observed results from the thermodynamic calculations point to some extremely important details. First, the influence of pH on this reaction is very high and very nonlinear, as an increase in one unit of the pH speeds up the reaction 10 times. This means that only one pH unit can be the difference between color free or high color product. Another important observation is the cascade effect of the initial aldehyde concentration on the reaction rate (second-order reaction), especially in the area of the products with high molecular weights – pentamers and above – which are the color bodies. Minute increases in the aldehyde concentration leads to huge changes in their rate of formation.

To reproduce the aldol condensation reaction in lab conditions, 100 ppm acetaldehyde solution was prepared in ethylene glycol at 35 °C and pH was adjusted to 9. After only 120 minutes at these conditions, visible color was discovered. The conditions are not the same as in the plant – the temperature is much lower and the concentration of acetaldehyde is much higher; however, the temperatures in the plant are much higher and if standard Arrhenius law is assumed for the temperature dependency of the rate constant, the reaction in the plant will be much faster, and this may compensate for the high concentration of the reactants. LC-MS analysis of the synthesized material revealed a series of aldol reaction products.

The most important information related to pretty much any color problem will come from the experimental UV-Vis data. Two samples were obtained from the plant, one with low color and another with high color. Their UV spectra and the difference spectrum are shown in Figure 15.5.

The difference spectrum appears to be structured, and therefore further investigation was conducted. A correlation analysis was performed, using the calculated spectra and the difference spectrum. The objective was to isolate the calculated spectra, that when summed match the difference spectrum of the real samples in the best possible way. The result is shown in Figure 15.6.

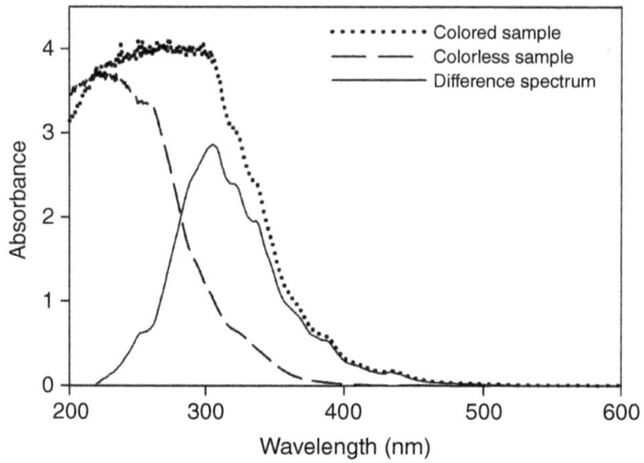

Figure 15.5 Low- and high-color plant samples – UV spectra.

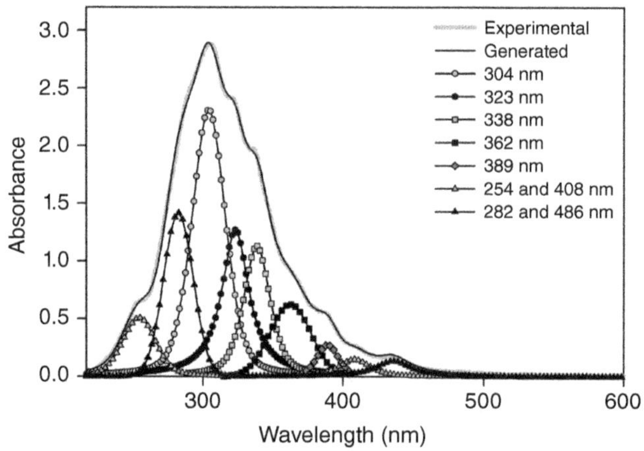

Figure 15.6 Fitted peaks from calculated spectra and difference spectrum from real samples.

A very good fit is observed, with many aldol condensation products present. In order to account for all theoretical peaks needed to reproduce the experimental spectrum, a new class of aldol condensation products was calculated: unsaturated aldehydes with terminal OH groups. These species account for the 304 and 338 nm peaks present in Figure 15.6. In summary, the UV peaks at 254 + 408 nm and 282 + 486 nm are due to nitro-type molecules ($NO_2C_xH_{(x+1)}O$, where x = 6 and 8). The 304 and 338 nm peaks arise from the conjugated aldehydes but with a terminal OH group ($HOC_xH_{(x+1)}O$, where x = 8 and 10). Finally, peaks at 323, 362, and 389 nm belong to the first class of molecules investigated utilizing computational modeling, unsaturated conjugated aldehydes ($C_xH_{(x+2)}O$, where x = 8, 10, and 12).

The methodology presented so far has shown a very high confidence about the validity of the aldol condensation reaction hypothesis. Chemometrics analysis of the process data confirmed the importance of process variables that are caustic related and therefore affect the pH of the process streams in the glycol system, which is consistent with the aldol chemistry. The process conditions as temperatures, residence times, and compositions are consistent with the aldol condensation chemistry. Thermodynamic calculations confirmed that the aldol condensation is favored in these conditions. A laboratory experiment was performed at conditions that tolerate aldol condensation and color was obtained after a residence times comparable to the residence times in the real plant. The color bodies in the synthesized color were confirmed to be aldol condensation products by LC/MS. Finally, there was a high degree of consistency between the UV spectrum of simulated aldol condensation product mixture and real case of color compounds present.

For final confirmation of the aldol condensation reaction hypothesis and the influence of the caustic on the color, a plant test was performed. The plant was stable at the beginning of the test. Four steps in the caustic flow to the EG reactor were performed, two increasing the caustic flow and two decreasing it.

The results show that the two increases in the EG reactor caustic flow led to two increases in the MEG column make flow color exactly as expected. The residence time between the change in the caustic flow and the change in the color is comparable to the residence times in the train. The plant test results are the final proof of the validity of the aldol reaction hypothesis. Since the pH (namely, the NaOH addition rate) at the reactor inlet was kept at values of 9 and below, no significant color problems have occurred in the plant.

The color problem-solving methodology presented above was quite successful in resolving not only the problem above but numerous other color issues in the past. This approach consists of using multiple techniques: chemometrics plant data analysis, process chemistry and side reactions, computational chemistry, laboratory and plant experiments, liquid chromatography/mass spectrometry (LC/MS) and UV/VIS spectral analysis.

Section IV

Problem-Solving Tools

16

Emphasis Is on Speed, Low Cost, Productivity, and Right Answer

The early stages of a problem-solving effort can be thrilling or completely nerve-racking depending on the seriousness of the problem, the team and stakeholder dynamics, and the ability to get "quick wins" early in the project. It is important to keep the team focused on the ultimate goal (What is Success?) while maintaining progress on the rate-limiting requirements to achieve that success. If the proper balance can be established, then team member engagement will be maintained at a high level while garnering trust from the stakeholders.

Although this book focuses on using scientific fundamentals during problem-solving efforts, the authors do acknowledge that there are other priorities that must be managed and sometimes take higher priority. Often determining the true scientific root cause of a problem can take several days or weeks and may require quite a bit of data collection. If the production unit is losing thousands to millions of dollars per day, then there is very little tolerance for the slow methodical approach of hypothesis gathering and testing. In such cases, it is often helpful to think of the production unit as an emergency room patient. The project team must first act as the team of medical professionals assigned to quickly determine the critical ailments of the patient and stabilize the vital signs. After this is accomplished, then the team can begin acting as specialists that can fully evaluate the patient's health and correct any of the underlying issues that led to the critical health event.

To put this in a manufacturing context, imagine you are asked to solve a chemical plant issue in which Product CJS is failing to meet quality specifications. The unit has already filled their available storage capacity with off-spec product, and they have no good material to ship to their customers. The next order must ship out in two days or it will be late. The last thing the stakeholders and plant personnel want to hear is that you and the team will spend the next three days in a

Fundamentals of Industrial Problem Solving: A Practitioner's Guide, First Edition.
Zdravko I. Stefanov, Eldad Herceg, Carla Schmidt, David M. Jacobson, Dana Livingston, J.P. Chauvel, Sunil Kumar Chaudhary, and Christopher Paul Christenson.

conference room brainstorming possible hypotheses and several more days testing them. This is definitely a situation that requires a quick, low-cost solution to stabilize the situation. Obviously, your team's first efforts should be focused on how to get in-specification Product CJS out to the customer as fast as you can while maintaining safety. Depending on the severity of the production interruption and the resources available, you may want to consider forming two teams: one to work on discovery and implementation of a quick fix and the other to work on the fundamental understanding and long-term solutions. Whether the choice is made to operate as one team or two, the team needs to be flexible and agile. This means streamlining communications for optimal effectiveness and ensuring that interpersonal relationships, politics, etc. do not interfere with the progress of the team.

With respect to streamlining communications, there are many formats available in today's digital world, but the reader will have to work within the framework of what is available and allowed within the company. Therefore, rather than to discuss the different communication and data-sharing platforms that could be used, it is better to highlight that expectations should be set within the team very early for what is communicated, when or at what frequency it is communicated, and to whom the communications should be addressed. For example, if the impurity issue in Product CJS mentioned earlier in this chapter could become a regulatory issue, then all communications will have to be carefully managed. This would mean that preliminary analytical results on the impurity should not be shared broadly until the proper replicates and data quality standards are verified. It may also mean that the legal department should be consulted to advise the team members on proper communication (content, means, and distribution) since these communications could be subpoenaed if legal or governmental action is ultimately taken. Even if the project does not become a regulatory or legal issue, careful communications are still very necessary. Any miscommunication or false communication that has to be recalled and revised will erode trust within the team and among the stakeholders and spread more confusion in a chaotic situation.

Another point on communications is that lack of communication is also a major pitfall. Often if nothing is communicated, then the assumption is that nothing is being done or no progress is being made. Whether it is recognized or not, stakeholders and team members will each have an expectation of when they should receive information. For example, if a team member is expecting to receive samples for analysis and they do not receive them by the expected time, they may assume that the samples are not coming that day and go home early. In contrast, if they were not informed ahead of time that the samples would arrive that day and they are caught off-guard, then the instrumentation may not be available to conduct the analysis. Both situations could significantly delay the project progress at a time when speed is the most important. Stakeholders are even more sensitive to not receiving timely communication. For example, the plant and company leaders are accountable to upper-level leaders for explaining what actions are being taken to correct the plant issue. They need information from the investigative team to be sure that the right resources are working on the problem and that a resolution is closer rather than farther away. Neglecting the task of making a proper communication plan is risky and may result in delays, rework, or extra stress.

17

Literature

The search of the literature is one of the first steps in the problem-solving process and its importance cannot be emphasized enough. Multiple problems in the authors practice have been solved at that step and doing proper literature search has saved huge amounts of resources measured in time and money.

In the current age, the amount of scientific information and the extent of scientific research is immense and is growing exponentially. Therefore, there is quite high probability that information about the problem at hand can be found somewhere in the literature and the rate-limiting step is to know where to look and how.

17.1 Why Literature Search

So why do literature search? Here are several very good reasons.

17.1.1 Problem Understanding

The first and most important reason is problem understanding. To solve any problem, it is absolutely necessary to properly understand the science behind that problem. Improper understanding of the fundamental science often leads to problems being "solved" again and again many times. The existing literature happens to be one of the most important sources of problem understanding.

17.1.2 Literature Search Is Economical

There are several benefits of literature search vs. other means of problem understanding. One benefit is that literature requires a dedicated effort to document

Fundamentals of Industrial Problem Solving: A Practitioner's Guide, First Edition.
Zdravko I. Stefanov, Eldad Herceg, Carla Schmidt, David M. Jacobson, Dana Livingston, J.P. Chauvel, Sunil Kumar Chaudhary, and Christopher Paul Christenson.

knowledge and as such it possesses intrinsic degree of reliability, greater to, for example, word of mouth or personal experiences.

Another benefit is that the literature can provide an answer quickly and without the necessity of running lengthy and expensive experiments. As it is often said, "A month in the lab can save you an hour in the library." The same goes about experimenting with the real plant.

Yet another benefit of literature search is that it provokes thought about the problem.

17.1.3 Lost Knowledge

Another reason to do the literature search is to uncover lost knowledge. Very often during a problem-solving project, it is encountered, that knowledge about the problem has existed since a long time. The particular problem happened to already be solved many years ago, but due to corporate memory loss and people turnover, no one remembers that. Just like the quote from The Lord of The Rings: "And some things that should not have been forgotten were lost. History became legend. Legend became myth. And for two and a half thousand years, the ring passed out of all knowledge." The same thing happens to plants, but it takes much shorter time than two-and-a-half thousand years. Usually takes about a decade.

17.2 How to Do Literature Search Properly

Experience shows that there are certain considerations when performing the literature search that usually lead to good results. The most important ones are listed in the following.

17.2.1 Sources

For a good literature search, it is very important to pick the right and the best sources. The best sources are usually peer-reviewed publications – books or journals, internal or external.

Peer-reviewed Internal Sources

The peer-reviewed internal sources are the first choice. If the organization possesses a centralized research archive (as most large companies do), this is the best source to work with first. All of the literature in a company centralized research archive is guaranteed to be very closely related to the plants and processes of the company. It is a great resource to look for what has been done before on the problem at hand, to understand the plant and the process and who worked on those.

Understanding who worked on the problem before, or knows the process solves an important networking problem if such exists, i.e. one does not know who to talk to about the problem at hand. The authors list in relevant reports provides immediate networking information and it may turn out the person of interest is just several doors down the hall.

Peer-reviewed External Sources

The peer-reviewed external sources are usually books or journal papers, less often conference presentations. They can also be an excellent source of information. If the organization has a subscription access to the large journal publishers, this should definitely be utilized. If a paywall is preventing access, then another tactic is to look for a copy of the paper translated in another language, or the author's web site. Sometimes even contacting the author directly can help. Another place to look for papers is again a colleague that was working on similar problem for many years and happens to keep a large archive of paper copies stashed in their office.

Internal Never Published/Paper-only Sources

These are usually technical reports that exist only on paper and are, as mentioned before, stashed in someone's office. These can be quite hard to find, but sometimes they are extremely useful. Finding these kinds of sources necessitates picking up the phone or going out of the office (bar a pandemic has choked the world for half a decade!) and talking to people. Apart from the interesting technical information that can be obtained, one can get a good idea what people already tried, and most importantly what they did not try (or even did never even think about trying).

Plant Libraries

Exploring the plant libraries is probably the hardest, as it requires going to the actual plant and digging into dusty, sometimes disorganized paper folders. The gold that lies within is usually the plant original design, including P&IDs, operating parameters, and specifications. It is amazing how often a plant slowly drifts away from its design intent over the years and ends up in a very different operating point. Bringing the plant back to the original design can be very helpful for solving a problem and the original design is in the library. However, with the current high people turnaround observed in industry, the people in the plant don't remember it or don't even know it does exist.

Patents

Patents can be an interesting source, but they should be used with great care. In a patent, everything that can be omitted to make the patent useless for a potential replication of the process will be omitted. Therefore, one must sift through the patent and make sure to ask questions about how close to reality what is described in the patent is and what is not there.

The Unpublished Works

Once in a while the topic of interest may be so obscure, that there is literally nothing published in the open literature. This, however, does not mean work has not been done. If there is a researcher that did any kind of work in this general area, it is worth contacting them and asking if they ever did anything more closely related to the problem at hand, but it never got published. It's a small chance, but sometimes it produces a quite useful result. A problem in our practice was solved exactly in this fashion, when a colleague found out a professor that did the kind of research he needed for solving the problem. Getting the information involved getting on a plane, flying to the other side of the world, meet the professor in person for a private consultation and get the information in the form of hand-written notes. All this complication still saved years of lab research work and helped start a plant that was due to start in several weeks and it could not be started without this research.

17.2.2 The Search Process

Regardless of the source, a good question to start with is "What am I looking for?" There are several ways to approach this. One is to directly look for the particular problem, for example "Sea water corrosion of stainless steel heat exchangers." If the topic is well researched and the problem is substantial, chances are that a direct hit can be obtained. If this direct approach fails to lead to a useful result, the next step is to search the phenomenon in general, such as "Sea water corrosion of stainless steel," "Sea water corrosion of heat exchangers," "Stainless steel corrosion in saline water," etc. The search can be expanded even more if necessary by just specifying the conditions at which the phenomenon occurs, such as "Stainless steel corrosion in hot salt water" and so on. It is important to explore all possible angles associated with the problem. Sometimes the important information is auxiliary to the main topic of a research paper and it won't be found with direct search.

Another important point is to obtain as much information as possible from initial observations in order to narrow down the search. For example, if a fouling problem is encountered, it is very helpful to analyze the foulant and determine what its composition is. Then a much more productive literature search can be performed, when several hypotheses can be formed about the potential mechanism causing the fouling. This loops back to the question "What am I looking for?" The more details can be added to that question, the better the results of the literature search will be.

It is important to understand that the literature search process is not a part of a linear sequence of steps in the problem-solving process. Usually, it is an iterative process of literature search – hypothesis generation – hypothesis testing – literature search, etc. In this iterative sequence, the hypothesis testing step will also often include literature search as a subroutine. Therefore, the literature search starts with the project start and it does not end until the problem is solved. There is continuous need for new information during the problem-solving process, and the literature search is one of the ways to acquire that information.

Another useful advice in the process of problem solving is to be patient. It is not very often that the important find will be on the first page of Google search results. One has to be patient and dig deep. A very interesting example from the authors practice involved troubleshooting of a pipeline being plugged by a mysterious solid material that formed very slowly and was quickly decomposing on atmospheric exposure. Initial search results in all of the standard sources led to nothing. Persevering in looking at more and more sources finally yielded a very useful result from a German paper from 1905 that has happened to be digitized by Google without translation, apparently from old photocopies of the original journal. The paper described the formation and properties of a compound that matched the properties of the observed plugging material almost identically. This led to rather quick root cause determination and the problem was eliminated. It is worth noting that the German researcher was studying a completely different chemistry and he discovered that compound by serendipity. However, being curious about it, he researched it and published the results as an interesting side product of his original work. It comes to show that sometimes the important result can be hiding in the literature, and it takes perseverance to discover it. On a side note, being capable to search in several different languages also can be very helpful. Google translate also helps, with caution.

17.2.3 Pitfalls

Massive Errors Propagated Throughout Literature

One of the most perilous pitfalls when searching the literature is believing everything that is in print. Choosing well-respected publications usually helps, but there is no guarantee. A particularly egregious example is provided in the book "Ignition!: An informal history of liquid rocket propellants" by John D. Clark, where it is described how the density of liquid fluorine, determined by James Dewar to be 1.108, was used, cited, and repeated in the scientific literature for about half a century. When some people working on liquid fluorine rocket engines at Aerojet doubted the figure, Scott Kilner from that organization went on to measure it and it turned out it was just above 1.54. This prompted several other independent studies that confirmed the new figure. Apparently, Dewar's reputation was large enough to not cause anyone to doubt his figure. This example shows that information even from the apparently most solid source can be quite substantially wrong. Never believe everything that is in print, always try to get information from multiple sources and always cross-check with fundamentals.

Source Age

There is a certain age of the source that seems to be optimal, all other factors excluded. This is usually information published about 5–10 years ago. This is current enough so that modern techniques and equipment are used to determine the published facts. At the same time, it is somewhat old, but not too old. Being somewhat "seasoned" is not terribly bad, due to the fact that the paper had the chance to be in publication for a while, so lots of people got the chance to read it, try to recreate the results, and find out if there is something wrong. There have been plenty of stories lately when some kind of breakthrough research is published in a very respectable journal, only to be found later to be a complete fabrication and retracted in a hurry. Therefore, do not get too excited about the newest stuff, just because it is hot off the press. Caveat emptor. In mature fields, it is sometimes helpful to go back to the foundational papers. These may be more general and were written by a brilliant scientist. Later papers may be much more incremental. You may be very interested in the increment, as well.

18

Scientific Method

The scientific method is a systematic, empirical approach of collecting data and knowledge with the goal of understanding observations, testing hypotheses, and answering questions. The goal of the scientific method is to prevent and/or counter bias, primarily conformational bias that affect people's thinking by using a systematic, hypothesis-driven approach of collecting the data. It is extensively used and is a critical part of the Diagnostics phase of the FPS process.

The practice of the scientific method at a very high level enables solving the problem instead of just putting a band-aid on the cut. It is critical to understand where we are currently in the problem-solving space and to make a plan to improve the step that is currently limiting our performance. Simply, we need to go through the process of hypotheses testing with fact, theory, and models if we are to achieve this goal. Challenging the current hypotheses is much more effective than just looking for data and opinions that support our current thinking.

There are many variations of the scientific method but all of them have some common traits that in general can be described via six steps. These steps are visually presented in Figure 18.1.

The question comes first. Before the hypothesis is stated, one has to clearly identify the question that needs to be studied. The question can be stated as simple as "Why do we experience shorter run times of our X equipment?" or "What is the reason behind experiencing slower cure times for our sealant?" Still, this is a necessary step since it will help us to focus on stating specific hypotheses relevant to the question that we are trying to answer.

Next step is to do a background research around the problem and the hypotheses relevant to the question. The goal of this step is to research the science to get a high-level understanding of the problem, as well as to inform us about what was done in the past in the area of our problem. In many instances, we will not be able

Fundamentals of Industrial Problem Solving: A Practitioner's Guide, First Edition.
Zdravko I. Stefanov, Eldad Herceg, Carla Schmidt, David M. Jacobson, Dana Livingston, J.P. Chauvel, Sunil Kumar Chaudhary, and Christopher Paul Christenson.

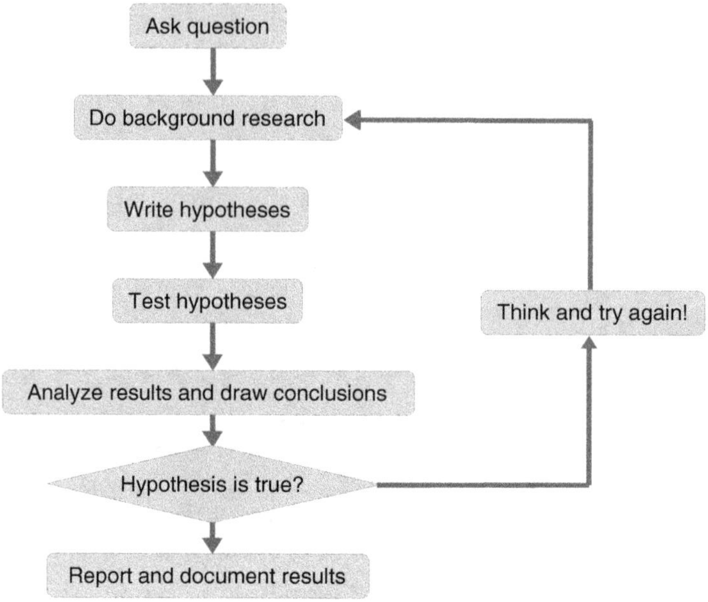

Figure 18.1 The steps of the scientific method.

to find the information that will exactly answer the question relevant to our problem, but the information will be useful and easily leveraged in most cases. The search can be done using library, internet search, through internal company's databases, etc. Our previous experiences show that this step is accelerating the whole problem-solving process and in many instances was the only step taken since the answer to the issue under investigation was found in the literature or by other background research means.

Next, we go to the hypothesis forming phase. Basically, in layman's terms, a hypothesis is an educated guess about how something works. It is a supposition or proposed explanation made on the basis of limited evidence as a starting point for further investigation. It is an attempt to answer the question stated in the first step of the scientific method with an explanation. Stated hypotheses should be clear and written in a simple language. Also, they have to be testable. To prove or disprove a hypothesis, one needs to be able to do an experiment and take measurements or make observations to see how important variables are related. Moreover, experiments should be repeatable, if necessary.

Hypothesis evaluation is the next step in scientific method process. Hypothesis testing can be done via first principles knowledge, mass balance, literature, data mining, fundamental predictive models, and powerful well-designed experiment.

Gathered data and facts are then evaluated against each stated hypothesis, and decision is made as to whether the data are refuting or supporting the hypothesis. One has to understand and accept that a theory can only be proven wrong by the data. A theory can never be proven right by the data. Basically, a theory is only right based on the available data at the time. A good theory is always vulnerable to new data that can disprove the theory tomorrow.

If none of the stated hypotheses survive the preponderance of the data and supplied evidence, one has to go back to the hypothesis-stating step, think hard, and come up with a new set of hypotheses to be tested. If a hypothesis or a small set of hypotheses survives, one has to draw a conclusion and make a decision whether these hypotheses are the root cause of the problem under the investigation. Finally, the results and work need to be documented for the future reference.

One of our colleagues suggested that the scientific method is practiced at several skill levels. He captured his thoughts in a spreadsheet on the Scientific Method (provided in Appendix). Don't try to jump to the top level in one step. Pick which step is most important to improve, now. Then focus on getting up one step till you are stable. Then pick the next focus for improving. This is a journey, not a sprint.

19

Process Data Analysis

The process data analysis is another important tool in the arsenal of tools used in plant problem solving. The analysis of the process data can lead to the generation of new hypotheses or it can be used to evaluate existing hypotheses. Whether used for hypothesis generation or evaluation, there is hardly any troubleshooting project where process data analysis is not involved.

It is important to note that it is of extreme importance to mate the process data analysis with process knowledge and fundamental process understanding. No matter how advanced the data analysis tool, it is pretty much impossible to get a meaningful analysis if the fundamentals of the process are not taken into consideration. If one doesn't know the fundamentals of the process, one will not be able to comprehend the importance of the data analysis results. The root cause of the problem may be obvious for the one with such knowledge and completely incomprehensible for another that lacks it.

In the following, we cover the main aspects of process data analysis.

19.1 Preprocessing

The data preprocessing is the most important step in any process data analysis and ignoring this step or doing the preprocessing poorly invalidates the analysis completely. The importance of proper preprocessing cannot be stressed enough. Even the most sophisticated data analysis algorithm is doomed to failure if bad data are fed to it – Garbage in – Garbage out.

19.1.1 Sampling

When one collects process data for analysis, the first thing to consider is the sampling interval. Plant historians offer many options and one can get data every

Fundamentals of Industrial Problem Solving: A Practitioner's Guide, First Edition.
Zdravko I. Stefanov, Eldad Herceg, Carla Schmidt, David M. Jacobson, Dana Livingston, J.P. Chauvel, Sunil Kumar Chaudhary, and Christopher Paul Christenson.

second, every minute, every hour. . . Also the data can be instantaneous or they can be the average of all measurements taken during the stated interval. There are two main questions one should ask. The first is "What is the time scale of the dynamics of my process?" The second question is "What is the actual frequency of the data stored in the historian?"

Process Dynamics Considerations

The sampling period needs to be appropriate with process dynamics time scale. One cannot, for example, expect to properly analyze a system that has dynamics in the order of seconds (a fast flow loop for example) if one collects only data that are from 10 minute averages. All of the important dynamics will be filtered by the averaging. Even if one collects discrete samples every 10 minutes, still the system dynamics is not properly sampled, and there will be aliasing issues, etc. According to Shannon's sampling theorem, in the ideal case, one can properly recover information from a signal if sampled at least twice as fast as the highest frequency component. This is, however, the ideal case. In practice, as a rule of thumb, one needs 5–10 samples per period to properly catch the dynamics. In another extreme, it does not make any sense to collect data from a system that has time constant of days every second. This is an overkill that makes the analysis very hard due to the humongous amount of data and provides very little additional information.

> ***Wisdom Tidbit – Data sampling frequency should be 5–10 times faster than the response rate of the system being analyzed.***

> ***Wisdom Tidbit – Data sampling frequency should be reduced where possible to manage the data sizes to be more easily processed.***

Historian Compression Considerations

The plant historians store tremendous amounts of data. To manage storage and maintenance cost, usually some clever tricks such as data compression are utilized. The most ubiquitous trick is to store data not at some predefined interval (say every minute) but on exception. What this means is that if the process variable does not change significantly, no new data points are recorded. If there is a large enough change, then the new data points are recorded. The change threshold is defined as certain % deviation from the span of the variable. And this is where the problems start. In many cases, the percentages are predefined values (say 2%) and are applied uniformly across all tags, regardless of the frequency of actual events in the signals. This can lead to severe under-sampling of process variables that measure properties that change at high frequency and are important at that frequency scale. Another fault is usually poorly defined span. If a flowmeter has an actual span of 0–200 GPM and the span in the historian is defined as 0–2000 GPM, then the same 2% exception deviation is 4 GPM in the first case and 40 GPM in the second. If that flowmeter feeds a chemical, flow rate of which needs to be regulated within ±5 GPM for ratio control, the historian will miss any of the changes and record a flat line.

Therefore, one should always inspect the data obtained from the historian. Ask for the actual data points for important variables. Usually a data historian request for "data at every minute" returns a linear interpolation of the actual data at every minute, not the real data. For important variables, do the due diligence and make sure you are getting the actual data, not an interpolated version that does not properly represent the process dynamics.

Transitions, Shutdowns, Missing Data

Usually when a process data set is been acquired, the first-pass check involves removing time periods where the data does not represent normal operating conditions. Such periods include grade transitions, plant shutdowns, and missing data. The grade transitions are usually accompanied by sufficiently large transients and can be a problem particularly when a steady-state modeling approach is used (which is the majority of cases). Plant shutdowns are the extreme of the grade transitions, so to speak, and the transients are highly nonlinear and discontinuous (feeds and utilities are shut off, etc.). Apart from the fact that these large dynamics are detrimental to steady-state modeling approaches, it also makes the data distribution anything but normal. This can lead to severe error particularly when any kind of statistical testing is performed that requires a normally distributed data. Many statistical tests assume normally distributed data and if the data are far from normally distributed, these test are not adequate. It is a common pitfall to use data that have not been cleaned and do not have a distribution approaching normal in such statistical testing, and it can lead to spectacularly wrong results.

Missing data is another problem usually present in industrial data sets. There has been quite a bit of research on how to deal with that particular aspect of data preprocessing. The easiest way is, of course, to remove any time periods where many of the variables have missing data. Or one can remove variables that have more than certain percentage of missing values. There are many other methods that involve interpolation or more sophisticated ways to fill in the gaps. The details can be found in the literature, but the important point is this: One needs to use common sense when making the decision which method to use. For example, let's assume there is a time period when something interesting is happening in the plant. If there are many variables of lesser importance that are missing data during that time, obviously it makes more sense to exclude the variables instead of removing the interesting time period. In another example, if an important variable is missing 80% of the data points, it does not make sense to keep it, as most of the information is gone. Interpolation can help filling gaps in the data, but it needs to be done with caution. The starting data set should have enough information already present to ensure we are not "assuming" too much about the data when interpolating. One should avoid interpolating large periods of missing data for this reason.

Data with Large Differences in Sampling Frequency

Sometimes there is a need to join process data and analytical data, and usually these two data sources have a large difference in sampling frequency. Process data is usually fast (on the order of seconds to minutes) and analytical (lab) data is usually slow (hours). This leaves large gaps in the analytical data that need to be resolved somehow.

Linearization and Other Transforms

In certain cases, it can be beneficial to linearize or otherwise transform certain variables. Most of the data analysis algorithms are linear and require a normally distributed data. Having variables in the data set that experience known nonlinear behavior is bound to lead to problems in the analysis. Therefore, it is a good practice to linearize such variables by taking the inverse transform of the original relationship. There are often fundamental principles that can be applied, carefully, to linearize a very nonlinear variable. This simplification is regularly very helpful. An example would be the kinetics of making a polyurethane foam. The temperature jumps 100–150 °C in less than a minute. The reaction rate changes very quickly also.

Lags and Boxing/Grouping

Another important issue in data preprocessing is the presence of lags. When large plants are analyzed, often there is quite significant lag between the quality variables measured at the plant exit and the rest of the plant. All kinds of inventories exist in plants (tanks, distillation towers [bottoms and trays, reflux drums], etc.) Each inventory adds up to the total lag, and on top of that, there is a filtering effect on any change passing through these inventories. Therefore, it is important to try and compensate for the lags as much as possible, shifting the variable timelines accordingly.

Bad Sensors

There are two types of sensor failures, one pretty obvious and the other not at all obvious and very sinister in nature. The first one is a flat-line failure leading to a constant, unchanging reading that looks exactly like this on a trend, a flat line. This fault is easy to detect and deal with. A related failure is a digital stepping between two readings, which looks unlikely, examples are triangular or square patterns. The second failure type is quite hard to detect, as it manifests itself in a signal from the sensor that is quite erroneous; however, it is not a flat line, and it has noise in it. This makes it extremely hard to detect. There are advanced approaches to detect this kind of sensor failure, but very often the failure is undetected until the associated process variable shows up as important in the following analysis and is scrutinized more closely. One particular example is a sensor board with aged electronic components that is exposed to the sunlight during certain hours in daytime. When this happens, the sensor goes "crazy" for several hours and then when the board gets back in the shade and cools off, it resumes normal operation. One way to catch this is to closely examine any variability that is observed on a ~24-hour period (diurnal variability) and see if it makes sense. There is lots of real diurnal variability in plants as lots of equipment that is in the open is exposed to diurnal temperature variations.

19.2 Univariate Methods for Data Analysis and Exploration

The univariate (each variable is analyzed without account for the interaction with other variables) are usually the first line of plant data analysis. They can be relatively simpler than the multivariate methods, but not always, as we will see below. In general though, they are more accessible by the plant engineers and require less specialized knowledge. Also, in many cases they are sufficient to troubleshoot a problem.

19.2.1 Trending

The trending of process variables is the simplest and most common method of univariate data analysis. It is used on a massive scale in plants by both operators and engineers. It is nothing more than a time-series plot of one or more process variables. If the proper time and process scale are used, it can be very informative about what is going on in a process, and it can be very useful in troubleshooting. For example, in our experience, it was possible to fix a distillation tower control problem with trending the key variables (steam flow, reflux flow, steam valve position, inventory levels, product and bottom flows and valve positions). This allowed the owner to identify huge oscillations of the control valves over the entire span and correspondingly huge oscillations of the process variables leading to very poor composition control. Realizing from the oscillation periods that the frequency was many times higher than the natural frequency of the process, it was possible to determine that all controllers were grossly overtuned (very aggressive) and were driving the tower to be completely unstable. Correcting the tuning parameters of the controllers allowed the owner to stabilize the tower within a few hours by manual tuning. So, in this example a simple trending was completely adequate to solve the problem. But, as already mentioned above, it is of great importance to know what fundamentally to look for in the trend.

19.2.2 Univariate (Shewhart) Control Charts

Control charts can be viewed as an advanced and statistically reinforced trending. They include, apart from the process variable data, one or two control limits (upper and lower). The control limits are usually the 95 or 99% confidence intervals of the assumed normal distribution of the process variable. Control charts can be quite informative and also alarm when a control limit is violated based on various rules. The usefulness comes from the ability to detect a drift or a change in a process variable that drives it outside its normal range. Univariate control charts can detect fouling, drift in measurements, drift in operating conditions, etc.

19.2.3 Statistical Distribution Analysis

Statistical distribution analysis can be very useful in troubleshooting. On one hand, it can provide useful information about the system. On another, it can help with using the right statistical tests if a hypothesis is to be tested. For example if a certain variable distribution changes from something close enough to normal to a distribution with a heavy tail, this can mean that a constraint of some type is being hit by the process and the relationship between the investigated variable, and the rest of the system is becoming nonlinear. A distribution that is chopped off on one side is even more interesting. If it is a process variable, it is definitely a hard constraint being hit. If it is a quality variable, it usually means that the lab is running repeated tests until they get a result that falls between the release specifications. Also, a simple multivariate analysis can be performed if the analysis tool supports the plotting of multiple distributions at the same time and when a certain bar on the histogram is selected in one distribution, it highlights the corresponding observations in the other distributions. For example, it can show that all samples with values close to the upper specification limit of the quality variable fall within the lower tail of certain temperature in the process.

19.2.4 Some Advanced Statistics Aspects

In certain cases, it may become important to use advanced or robust statistics, particularly when there are lots of outliers in the data. When this happens, using the standard deviation and the sample average as estimators of scale and location can bring very erroneous results. A small number of outliers can heavily distort these estimators. It may not look as a big difference when the data histogram is plotted, but it can be the difference when two different periods in time are compared and one asks if there was a statistical difference between them. In this case, it may become necessary to use advanced tools that employ robust estimators of scale (MAD, biweight midvariance) and location (median). The help of qualified statistician should be employed if such a need arises.

19.3 Multivariate Methods for Data Analysis and Exploration

The multivariate data analysis methods provide significant advantage over the univariate methods mainly because they take into account the interactions between the variables. Explaining the interactions between the process variables is in many cases of great importance. In a typical plant, there are many interactions, and a multivariate analysis can shed light on an important phenomenon where a univariate analysis will see nothing.

19.3.1 Principal Component Analysis

The principal component analysis (PCA) is one of the most important linear multivariate analysis methods. There are many aspects of plant data analysis where PCA can be used to great advantage.

Here, we will discuss the general principle behind PCA without resorting to equations and linear algebra. There is plenty of literature on the subject and the interested reader can investigate independently. Here the idea is to convey the PCA concept as simply as possible and help engineers to decide when to use it. For detailed description, a great reference is (Beebe et al. 1998).

In a typical plant, as already mentioned there are many measured variables, usually in the order of thousands. However, many of these variables are not independent and their variability is driven by a much smaller number of independent factors. For example, changes in plant production rate and recipe can drive changes in many variables. In a smaller scale (for example, a distillation tower), reflux rate changes on a distillation tower will drive changes in steam flow rate, cooling water flow rate, tray temperatures (profile shift), product and distillate flow rates, etc. In general, the number of independent factors is about two orders of magnitude smaller compared to the number of the measured variables. If there was an approach to isolate the independent factors, this would greatly simplify any analysis and troubleshooting. Such approaches exist and the PCA is one of the most popular. These independent factors are also known as principal components, ergo the name PCA.

The principle behind PCA can be explained in Figure 19.1. For simplicity and to retain the possibility to graph the data set, only three variables are used. The data points are represented by the black circles.

One can see that the data points form sort of an elongated cloud, or in other words, there is more or less correlation between the variables in the different directions, and they are not independent.

When one observes the data cloud, one can see that the cloud possesses a dominant direction of elongation. The elongation or variability in that direction is driven by the most significant factor or source of variation in the data set, the first principal component. If a vector is passed along that direction and its size is the distance of the data along that direction, one has isolated the first principal component. In a similar fashion, once the first PC is determined, then the next (orthogonal to the first) most dominant direction of variability is found and the second PC is extracted. This process continues until a new coordinate system is created, which is made up from the principal components.

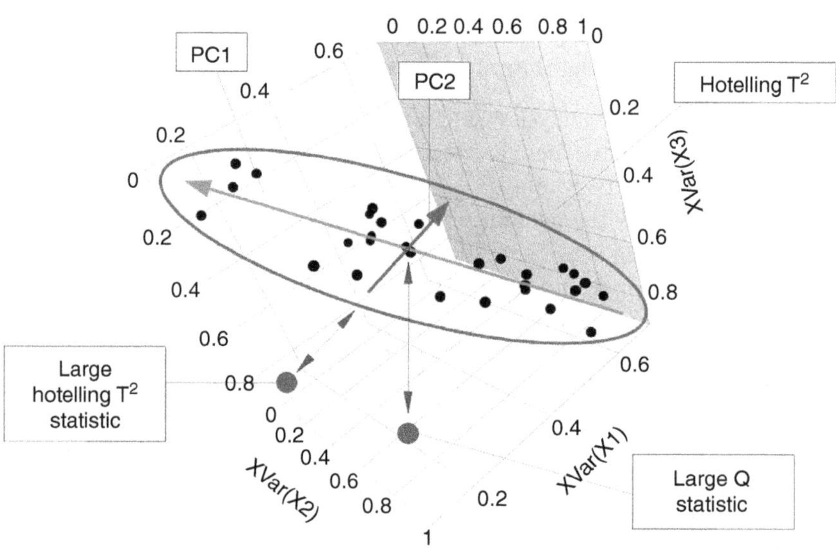

Figure 19.1 Principal component analysis. (*See insert for colour representation of the figure.*)

The maximum number of PCs that can be extracted is the same as the number of the original variables in the data set. However, this is never done, as the point is to extract only the major sources of variation. Everything else is noise. For that purpose, there are statistics that show how much of the original variation is explained by the selected number of PCs and one can stop extracting when sufficient percentage of the original variation is explained, and/or extracting more principal components does not increase the amount of explained variation.

After the principal components are extracted, one has in principle a linear model that describes the original data. Now the great advantages of PCA can be exploited as the fact that we have a linear model of the original data that gives us many possibilities for analysis. We can obtain new data points from the plant, feed them to the model (project), and see how they fit. There are two major statistics that provide great insight about the data point projected to the model. One is the Hotelling T^2 statistic and the other is the Q^2 or distance-to-the-model statistic.

The Hotelling T^2 statistic is represented by the ellipse in the plane PC1/PC2 (Figure 19.1), and it is the multivariate version of the 99 or 95% confidence limit of the normal distribution. Any observation outside the limit does not belong to that multivariate normal distribution with the corresponding confidence. This way one can detect outliers immediately, in the multivariate sense. A multivariate outlier may be of importance for the problem at hand and a univariate control limit will not catch it. A data point with high T^2 usually represent abnormal operating condition where the correlations between the variables are preserved. In other words, everything interacts as supposed by the physics/chemistry of the process, but the process operates away from its normal operating point.

The Q^2 or distance-to-the-model measures the distance between the multivariate model and the data point. As seen in the figure, a data point may have zero T^2 but quite large Q^2. What this means is that the correlation structure between the variables is somehow quite unexpected, even though the average values of the variables are not away from normal. This may indicate broken sensors, broken valves/actuators, changes in process chemistry, etc.

The fact that we have a linear model provides another very important diagnostic tool, the contribution plots. As the model is linear, one can swap between the original coordinate system and the coordinate system of the principal components. This gives the possibility to assign contributions to the statistics (Q^2 and T^2) of every data point. The contributions give quantitative and qualitative information of how each of the original variables contributes to a given value of the statistic in questions. For example, if a data point has a large T^2 statistic, it is possible to ascertain which process variables contribute to that deviation. This makes the contribution plots an invaluable diagnostics tool.

Another very important advantage of PCA and the linear model is that there are extremely well-optimized linear algebra subroutines that can solve the PCA decomposition problem in extremely short times. In a typical example, a plant data set of several hundred variables and about hundred thousand observations is decomposed in the order of 30–60 seconds on an 8 Core 2.9 GHz Intel Xeon workstation with 16 GB RAM. In other words, nowadays even a pretty standard computer configuration available to the regular customer is capable of doing the necessary math in very short times even for such large data sets.

> ***Wisdom Tidbit – Computing power is increasing exponentially and math heavy data analysis is becoming more and more accessible. It is reasonable to assume that you can now do a heavy math problem with standard resources instead of dismissing it beforehand.***

In summary, PCA provides the capability to do plant data analysis on data sets with very large number of data points, rapidly, with extremely well-developed level of diagnostics. It has been used extensively in practice (the authors and many others as indicated by the extensive literature evidence) and is highly recommended to become a standard tool in the toolset of every engineer or troubleshooter.

19.3.2 Partial Least Squares

Partial least squares or PLS is a method for data analysis that is very close to PCA; however, in the case of PLS, one builds a linear regression model between a data set X (usually process variables) and another data set Y (usually quality variables). When a regression model is needed and multiple least squares (MLS) is used, there is a major problem if the variables in the X data set are not independent. In that case, severe mathematical difficulties can be experienced that makes the generation of the model impossible or the produced model is of very poor quality.

As we already discussed above, in the case of PCA, these difficulties are removed as the resulting PCs are orthogonal, and therefore there is no problem with collinearity. So in the case of PLS, the regression is between the principal components of the X and Y spaces. In other words, the PLS model will rotate the principal component coordinate systems in X and Y until X explains the maximum amount of variation in Y.

The troubleshooting power of the PLS regression lies in the fact that it can extract only the important sources of variation in the X variable space that are capable to explain the variation in the Y variable space. It does that with the same advantages that PCA has – it can deal with data sets with very large number of data points, rapidly, with extremely well-developed level of diagnostics. Also, just like PCA, it has been used extensively in the troubleshooting practice of the authors and others (Chiang et al. 2001).

19.3.3 Some Batch Process Considerations

The batch processes data analysis poses certain challenges that deserve special attention. In the case of continuous processes, all of the problems listed in Section 19.1 of this chapter are of course present. In the case of batch processes, there are some additional issues.

Lack of Steady State

In the case of continuous processes, one can usually define some type of pseudo steady state, which makes possible the use of steady-state models that do not include time as an input. It is not often realized that this is not true in the case of batch processes. In a batch reactor, even if all parameters are the same, the time spent in the reactor impacts the properties of the product. Therefore, all batch models are by necessity dynamic, one way or another.

Batch Data Is 3D

The time within a batch is an important parameter, as batch processes are dynamic and have no steady state. So, time along the batch needs to be included in the data set. The time then becomes a third dimension, along with the other two (variables and observations, or batches). So intrinsically, a batch process data set is a parallelogram with three dimensions and a ragged edge. This poses a problem as the usual methods for data regression or analysis that involve linear algebra can deal only with 2D matrices. To circumvent this issue, the 3D data set is usually unfolded in variable or observation direction (Figure 19.2). Each unfolding is better for analyzing certain variability (batch-to-batch or in-batch). Each of the unfolding methods is therefore a compromise and a conscious decision should be made which one is to be used.

The Batch Durations Are Usually Not the Same Among the Different Batches

It is often the case, except in cases where it is enforced by the control system, that the durations of the batches are different. This has serious implications, as the batch data parallelogram described above has a "ragged edge," and it is not possible to unfold it in observation direction to model the batch-to-batch variability. Apart from this, there is the problem of information loss in the data set toward the ends of the batches as the data starts to disappear as the batches become longer and longer until there is usually only one batch left at the longest processing time. If there is something important happening right towards the end of the batches, it is not possible to analyze the data due to huge number of missing values. There are ways to deal with this problem and the batches can be either truncated at the length of the shortest batch, squeezed, or expanded by interpolating the batch

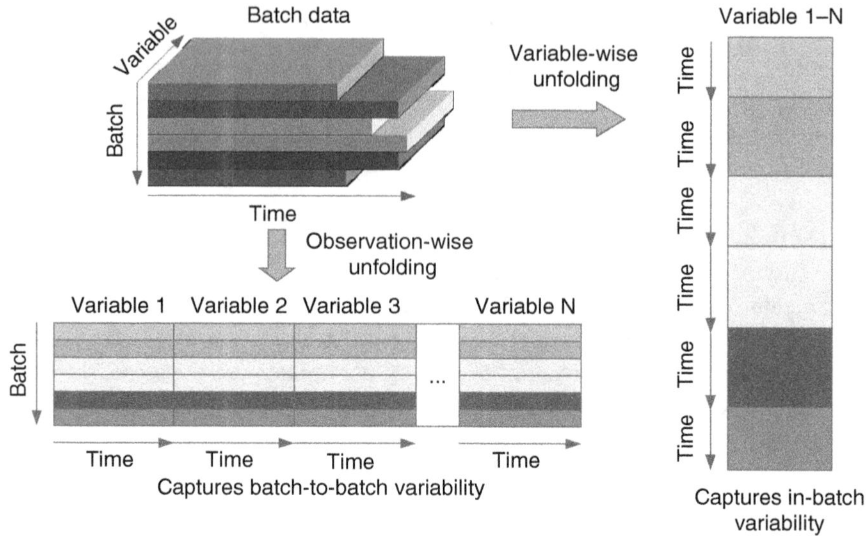

Figure 19.2 Batch unfolding methods.

length to a predefined average length batch timeline (Linear Time Scaling), using warping methods to match features in the batch trajectory in time (variety of Time Warping methods), etc. Unfortunately, each of these methods is a compromise that removes information from the data set (the last part of the timeline is discarded, or the timeline is shifted artificially). Artificial timeline shift is particularly insidious as it moves specific events to new times, which can be of huge importance for analysis of batch processes where it is not only important what happens, but also when. Knowledge of the process fundamentals may help you determine which method for alignment to select in order to introduce the least amount of distortion in the data.

19.4 Example of Plant Data Analysis

This example covers data analysis done as part of the color problem work, described in Chapter 15.

19.4.1 Multivariate Analysis of the Plant Data

An extended amount of process and color data was obtained for this analysis, encompassing over two years of time. The process data contains 286 process

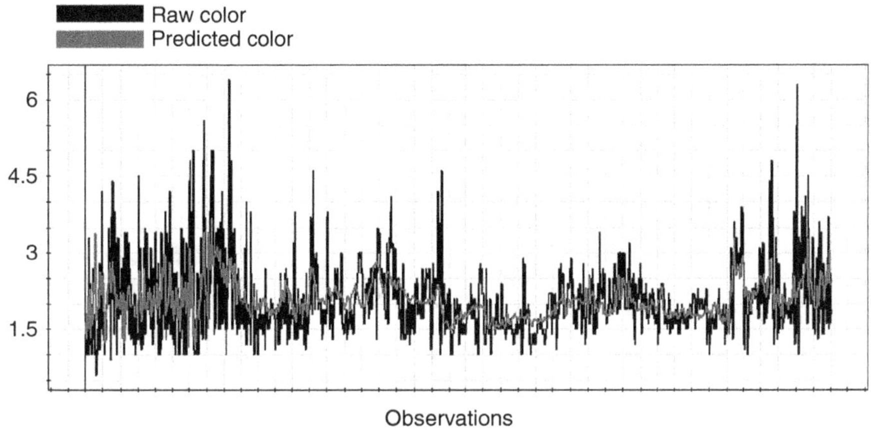

Figure 19.3 Observed vs. predicted color.

variables and the color data contains two variables – the MEG make flow raw color and the MEG make flow filtered color. The filtered color data was calculated using a first-order filter with time constant of two days.

A preliminary PLS model was built to gain a general data set understanding and to remove outliers in the data. The observed vs. predicted color are shown in Figure 19.3.

The model quality is satisfactory, keeping in mind that the color data has low signal-to-noise ratio. This is because the standard deviation of the color analyzer is 0.3 color units. What is important here is that the predicted color follows the trend of the raw color, even if the high-frequency variation is not captured by the model. The first three important variables in this model are the UV absorptions of an online UV analyzer mounted on the product make stream. The wavelengths selected by the model are 220, 250, and 275 nm.

The selection of the UV analyzer readings is not surprising, but these variables are not actual handles, i.e. cannot be directly manipulated. Therefore, all UV analyzer readings were excluded from further modeling, and the model was rebuilt with the rest of the process variables.

A new model was built by excluding the UV analyzer readings from the calibration set. The model is of similar quality as the one that includes the UV analyzer variables, see Figure 19.4.

The data set of this model was reduced to 24 variables. The top most important variables are listed in Table 19.1. They are sorted by their variable importance for prediction metric (VIP).

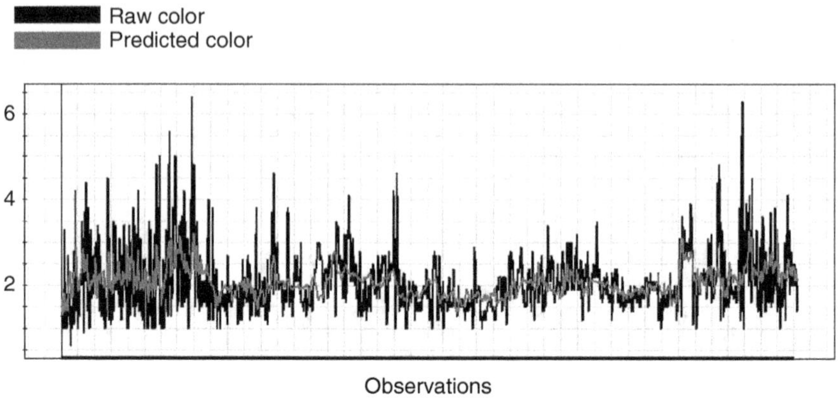

Figure 19.4 Observed vs. predicted color without UV readings.

Table 19.1 VIP metric values for the most important variables.

Variable	VIP
Ethylene Glycol Reactor Caustic Feed	1.81
Ethylene Oxide Reactor Inlet Dichloroethane	1.52
Downstream Location 1 pH	1.36
Ethylene Oxide Reactor Inlet Ethylene Oxide	1.24
Distillation Column Overhead Flow	1.12
Second Column Overhead Flow	1.09
Ethylene Oxide Scrubber Overhead Ethylene	1.09
Downstream Location 2 pH	1.07
Ethylene Oxide Reactor Pump Ditch Flow	1.00

To validate the model, the latest process data was left out of the calibration set. After the model was built, the color of this set was predicted using the model with the result shown on Figure 19.5.

The prediction of the validation set is of good quality. A major color event was predicted very well. This confirms the quality of the model.

It is of great importance here to isolate the cause of this problem, as it is well predicted and well separated from the rest of the data. Using the calculated variable contributions, three main contributors were discovered, see Figure 19.6.

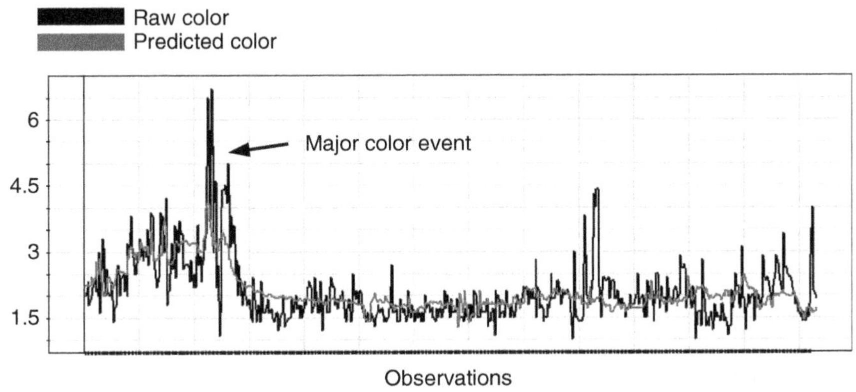

Figure 19.5 Observed vs. predicted color – validation set.

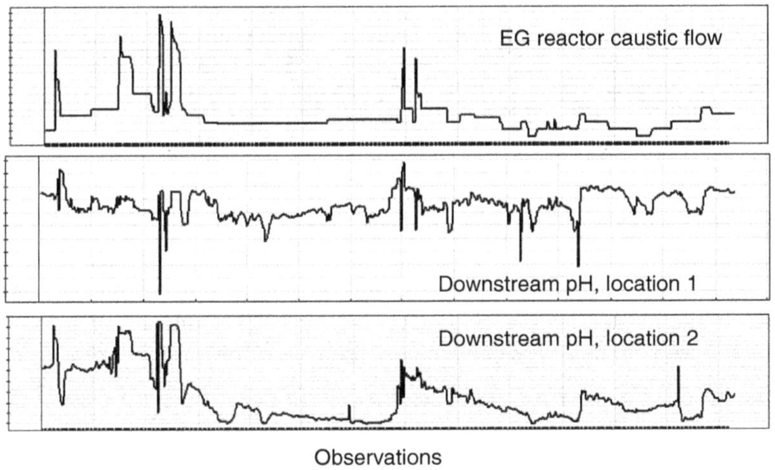

Figure 19.6 Main contributors to the major color event.

It is observed that all three variables are directly related to the amount of caustic present in the system. Another very clean indication of the caustic influence on the color is found in the calibration set, see Figure 19.7.

The peak in the ellipse on the plot clearly matches a high EG reactor caustic flow period (top small plot above) and downstream high pH period.

19.4.2 Summary of Data Analysis Results

The data analysis shows that the main process variables statistically related to the color are caustic related. The top candidate is the caustic flow to the EG reactor.

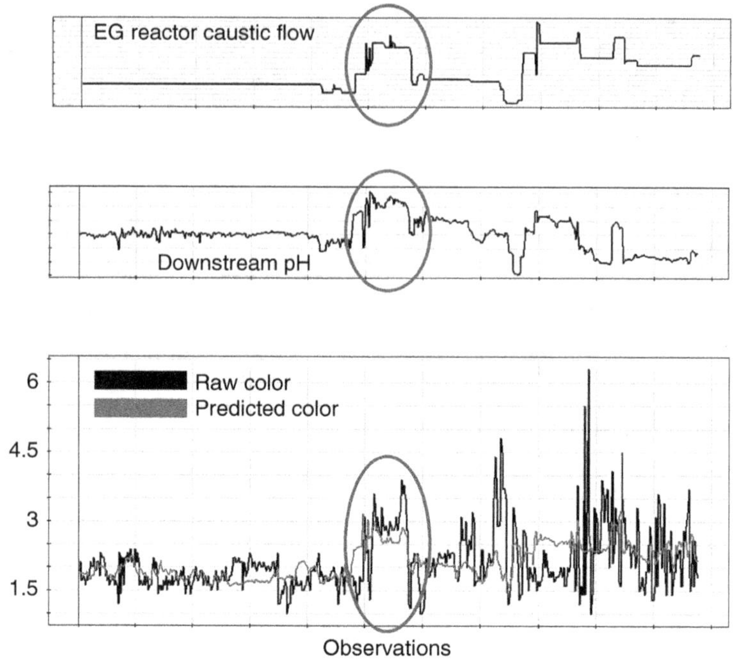

Figure 19.7 Main contributors to another particular color event.

Another variable that seems important is the EG reactor discharge pump flow, which points to the residence time in the glycol system. This could also point to a higher concentration of aldehydes being fed to the system due to a higher pure EO production.

References

Beebe, K.R., Pell, R.J., and Seasholtz, M.B. (1998). *Chemometrics: A Practical Guide*. New York: Wiley.

Chiang, L.H., Braatz, R.D., and Russel, E.L. (2001). *Fault Detection and Diagnosis in Industrial Systems*. London: Springer.

20

First Principles Modeling

In the past 15–20 years, modeling that relies on fast computers, in particular, has moved to extensive practical industrial applications. The fastest developing areas are materials modeling, including molecular simulations, large-scale process modeling and optimizations, computational fluid dynamics, statistical modeling, and data mining. As computer speeds and memory, as well as software technology, improved, modeling came to be more practical to apply in fast-paced industrial process problem solving.

One of the most important roles of modeling is to provide understanding and to generate rules for operations. These rules sometimes look simple, other times more complicated – with the common theme that we cannot deduce easily from available data. Fundamental or mechanistic modeling uses general equations of engineering or chemistry and builds models that are usually more complicated than a single equation; many times the equations are differential, partial differential, integral, or combinations of those. They need sophisticated computer algorithms to solve for generated information in formats that correspond to measured quantities.

Modeling in general provides great time savings by accelerating experimentation, as model-guided experiments may be carried out. In addition, developed models contain large amount of knowledge in a compact, reusable form. Consequently, knowledge is preserved so that it can be reused to resolve future problems. It is true for every model that they are built on experimental data. Experimental data are used either for model development (meaning to develop modeling equations), adjusting parameters of existing models/equations (parameter fitting), or model validation studies.

As models represent available knowledge, their form, information content, and range of validity may highly depend on the developers, and available experimental data. A model of a process will not contain more science than what the developers know. Developers means the possessors of the rules/knowledge,

Fundamentals of Industrial Problem Solving: A Practitioner's Guide, First Edition.
Zdravko I. Stefanov, Eldad Herceg, Carla Schmidt, David M. Jacobson, Dana Livingston, J.P. Chauvel, Sunil Kumar Chaudhary, and Christopher Paul Christenson.

approximations used, availability of accurate and efficient computer algorithms, and the people who program those into a software package. Because of this, models may not be unique, but considered to be valid until experimental data surface that cannot be explained by the current models. When models break, either re-calibration becomes necessary or a new model needs to be developed.

Thirty years ago, there was a shortage of chemicals, and customers would pay premium prices for any chemical that worked at all. Trial and error was used with success to develop new chemistry and to optimize/troubleshoot processes. Today, however, the trend has shifted due to increased competition. In addition, chemicals produced today must be manufactured in a way that minimizes down-time, pollution, and energy costs. Trial and error has become too costly and time-consuming to improve or design manufacturing processes and products. Effective use of modeling creates a competitive advantage at the marketplace. Modeling in the past ten years has been firmly established in the chemical industry. The main use is to speed up process and product development, and guide process trouble shooting by providing rational, mechanistic hypothesis.

Rational, mechanistic hypothesis building is a strategic approach to problem solving as it provides the base for better decision-making in creating long-lasting solutions. Design rules are transferred to and implemented in computer models that can be reused, so developed knowledge is preserved. While descriptive models are good and suitable for process optimizations within the limited range of process parameters, mechanistic models have the ability to predict outside of those with higher confidence. One of the most useful features of modeling is to eliminate false hypotheses.

Modeling, and use of modeling in the industry, is generally different from in an academic research environment. In the industry, the time to solve a problem is very limited, consequently multidisciplinary approach, and team work is a must. When communicating results, visualization of those is very important as decisions are made generally by management, at positions possibly far away from the model developer. While in the academia, full resolution of problems is preferred, in industrial settings that approach may not be afforded. Many times, it is enough to provide a partial resolution of a problem to eliminate false hypothesis, which in turn greatly helps to drive teams toward finding root causes of a problem.

20.1 Selecting the Right Theoretical Modeling Approach

First, we will describe levels of modeling in terms of complexity, and length scales. We will discuss how to choose between various available modeling techniques. In this chapter, we will concentrate on fundamental or mechanistic modeling of chemistry, and their applications on various fields of interest to the chemical industry. We will show a number of examples to illustrate how modeling can be utilized effectively.

When you ask various subject matter experts about the modeling approach they would use to resolve a problem, they most probably will have different answers depending on their own project experience, and field of expertise. However, the problem itself is very objective in this regard; it does not care about the available experts and their modeling approach. Consequently, where and how modeling can help will depend on the problem, so we need to be ready to adapt both in the approach and in personnel. How do we decide about what kind of modeling we need?

- **Define the goal of the model.** When we want to describe, control, or detect changes in a process, the appropriate modeling technique is phenomenological, chemometrics, or statistical modeling. If we need to design or predict outside the range of known operations/experiments, then mechanistic/fundamental modeling would be the dominant approach.
- **Industrial applications and needs may drive model selection.** Considering industrial process problem solving, most probably both approaches are necessary to various extents depending on the project. On the mechanistic side, we may not need the detailed mechanistic model, but we will need mechanistic thinking. Without mechanistic thinking, it is easy to overlook potential bad consequences of a process parameter change. The variable of interest may improve, but many other, currently not investigated parameters/properties may change for the worst. Fundamental thinking/models help to analyze these situations.
- **Availability of resource may drive the modeling decision.** We may have a very clear idea of what type of modeling would be necessary for a problem; however, the expertise may not be available either within the company or because of existing project load of the expert.
- **Lack of knowledge may drive modeling decisions.** At the beginning of a problem-solving project, we probably don't know the root cause and potential solution to test. We have to be ready to adapt to the path that the problem solution dictates. As we work with chemical processes, it is pretty much guaranteed that some form of materials modeling will be necessary. These may involve computing spectral properties to identify components of a complex mixture, predict physical properties, reactivity. For equipment design and sizing, it is very probable that fluid dynamics modeling will help. For so-called wicked important (not resource limited) problems, where single solution may not be identified easily or does not exist, modeling experts from various fields are advantageous to be present right in the beginning. They can work in parallel for analyzing various aspects, hypotheses, potential root causes. In these situations, disqualification of hypotheses (dead-ends) is as valuable as validation. This is where modeling can excel.

20.2 Fundamental Models

Let's see what we mean about fundamental models. Predictive, mechanistic models include fundamental understanding of all sides of the problem. In chemistry, it means knowledge of relevant chemical species, reactions, kinetics, physical properties. On the engineering side, it means the knowledge of relevant process conditions, and all the associated engineering issues and models. These models, when they are built well, will predict the most important trends. Validations and appropriate adjustments are made to describe current conditions. As the underlying chemistry and engineering model's validity are generally broader than current process conditions, fundamental models can be used with high confidence outside of the range of development.

Sometimes these fundamental models are considered as competitive to phenomenological or statistical models; however, we should consider them as complementary. They are used for different purpose. As we cannot implement complex chemistry and engineering equations into process control logic directly, it can be done with a simple, function-based phenomenological model. Statistical models have great use in finding important control parameters in the haystack of large number of process variables.

The development time of fundamental models is generally longer than that of statistical models. However, the tools, available computer codes, and databases that collect physical and chemical property information greatly accelerated the development of mechanistic models, so they became practical for fast-paced problem solving. The speed arises not only from the increase in computer speeds but also from distributed computing, data management, storage, and retrieval through the more widespread use of databases.

For models used in the chemical industry, we may distinguish a number of distinct levels associated with various length scales (Figure 20.1). The largest level of model mentioned here is plant-scale, which may involve a single or multiple plants in one location. However, the largest length-scale that modeling in the chemical industry practices is supply chain modeling. This is associated with modeling of scheduling with our suppliers, customers, and internally between purchasing, production, and distribution. Most process engineers are familiar with plant-level models, as they exist for many of the industrial processes. We use those for testing how a change of process variables affects others and to identify equipment limitations. In the plant-level models, the details of many pieces of equipment and physical properties of the materials are included. Because of the complexity of unit operations, many times a very approximate, simplified model is included. However, recent trends are such that more and more details of engineering equations are put in as increased computer speeds allow that. The next lower level, what every engineer is familiar with are continuum models handling

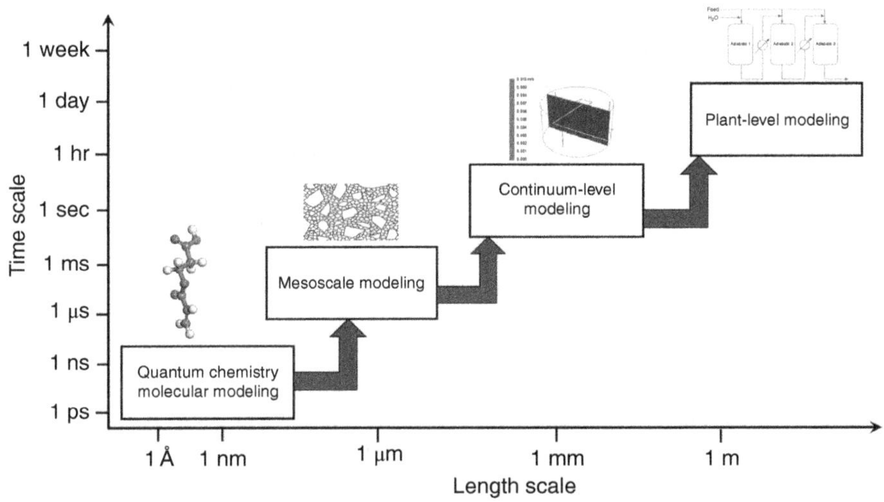

Figure 20.1 Levels of fundamental modeling (multiple length-scales are present).

length-scales greater than a few hundred microns. These include large-scale reactor models, which maybe simple approximations, or more detailed fluid dynamics models. Molecular models that consider electrons of an atom have length scales shorter than a few hundred Angstroms (n*0.01 μm). Meso-scale or coarse-grain models are in between the atomistic and continuum scales. The borders among them are somewhat fuzzy, and evolve in time, as increased computer speeds allow the computation of more and more details.

Models developed for different length-scales may be connected with appropriate techniques. For example – results of molecular modeling, with applications of rules of statistical mechanics, are connected to continuum models, including CFD. Interactions between molecules can be translated to information necessary for meso-scale models, so as the results of meso-scale models (viscosity, interfacial tension) to CFD simulations. Interfacing between the length-scales generally is not done within a single simulation code, especially in the industry. However, a number of academic efforts have been made in this direction – called multi-scale modeling. Probably the greatest success comes from Yannis G. Kevrekidis group, Princeton, Chem. Eng. Dept. (https://scholar.google.com/citations?user=N4NN1xwAAAAJ&hl=en; and https://engineering.jhu.edu/chembe/faculty/yannis-kevrekidis/). The technique they developed has been used for reactor modeling, modeling economy, ecology, weather, and many other fields.

Let's see where these different levels of modeling can be applied among the fields of our interest.

- **Molecular modeling** is widely used in chemical engineering, where thermochemical information, including heats of formation, heat capacities, phase-partitioning information, is utilized. Both qualitative and quantitative understanding of reactivities are used for predicting not only main product formations but also by-products and trace amounts of chemicals that may be even hard to detect, but can cause color or odor problems. In analytical chemistry, molecular modeling can provide spectral information, partitioning for separation, reactivities. Modeling of polymer structure helps materials science to elucidate most probable points of cross-linking, or oxidative break of polymer chains. One of the first success stories of molecular modeling in the 1980s was to predict toxicity and environmental fate of chemicals.
- **Meso-scale modeling** is mainly used in materials science and interfacial science. In materials science, predicting phase equilibria, viscosity, stresses, and Young's modulus are examples. In interfacial science, meso-scale computations are used for modeling molecular assemblies, surfactancy, which are associations developed between large numbers of molecules. Their appearance greatly modifies materials properties, such as viscosity, toughness, ductility.
- **Macro-scale modeling** is the most widespread technique in the industry. This area includes classical reactor, distillation, separations modeling, and computational fluid dynamics modeling. One of the more recent trends is to use the results of molecular and meso-scale models as an alternative to experimental data in continuum models. When molecular-level models are accurate enough, they greatly speed up problem solutions as time-consuming experimentation can be reduced. In the area of computational fluid dynamics, more and more details of chemistry and phase equilibria are included. In the academia, these approaches are broadly used; every major chemical engineering department has a number of faculty members versed in molecular, meso-scale, and computation fluid dynamics modeling, and collaboration between the experts is considered to be a must.
- **Plant-scale modeling** is used for integrating a variety of process equipment to find out equipment sizing to create a desired amount of product. This includes reactors, storage tanks, pumps, separation equipment, etc., and in addition process control logic. Every one of these needs an appropriate model to account for physical property, momentum and composition changes of the flow of materials. The sub-models may be simple, other times more complicated. Usually it is possible to include very detailed sub-models which allow more accurate predictions. As computer speeds increased dramatically, the recent trend is to put in more and more details, which is especially true for chemical mechanisms.

20.3 Examples of Theoretical Modeling Application

First principles modeling tools have been used extensively in the past. They have been used either as a sole technique to tackle the issues or as a part of a broader effort as one of techniques used throughout the problem solving process. These tools have been proven as a very valuable addition in our "problem solving kit" to attack technically challenging high-value problems. In this section, we will briefly discuss the role and impact of theoretical modeling techniques in solving some of the most challenging and frequent problem in the chemical processes (fouling, trace chemistry, and catalyst issues). An example for trace chemistry issue and of how the first principles modeling techniques were used to aid in solving it was presented in Chapter 15. We will not discuss these here further. In this section, we will present examples for fouling and catalyst issues and discuss the role of first principles modeling in the problem-solving process that we follow.

Fouling issues are one of the most common problems encountered in the chemical industry. At the same time, they are extremely challenging and difficult to solve due to complexity of the physico-chemical processes and engineering aspects leading to their manifestation. As an example here, we will discuss how the first principles models were used within the problem-solving process to attack and solve a decades-old fouling problem within a plant (Wright et al. 2008). A high frequency of plugging incidents in a heavies unit, used to separate catalyst and heavies from a lighter product stream, disrupted and limited sales for several products. With the impacted plant only operational for periods of three to nine days followed by plant shutdowns of two to three days for cleanup, each incident cost the business critical pounds for one of its primary products. Unfortunately, plugging was not a new occurrence for the plant, as the heavies unit had experienced severe plugging episodes in the past. Several previous efforts in the plant had attempted to resolve the plugging problem over the years; however, those efforts were unsuccessful. Market demands and the earlier lack of success led the business eventually to form a multi-discipline team of chemists and engineers to understand the fundamentals of plugging. The objective for the team was to use this fundamental understanding to devise chemical, engineering, and operational strategies to mitigate the effect of heavies formation, so the plant could operate without plugging for a minimum of five months.

The role of theoretical techniques in this particular project was to use the characterization results obtained from numerous analytical studies to develop a chemical mechanism for heavies formation based on the actual process conditions. The tars were determined to be organometallic, with the inorganic part of the molecule coming from the catalyst used in the process. The structures of the ligands were confirmed by isolating the ligands with preparative LC methods. The isolated materials were then characterized by C-13 nuclear magnetic resonance (NMR) and single-crystal X-ray diffraction (XRD). This characterization work led

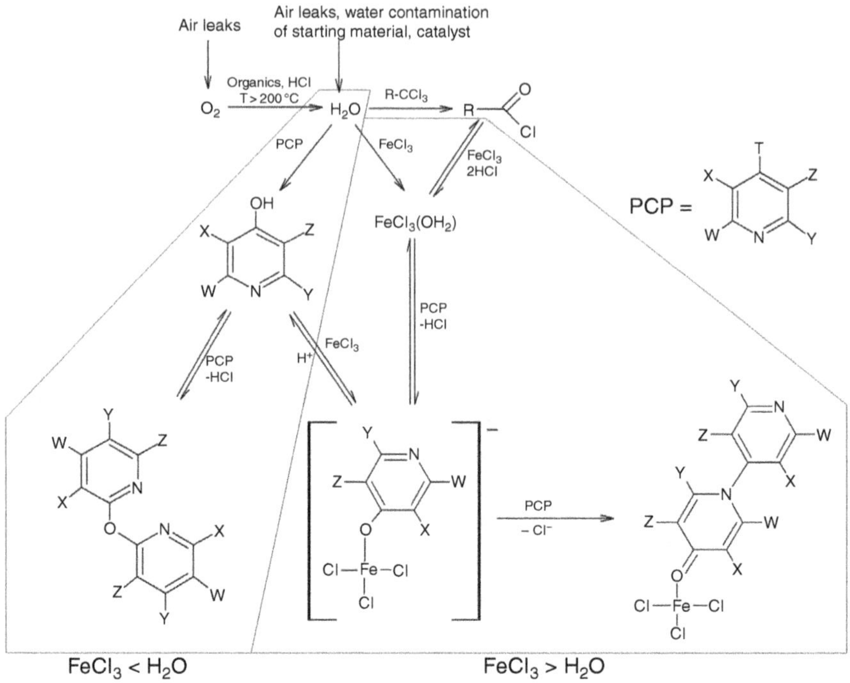

Figure 20.2 Chemical mechanism for tar formation.

to the development of a chemical mechanism for their formation using first principles models and showed that oxygen and the catalyst was embodied in the two tar molecules (Figure 20.2).

The mechanism, proven experimentally in the lab with water-spiked process samples, supported the proposed mechanism and showed that the catalyst was a carrier of oxygen in the process. Oxygen in any form, whether it be in the form of water, air, or an organic or inorganic oxygenate, was an abnormal reactant in the process chemistries and could only be introduced through a combination of feedstock impurities, vacuum leaks, and/or wet maintenance procedures. The mechanism that was proposed showed that once oxygen and water entered the process, they reacted with the catalyst and the process feeds to produce tars (or tar precursors that led to tars) that plugged the "heavies" unit. The mechanism for tar formation focused the team's efforts on identifying oxygen (water and air) incursion sources and removing those sources from the process. Once this was done, the decades' old Tar Wars' battle was won in less than three months by addressing the underlying cause of plugging. The team's goal of 5 months was far exceeded, as the plant has experienced one year of "plug-free" running (Figure 20.3).

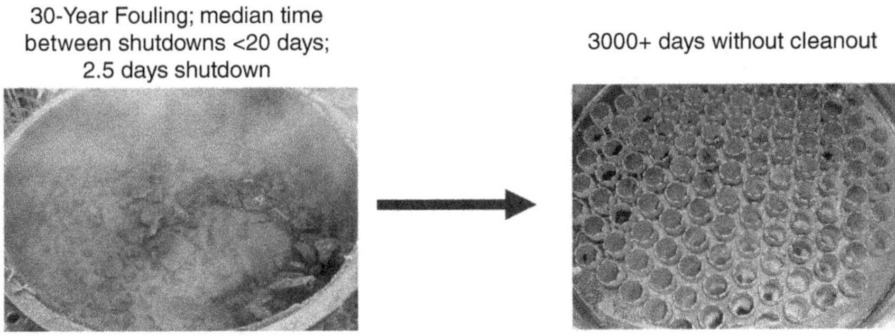

Figure 20.3 Before and after pictures of the fouled equipment.

Next, we will discuss dynamic modeling to study reversible poisoning of a catalytic bed (Somasi et al. 2013). Catalyst poisoning in tubular reactors is a common problem in many chemical processes using a variety of catalysts including zeolites and transition metals. A commercial reactor for the hydrolysis of an organic feed suffered a dramatic reduction in catalyst lifetime, and this was found to be correlated to an increase in the concentration of ammonia from 30 ppm historically to 100 ppm. This led us to hypothesize that reversible ammonia poisoning of the zeolite catalyst might be responsible for the complex dynamic behavior that had long been observed in this system and the change in ammonia levels might be the cause of the observations concerning catalyst life. A dynamic model of the reactor system for this highly exothermic reaction has been developed to evaluate this hypothesis and explore alternate reactor operation and design options to manage the varying NH_3 levels observed in the feed.

Computational chemistry modeling methods were first used to calculate the Gibbs energy of ammonia adsorption on the catalyst surface. Figure 20.4

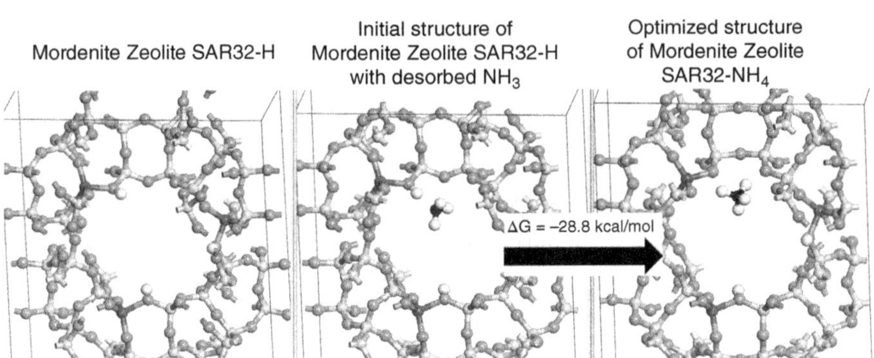

Figure 20.4 Adsorption of ammonia in the SAR32 mordenite.

visually represents the adsorption process of ammonia on the SAR32 mordenite. As noted in the figure, ammonia adsorption in SAR32 is thermodynamically favorable, and the reaction proceeds with a proton transfer from the SAR32 acidic site to ammonia to produce a negatively charged zeolite framework that is neutralized with an ammonium cation. Since this is in essence, an acid–base neutralization reaction, it proceeds with virtually no activation energy (Ea). The absence of Ea and a negative ΔG value indicate that ammonia adsorption in the zeolite catalyst is a very fast process under the reactor conditions. Obviously, this leads to deactivation of catalytically active sites and deterioration in the catalyst performance. Modeling was then used to estimate activation energies for ammonia desorption in the presence of water and HCl since they are always present in the reaction system. It was found that in the absence of water and HCl, activation energy is ~35 kcal/mol, which matches well with the values measured and estimated by previous work (Parrillo and Gorte 1993; Kyrlidis et al. 1995). Water lowers Ea to ~28 kcal/mol, while HCl lowers it still further to ~23 kcal/mol.

Next, equilibrium constant was estimated from the miniplant (laboratory reactor). It was found that the Gibbs energy estimated using the miniplant data is bracketed by the theoretically estimated Gibbs energy in the presence of water and HCl, which implies that the results of the poisoning study are quite reasonable as seen in Figure 20.5. Therefore, this estimate of the adsorption equilibrium constant has been used for the reactor modeling.

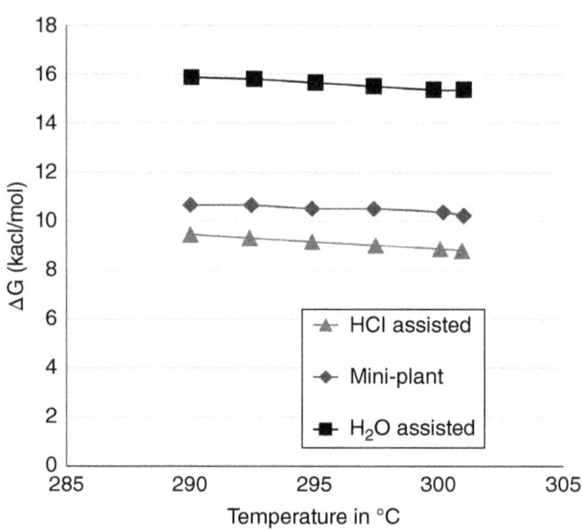

Figure 20.5 Estimated Gibbs energy for NH_3 adsorption from theoretical calculations and from equilibrium constant estimated from miniplant.

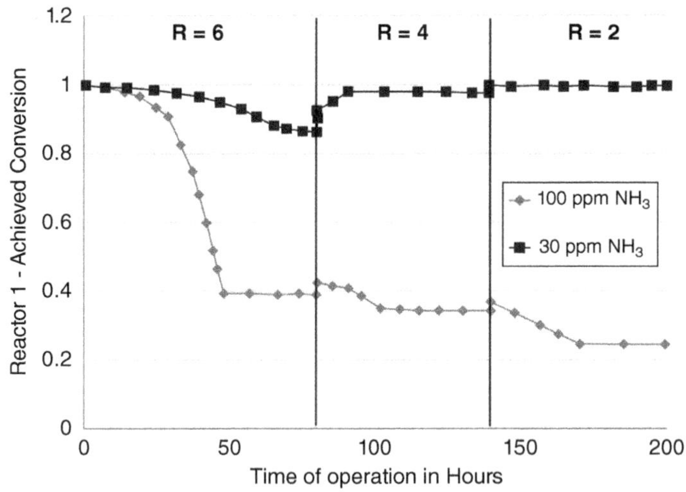

Figure 20.6 Effect of water/primary reactant ratio on different NH_3-level feeds.

Subsequent dynamic models, built by using the estimated equilibrium adsorption constants, showed that an ammonia levels of 100 ppm will decrease the catalyst performance in a shell-and-tube reactor significantly compared with 30 ppm. More importantly, the model shows that changing the water to primary reactant molar ratio allows the reactor to run successfully despite irreversible deactivation when NH_3 levels are on the order of 30 ppm, but not for 100 ppm. This is shown in a graphical form in Figure 20.6.

Developed models also showed that steaming can be used to extend useful catalyst life by stripping adsorbed NH_3. For the shell-and-tube reactor, increased duration of steaming is required for regeneration, whereas for the adiabatic reactor, increased steam temperature can be substituted for the duration of steaming. In both cases, steaming will cause a temporary increase in conversion. Simulations of an alternate, adiabatic design (Figure 20.7) reveals that the heat removal capability of the shell and tube reactor, originally designed to allow better temperature control, hurts the performance in the presence of NH_3 by removing heat that would otherwise aid in the desorption of NH_3 and retention of catalyst activity. In this new design, the primary reactant is split between the three adiabatic reactors – and is shown to be much more robust toward varying NH_3 levels. The model showed very little difference in reactor operation for the two different levels of NH_3. This can be a significant advantage of this reactor design compared to the system comprised of a shell-and-tube reactor followed by a polishing adiabatic reactor. For this system, keeping reaction heat in the system provides the advantage of driving NH_3 desorption, making the adiabatic system more robust to this specific mode of catalyst deactivation.

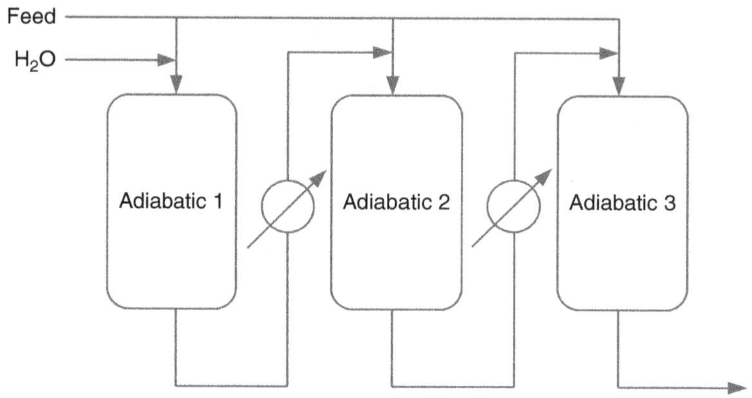

Figure 20.7 Adiabatic reactor scheme.

References

Kyrlidis, A., S J Cook; A K Chakraborty; A T Bell (1995). Electronic structure calculations of ammonia adsorption in H-ZSM-5 zeolites. *J. Phys. Chem.* 99: 1505–1515. doi:https://doi.org/10.1021/j100005a022.

Parrillo, D. J. and Gorte, R. J. (1993). Characterization of acidity in H-ZSM-5, H-ZSM-12, H-mordenite, and H-Y using microcalorimetry. *J. Phys. Chem.* 97: 8786–8792. doi:https://doi.org/10.1021/j100136a023.

Somasi, S., Witt P.M.; Calverley E.M.; Livingston D.A.; Herceg E. (2013). Dynamic modeling to study reversible poisoning of a catalytic bed. *Ind. Eng. Chem. Res.* 2013, 52:15330–15341. doi:https://doi.org/10.1021/ie400616b.

Wright, L.G., Bell, B.M., Paul Chauvel, J. et al. (2008). Tar wars in a complex chemical plant: a fundamentals approach to problem solving. *AIChE session titled "Plant Troubleshooting and Problem Solving: The Chemist and Engineer Working Together." AIChE Spring Meeting and Global Congress on Process Safety*, New Orleans, LA (April 2008).

21

Analytical

In the chemical industry, there are many issues related to unintended chemical reactions such as corrosion, fouling, and trace chemistry involving raw material impurities. These reactions can lead to significant profit losses due to poor product quality, reduced catalyst efficiencies, loss of primary containment, etc. The primary tool in gaining fundamental understanding of these issues is analytical chemistry. In this chapter, we will explore how to apply the tools of the analytical chemist, from simple test methods to sophisticated instrumentation, to narrow down the hypothesis matrix and focus the problem-solving team on the right avenues of investigation. This chapter does not, however, explore the breadth of analytical techniques and their actual application to sample analysis. The reader is recommended to refer to analytical chemistry texts for such information.

21.1 Definition of Analytical Chemistry and Data

The words "analytical" or "analytics" are found in high frequency recently in the manufacturing publications and in popular media. However, for the purposes of this section of the book, analytical data only refers to chemical analysis data acquired for the purpose of identifying qualitative and/or quantitative chemical information on a chemical sample. These analyses can be quite simple such as the sensory methods used by people who cook or bake, brew beer, ferment yogurt, or make candy or jelly at home in which changes are observed in color, clarity, smell, taste, sound, or viscosity. Some of the same observational methods are used in the chemical industry. For example, visual inspection is used to verify product clarity and ensure there is no foreign matter in most pharmaceutical- and food-grade products. Smell is often a leading indicator of whether a wastewater treatment plant is operating properly or not, much to the disgust of the plant's neighbors.

Fundamentals of Industrial Problem Solving: A Practitioner's Guide, First Edition.
Zdravko I. Stefanov, Eldad Herceg, Carla Schmidt, David M. Jacobson, Dana Livingston, J.P. Chauvel, Sunil Kumar Chaudhary, and Christopher Paul Christenson.

While these techniques may be inexpensive and quick, they are not without their pitfalls. First, in the chemical industry, there are generally safety concerns when using taste, smell, or touch to characterize process samples. In addition, any method that relies on human senses will have a certain level of subjectivity since every person's senses are slightly different. This can make it difficult to use observational data in killing hypotheses. (Note: This is why the use of photos and video to capture observations is highly encouraged whenever practical.)

Advancements in analytical techniques have been developed to overcome subjective observational methods. Instruments that measure and quantify color and clarity are one example. Many instrumental analytical techniques that are used every day in production units to verify that the plant is operating properly include gas chromatography, titrations, infrared spectrophotometry, etc. These techniques can be either conducted in a lab using samples that were collected manually in a jar or gas cylinder or can be conducted continuously online without automated sample collection. The data from these methods are added to the process control system and automated adjustments are made to operating parameters based on the input from these analyses. These data are often quite valuable in determining what has changed in a chemical process and what has not changed since there is an established baseline of historical data to compare current performance against.

There exists another class of analytical instrumentation that is even further advanced. For example, there are highly sophisticated methods such as gas chromatographs that can measure impurities to the part-per-billion concentration level. There are electron microscopy instruments that can pinpoint the location and quantity of precious metals on catalytic surfaces. While these methods push beyond what the human senses are capable of accomplishing, they come at an extremely high cost (instrument purchase, continual maintenance, cost per test, and skilled personnel to operate them). Most of these techniques are only used by large companies or universities. As a consequence they are not available in every problem-solving situation.

Therefore, it is important to strike a balance between using the existing analytical data (from observations and daily/weekly quality control data) and specially requested analytical tests to further elucidate what has changed in the process. The balance will be a judgment call by the problem-solving team. Their decision will depend on the severity of the problem (how much it is costing the company) and the cost of acquiring extra analyses (cost of higher sampling frequency, sending samples to external labs for testing, etc.). This is analogous to the medical industry when a physician will request testing to understand the health of the patient. However, a doctor will not request five or more tests without a reason to suspect that it will provide valuable data to help assess and/or diagnose the patient. This is done for two primary reasons: the cost of the tests are substantial and most tests lead to higher stress within the patient. In the same way, problem solvers should be careful to balance the desire for data with the costs for the tests and the possible stress on the project team to get the samples and complete the analyses. Time should only be invested in the analyses necessary to test the highest priority hypotheses. The reader is cautioned to not fall into the trap of getting data just because it is possible. Any effort spent on paths that are not critical to testing the highest priority hypotheses will only lead to more stress and confusion.

21.2 The Relationship Between Hypotheses, DQOs, and Analytical Data

Consequently, the ideal scenario is to write the hypotheses first and prioritize them before beginning any analyses. Then the project team should determine what data is needed to kill the hypotheses. Next, the team should write a brief Data Quality Objective specifically for the analysis. This ideal process is outlined in Figure 21.1.

The above process is mentioned as being ideal, but it is important to recognize that in reality, it is rarely as simple or as linear as Figure 21.1 would indicate. As will be highlighted in subsequent examples, there are frequent situations that call for reiteration in the analytical DQO process. In fact, sometimes the analytical capability dictates whether a hypothesis can be killed or not.

The analytical DQO will be different from the DQO for the project that was discussed in earlier chapters. However, it will still have the same three questions about success:

- What is success (for the analytical and/or lab work being requested)?
- What is necessary for success?
- What is rate limiting?

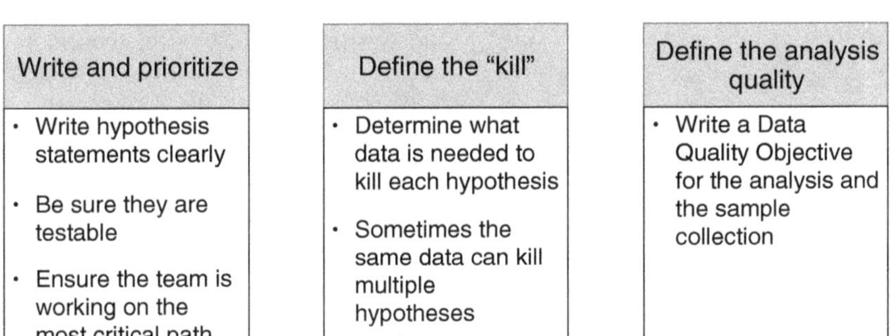

Figure 21.1 Ideal process for analytical data collection to test hypotheses.

In addition, the analytical DQO should go into more detail about the actual data quality and quantity needed to make a decision on the hypothesis. The next section will highlight the many possible considerations that could be included in the analytical DQO. The reader will have to decide which components are most critical for a given situation. Care should be taken to balance between getting the needed agreement between the parties involved and overburdening the team with too much information and paperwork that it becomes a hindrance to making actual progress on the work.

There are some situations in which a problem-solving team will believe it is necessary to have the analytical testing done before they can begin formulating hypotheses. For example, if a black tarry substance is found plugging the inlet of a pump, the team will naturally want to have some clue as to the composition of the tar prior to forming hypotheses. The reader is cautioned to avoid this pitfall and to stress with the team that even in these instances a hypothesis is implied and should be written down before testing begins. Examples of implied hypotheses are given in Table 21.1.

There are several advantages of writing down the hypotheses before beginning lab work or even sample collection. The act of writing down the hypotheses

Table 21.1 Examples of implied hypotheses.

Situation	Questions	Implied hypotheses
Foreign material observed in pump inlet	What is it? Could it be a pump seal or gasket a polymer rust biological growth a partially dissolved rubber glove?	The pump seal material is incompatible with the process stream and a portion of the seal has dislodged and been recirculated back into the pump inlet The raw material is polymerizing in the unit op before the pump due to high temp, low inhibitor, cross-contamination, etc. Fine metal particulates are acting as a catalyst for the polymerization of the monomer, which fouls the heat exchanger and flakes off as thin sheets into the pump suction There is insufficient biocide to prevent biological growth in the aqueous recycle stream, and it sticks to the inlet pump strainers Someone left a rubber glove in the piping during the recent maintenance activity.

Table 21.1 (Continued)

Situation	Questions	Implied hypotheses
Product quality failure	What has changed in the raw materials? Compare recent raw material retains to historical.	There is a new impurity or increase in concentration of existing impurities in the raw material from Vendor A or Batch 123 that is causing the product failure
Frequent filter plugging	What has changed in the raw materials? Compare recent raw material retains to historical ones	The particle size (or distribution) in Vendor B material is too large, and there is insufficient time in the dilution step to dissolve all the raw material which leads to filter plugging There are insoluble species in the raw material
Frequent high pressure alarms in the distillation tower after August 5th 5 p.m.	What has changed in the tower feed streams? Could it be an increase in water?	An emulsion in the dewatering tank caused by washing activities on August 5th led to poor liquid interface detection and periodic increases of water above 2% in the distillation feed and the higher water concentration has exceeded the tower's design capacity for vapor loading

causes the team to make the hypotheses more specific. The higher specificity a hypothesis has, the higher the likelihood of killing it with the appropriate data. Also, the person who is asked to perform the testing will be more engaged and attentive to the request if they understand why the request is being made and how the data will be used. This knowledge will enable them to make more informed decisions on how to analyze the samples and to make the proper observations during the analysis. For example, when a request for metals composition by inductively coupled plasma (ICP) of a liquid sample is made, the analyst has to understand if the request is for total metals or only dissolved metals. This guides their sample preparation decision: to shake the sample or filter it prior to removing an aliquot for analysis. Furthermore, a discussion about the hypotheses prior to beginning lab work prevents the generation of data that is not directly related to any of the hypotheses (wasted time, effort, and sample).

Once the hypotheses are properly understood, writing an analytical DQO ensures that the data that are generated are of the appropriate quality to be used to kill the hypothesis. If the wrong data or poor quality data are used to test a hypothesis, then this could lead to a premature or false killing of a hypothesis, thus delaying resolution of the problem. Alternatively, erroneous data could lead the project team to falsely believe that a hypothesis is still valid and cause them to spend more time testing it or implementing measures to prevent that perceived root cause. Either situation can be incredibly costly to the problem-solving process in both time and money. This is why having written hypotheses and an analytical DQO prior to beginning any lab work is highly recommended.

A useful exercise is to write down the consequence matrix for the analytical data. Consider that the analytical answers are a simple yes or no. Each can be true or false.

This makes a nice 2 × 2 matrix:

	True	False
Yes	True positive	False positive
No	True negative	False negative

For each of these cases, write down the consequence of each case. How will you deal with the result? How long will it take to detect errors? What are the consequences of the results correct or incorrect? Further, will the analytical results require notification of regulators? How long do you have to make that notification?

We recommend that all of this is carefully considered and documented before the analytical measurement is made.

These considerations will be important in determining the required analytical precision/accuracy and whether the analytical method requires validation before measurements are made.

21.3 Preparing an Analytical or Experimentation DQO

The United States Environmental Protection Agency (EPA) began using DQOs in the mid-1980s. The agency recognized that a significant time and effort was being expended on collecting environmental samples and analyses. However, some of these data were being collected without a specific goal in mind and the quality of some of the data was insufficient to use for decision-making. It was also recognized that there were situations where extremely high-quality data were being collected to make decisions that could have been made with lower quality data, leading to saving in time and money for data collection and analysis. Therefore, the agency implemented the DQO as a necessary front-end step before beginning a major project. The 1986 version of the EPA's DQO was designed to facilitate a discussion and agreement around clear statements of the following:

The decision to be made

- Why environmental data are needed and how they will be used
- Time and resource constraints on the data collection
- Descriptions of the environmental data to be collected
- Specifications regarding the domain of the decision
- The calculations, statistical or otherwise, that will be performed on the data in order to arrive at a result.

"Development of Data Quality Objectives: Description of Stages I and I," by the Quality Assurance Management Staff, ORD, U. S. Environmental Protection Agency, Washington, DC 20460.

The EPA's guidance document on writing DQOs is a rather lengthy document designed for getting the optimal balance of between data quality and cost prior to beginning a large-scale data collection effort. The process was designed to be an iterative, top-down approach starting from the decision-makers defining the decision to be made and how the data will be used and ending with the technical experts that will oversee the data collection and analysis. This process is a bit too cumbersome for use on smaller magnitude projects or projects that have an extremely compressed timeline, such as when a manufacturing unit is in crisis. Therefore, the reader is encouraged to tailor the length and detail in the analytical DQO to match the needs of the project and the criticality of the hypothesis to be tested or the decision to be made. Generally, the statements above from the EPA's version of the DQO are a good place to start in crafting the analytical DQO and the depth of the detail should be modified to the appropriate level. If, for example, the data will be used to determine whether to execute an unscheduled shutdown and maintenance activity of a chemical process, then there will be an economic and safety impact if the decision is made incorrectly. Therefore, care should be taken to ensure that the correct data are collected and that they have the appropriate level of quality. If the data will be used, in conjunction with other process data, to determine the outcome of an incremental process change such as changing the distillation outlet temperature by 2°, then the data quality may not need to be very stringent.

One useful exercise when preparing the analytical or experimental DQO is to predict what the data will have to look like in order to make a decision or to invalidate a hypothesis. This can be helpful to force the team to visualize what the data could look like and how they will use it. For example, a common occurrence in project teams is to ask for compositional data on a process stream that has never been analyzed before. However, if there is no baseline to which to compare that data, no decision can be made with a single set of data. Has the composition changed or is it consistent with historical data? Is the process variation such that one sample set is representative of the stream? Are samples needed three times per day over the next week in order to get an appropriate representation of the stream composition and its variation? By requiring discussion on how the data will have to look in order to be able to make a decision or test a hypothesis, it encourages the team to visualize all the way through to the end of the process. This typically causes teams to explore the possibilities and better define the specifics of how the data should be collected and reduces the probability of missing critical pieces of data that are needed. In addition, it reduces the likelihood of the team getting the data and then being unable to decide whether they can be used to kill or validate a hypothesis. There are frequent situations when data will be collected that invalidate a hypothesis that is strongly held by influential members of the team. Those members will want to reiterate the data collection over and over again because they are unwilling to accept that the hypothesis is false. A powerful way to avoid this situation is to get agreement ahead of the data collection on what the data have to look like in order to deem the hypothesis is false. Another advantage of predicting the data ahead of time is it can help the team to avoid falling into the trap of being led by the data, i.e. formulating theories that match the data rather than basing them on the fundamentals of science and technology. For example, iron corrosion products are often hypothesized to be the root cause of yellow discoloration in a chemical products. Investigative teams will then ask for iron composition data to test that hypothesis. If no iron is seen at the 1 ppm level, is this sufficient to invalidate the hypothesis or does the detection limit need to be 100 ppb or even 1 ppb? Does it matter whether the iron is present as a particulate in the sample or does it have to be dissolved in the liquid stream? If solids can cause discoloration, is the method of sample extraction appropriate to collect a representative sample of those solids? If these questions can be answered ahead of time, then the appropriate sampling and analytical techniques can be selected to collect the test data.

Another good practice is to challenge the assumption that new data are needed to make the decision. Most chemical processes already have significant quantities of data collected every day using process instrumentation (temperature, pressure, flow, etc.) and compositional data from analytical instruments. Sometimes a project team will become blind to using their existing data because they have never thought of using their data in novel ways before. In addition, some people are risk averse and wish to have more data to increase their confidence in making a decision. If they happen to make the wrong decision, they want to use the new data as a defense to show that they made every effort to verify their decision before proceeding. There are certainly occasions where collecting more data is the correct approach; however, the existing data may be sufficient to make the decision and will not require more time and resources to collect it.

Rapid problem solving requires carefully considered risk in decision-making. If you wish to make decisions at the 99% confidence limit, you will need a lot of data (usually) and it will take a lot of time. First, you will have to develop and validate the method. . .

You don't have that much time. Consider fully the consequences of errors. If the error causes much damage and/or takes a long time to detect, the confidence has to increase. If the consequence is small and the error will be found soon, lower the required confidence and gain speed; 80–90% confidence is often about the right trade-off. Beware of the errors that make things go bump in the night.

There are a number of other details that should be considered when preparing an analytical DQO. A list of possible considerations is given below as an example, but it is not an exhaustive list. The reader and/or the project team will have to select the level of detail needed for each situation.

21.3.1 What Are the Best Means of Collecting the Sample?

- Are there two phases in the sample? Solids in a gas or liquid sample? Liquids in a gas stream or vice versa?
- Does the sample need to be preserved? Is refrigeration needed? Does the sample need to be analyzed while still warm to prevent solidification/precipitation?
- Does the sample need to be protected from ambient light? If so, is collection in an amber bottle or a foil-wrapped bottle needed?
- Does the sample need to be protected from air exposure? Zero or minimal headspace in the collection vial? Handling in a purged or dry box enclosure?
- How much volume or quantity of sample is needed to conduct the required testing?
- Will the sample be collected after shutting down the process? If so, will the process of clearing the chemicals from the system alter the sample? Will the sample be exposed to a water flush or exposed to air and humidity?
- Is there a sample that needs to be collected to use as a basis for comparison, perhaps from another train or similar manufacturing unit?
- Should the samples be collected in glass, plastic, or metal vessels? Are there enough of these on hand to accomplish the data collection?
- What are the sample points that will be used? Document these well with descriptions and photos if possible. How should the samples be labeled to remove ambiguity?
- Will the sample change after collection? Changes of temperature or pressure resulting in precipitation, dissolution, condensation, etc.
- Is chain of custody needed for these samples, typically needed for regulatory or litigation situations?
- How should the samples be stored if not analyzed immediately?

21.3.2 What Are the Best Ways to Prepare the Samples?

- Which analyses must be run before others to prevent contamination or introduction of air, water, etc. causing changes in the sample?
- If the amount of sample is limited, which tests need to be conducted first? Conduct nondestructive testing or highest priority analyses first.
- Should the sample be mixed, blended, ground, centrifuged, filtered, etc., prior to removing an aliquot for analysis?
- Should it be heated or cooled to create a homogeneous sample?
- What are the acceptable solvents for dissolving the sample?
- What are the safety concerns that the analyst should be aware of before working with the samples?
- Are sample purification techniques needed prior to analysis?

21.3.3 What Are the Key Considerations for the Analysis?

- What standards are needed for comparison?
- What is the applicable concentration range?
- What limits of detection and quantitation are needed?
- What precision and accuracy are needed? Does the sample need to be preconcentrated to improve detection limits?
- How will the data be recorded? Are photos or video needed?
- How will the data be further treated after analysis? Averaging, statistical analysis, modeling, etc.
- How will the data be stored and communicated to the project team?

Once the sample and data quality requirements are defined, then the team can determine which analytical techniques best match the requirements. This does not mean that only the most expensive techniques should be considered. For example, when an experienced analytical chemist is presented with a dark-colored particulate in a solid or liquid sample in a clear glass jar and asked to identify it, they will often hold a strong magnet (magnetic stir bar) to the outside of the sample container to see if the particles are magnetic or not. If the particles move with the magnet, then there is certainly iron present in the sample and it is likely rust. Magnifying glasses or small digital microscopes such as those that can be purchased for children to use at home can be helpful for determining whether a haze in a product is a suspended droplet/emulsion, an insoluble particle, or biological material. A simple pocket laser will detect low concentrations of particles and droplets. Simple applications of pH paper can be used to determine if conditions are appropriate for corrosion or other reactions. These tools generally cost less than US$100 and do not require great expertise to use. However, if the team determines that higher end analytical techniques are needed, then the DQO can be used to engage the analytical experts in determining what the best techniques are that will fit the requirements.

21.4 How Analytical Data Are Used to Solve Problems

As mentioned before, analytical data should be used to test hypotheses that have been generated from a fundamental understanding of the chemical process. This method is preferred over the empirical method of looking at existing data and then trying to generate a theory to fit the data. This is not to say that the reader should blindly form hypotheses without ever looking at the data. However, the reader should be careful to examine data with the fundamentals and a hypothesis in mind, an implied hypothesis, to train the brain to ask questions first and then look for answers rather than the other way around.

At the simplest level, analytical data can be used to answer the questions of (i) what is it? and/or (ii) how much is there? However, the data can be applied in highly complex ways such as to complete mass and energy balances, to identify changes in the raw materials or process streams, to control unit operations such as distillation towers, to identify the components of solids contributing to fouling, polymerization, sludge formation, etc. The greatest value in using analytical data is when the data are used to refute a whole family of hypotheses at once. If when using the hypothesis onion, it is identified that several hypotheses can be tested with a single data set, then emphasis should be put on collecting that data first and ensuring the data quality is appropriate for those determinations. For example, in Table 21.1, a situation was proposed in which a foreign material was observed in the pump inlet. By determining the identity of the solid using a quick infrared analysis, the team may identify that the material is composed of a fluorinated polymer. This finding refutes hypotheses related to rust, biological growth, and rubber gloves. If the gaskets and seals in the process are not made of fluorinated materials, then those hypotheses can be refuted as well. This helps focus the team on a much narrower scope of hypotheses and sources for the material. For example, it might be hypothesized that the wrong seal material was substituted into the process or that the wrong seal material was ordered by mistake. By examining the hypothesis onion and looking for tests that can be used to test the highest number of hypotheses, the team can ensure the fastest progress in gaining understanding and ultimately identifying the root causes.

Another example of the power of analytical chemistry in solving problems is when it is used to create a mass balance. The process of creating a mass balance can not only kill many hypotheses but can also reveal many more possible hypotheses that would not otherwise be proposed. If, for example, a feed(s) to a unit operation contains <10 lb/hour of a sulfur-containing impurity and more than that can be found in the outlet streams, then there are several hypotheses that can be proposed to explain that phenomenon. Possible hypotheses include:

- The analytical methods are missing some sulfur species in the analyzed streams due to interferences or the methods were not designed to detect them.
- There is a stream (possibly leak or dissolved gas) that is not included in the mass balance.
- The samples that are being analyzed are not representative of the total flow (two-phase flow).
- Chemistry is occurring in the unit operation that transformed the sulfur species into a component that can be detected in the outlet streams. Ex: The analytical method requires filtration of the sample before analysis, and this removed some sulfur-containing species in the feed stream, but the unit operation reacts those solids into a component that is then soluble and detectable in the outlet stream.
- Error on the part of the analyst such as not shaking the sample as required in the procedure or allowing volatile species to evaporate in the feed samples, etc.

By determining when the analytical data departs from the expected plant performance, it is easy to see that there are phenomena at play, which are not expected. Further examining these discrepancies will typically lead to identification of the root cause(s) of the issue.

Mass balances do not have to be perfect to be useful. Often a mass balance that is correct within an order of magnitude can help guide the project team to quickly narrow their focus or to challenge their assumptions. For example, a bed of catalyst material is fouled every three months to the point where the differential pressure across the bed becomes too high for pump. The team hypothesized that the fouling material may be corrosion products from the upstream equipment, which is made of Monel, a nickel and copper alloy. Some of the team members believe that the amount of corrosion products observed is too large to be coming from the Monel equipment because they would be able to detect the thinning wall thickness in the ultrasonic scans of the equipment. The team begins with a crude mass balance to understand the order of magnitude of the problem. They need these few key pieces of data:

- An estimate of the foulant weight. This could be obtained by weighing the material prior to loading and then again upon removal. (Ex: 1000 lb of foulant.)
- A rough composition of the material. Elemental composition is often used for these purposes from techniques such as elemental combustion (yields CHNOS percentages) and X-ray fluorescence (yields metallic element percentages). (Ex: 50% C, 25% Ni, 12 % Cu, less than 10% of all other elements combined.)

This quickly leads to the calculation that ~250 lb of Ni and 120 lb of Cu would have to be lost from the upstream equipment every three month in order for it to be the source of the metal portion of the fouling deposits. Then using the internal surface area of that equipment, an estimate of the thickness loss could be calculated and compared to the sensitivity of the ultrasonic instrumentation to determine if the hypothesis is viable or refuted.

To further estimate the size of the source of the carbon in the deposit, an estimate of how much of the process flow contributed to the fouling accumulation is needed. In general, if the team cannot agree on one assumption, it is useful to do the calculation at the two extremes to define the boundaries of the problem. In this example, the deposition of the material is assumed to be continuous from the beginning of bed use until the end, and the average flow rate over the 90-day period of bed use was 10,000 lb/hour. This means that 500 lb of carbon were deposited from a total of 21 million lb of flow. Assuming 100% of the carbon impurity that enters the catalyst bed is accumulated within the bed, then the team would have to look for an impurity at ~23 part-per-million (ppm) concentration. If the impurity is hypothesized to be a hydrocarbon such as dodecane (elemental composition = 84.7% C), then the team is looking for 27 ppm dodecane in the process stream. If the impurity is hypothesized to be a polyacrylic acid polymer (elemental composition = 50% C), then the team is looking for 46 ppm of this material in the process stream. If accumulation rates within the bed are lower than 100%, then the concentrations of the impurities that the team is looking for will increase. This example highlights how mass balance data can be used to guide the hypotheses, as well as further analytical studies.

21.5 Conclusion

Analytical data are an important tool in determining the root causes of problems in the chemical industry. It is the project team's responsibility to determine what hypotheses need to be tested with analytical data and what the requirements are for these data to be useful in determining whether the hypothesis can be killed or not. An analytical DQO can be very helpful in striking the appropriate balance between data quality, cost, and speed. Some problems can be solved quickly, or greatly reduced in investigative scope, with the simplest analytical tools. Other problems will require more sophisticated analytical techniques and will require a higher level of expertise to operate the techniques. If higher level analytical expertise is needed, then the analytical DQO can be used to engage those experts to find the optimum solution that fits the analysis and timeline requirements. The most effective use of analytical chemistry is realized when it is used to kill multiple hypotheses at once. This helps narrow the focus of the investigation quickly and leads to faster identification of the root cause(s).

22

Engineering Disciplines Examples

A common property of the problems solved by FPS is that they are highly complex and multivariable. Therefore, FPS usually relies on a multitude of other engineering disciplines. These include Reaction Engineering, Fluid Mechanics and Mixing, Computational Fluid Dynamics (CFD), Separations, Solids Processing, Reactive Chemicals, Heat Transfer, Process Analytical, Process Control, etc. In this chapter, we discuss several engineering disciplines and provide some examples of the application of these disciplines to various problems encountered by the authors.

22.1 Typical Engineering Disciplines in FPS

22.1.1 Reaction Engineering

The Reaction Engineering discipline is obviously one of the most important, the chemical industry involving after all, chemical reactors as the primary unit operation. As Reaction Engineering studies the chemical reactors in general, it can not only help with solving problems with chemical reactors, designed to be chemical reactors, but also with other equipment, not designed to be specifically a reactor. Serious issues with side chemistry products stemming from various problems inside the reactors can be resolved by the Reaction Engineering discipline. It is also extremely important to involve Reaction Engineering experts in the unfortunate cases of uncontrolled reactions.

22.1.2 Separations

Separations is just as an important discipline, as virtually in all cases an industrial reactor is followed by a separation train. The separation train has to perform at the top of its capacity all the time by both satisfying the production demand and the

Fundamentals of Industrial Problem Solving: A Practitioner's Guide, First Edition.
Zdravko I. Stefanov, Eldad Herceg, Carla Schmidt, David M. Jacobson, Dana Livingston, J.P. Chauvel, Sunil Kumar Chaudhary, and Christopher Paul Christenson.

quality specifications. In many cases, the separation train can be a major source of trouble either due to problems internal to the separations train or when being fed streams that have changed impurity profile that makes it impossible for the separation train to do its job. Isomer separations are usually the most sensitive examples due to the inherent problems related to that system (the isomers having very similar physical properties). Separation trains are regularly reactors that haven't been studied well or at all. Many times these trains produce substantial amounts of fouling, for example.

22.1.3 Fluid Mechanics and Mixing (FM&M)

Fluid Mechanics and Mixing goes hand in hand with Reaction Engineering, when it comes to chemical reactor performance. The classical CSTR design assumes perfect mixing, but this is not always the case in practice. Fluid Mechanics and Mixing can help resolve many problems stemming from poor mixing in chemical reactors. Apart from chemical reactor problems, there are other applications, where fluid mechanics is involved. One example is lubrication systems often found in pumps and compressors. Computation Fluid Dynamics (CFD) is one of the tools in the FM&M arsenal.

22.1.4 Solids Processing

Many products made in the chemical industry are solids, such as pellets, crystals, powders, etc. Handling them can be as challenging or more challenging compared to handling liquids. Improperly designed solid-handling equipment can lead to fouling, agglomeration, caking, and pluggages that can paralyze the entire plant. Another issue can be mixing and heat transfer, particularly when a sensitive solid material is heated – localized pockets of high temperature due to poor heat transfer can lead to product destruction or even worse consequences in case of a decomposition. Finally, dust explosions are a very serious danger related to improper solids handling.

22.1.5 Reactive Chemicals

The Reactive Chemicals discipline deals with the dangers related to uncontrolled energy release from chemical reactions. Many plant problems can be due to reactive chemistry, which includes decompositions, deflagrations, vapor explosions, loss of containment due to corrosion, etc. Therefore, Reactive Chemicals gets involved every time there is a change in the process to make sure there will be no inadvertent chemistry that can make the system unstable/runaway and or corrode/destroy the plant. Reactive Chemicals is the discipline that helps to keep the plants safe in daily operation, and even though it is often in the background, it is one of the most important disciplines.

22.1.6 Heat Transfer

As many chemical processes either generate heat or require heat to proceed, heat transfer is very important and many problems may arise from improper heat transfer. Fouling of heat exchangers, for example, can impede heat transfer, lead to increased operating temperatures and even equipment damage. Many separation trains rely on distillation to achieve their goal and distillation is heat driven. Therefore, heat transfer problems in the separation train can easily lead to a complete plant shutdowns.

22.1.7 Process Control

It has been a long time since chemical plants could run without automatic control. There is no way to run a modern chemical plant without automatic process control. There is hardly any process that does not need it. Inventory control, feed flows and ratios, temperature control, composition control are just some major examples. Many processes are not inherently stable either, and there is no way to run them at the desired operating conditions without automatic control. Therefore, poor process control can paralyze a plant. Many times, the process control in the plants is poorly understood and maintained, and this usually leads to myriad of problems.

22.1.8 Process Analytical

One cannot control what one cannot measure. So, in-line process sensors are a prerequisite for a good process control. Process analytical usually deals with the higher level measurements that require a process analyzer, but that discipline can also offer improved basic measurements such as flows, temperatures, etc. Process analytical can enable the solutions of many problems by enabling improved process control via more frequent, more representative and higher quality measurements of various process variables.

22.2 Examples

22.2.1 Unwanted Material Messes Up a Separation Train

In this example, a plant was experiencing high amounts of unwanted material that was causing significant problems in the separation section of the plant. The material was tracked down by taking samples of various streams to develop an impurity mass balance through the process. It was determined that the problem species was mostly produced in the reactor, i.e. it was a side product.

The conditions required to make the side product were high residence time and poor mixing. The reactor design was inspected, and a hypothesis was generated that the current reactor internals were not optimal and had potential stagnation zones (high residence time and poor mixing). A classic approach to resolve this hypothesis would be a tracer test and measurement of the RTD (Residence Time Distribution). In this case, however, tracer test was not possible due to various issues (highly toxic and corrosive chemistry, safety concerns with sample collecting, product contamination, etc.). To evaluate the hypothesis without an in-process experiment, the alternative approach was to develop a CFD model of the reactor. With the help of the CFD model, it was confirmed that the current reactor design produced an RTD with a long tail that was likely to produce significant amount of the side product. Then, several different modified designs were proposed and quickly evaluated via the CFD model. In this way, a much-improved design was identified, which required minimal capital investment. The design was implemented and the amount of the side product decreased drastically, leading to significant improvement of the plant performance. In this case, Reaction Engineering, Fluid Mechanics and Mixing, and CFD were used in tandem to solve the problem.

22.2.2 Isomer Tower Suffers from Poor Performance

In this case, a packed tower that separated two isomers was suffering from poor performance compared with a historical baseline performance for no apparent reason. It was a long-term issue, and it was assumed that this was the new "normal." When the business climate changed and the plant needed to make much more of one of the isomers, the tower could not deliver the new capacity demand that had previously been possible. The FPS team reached out to several experts in separations. The first step was to build a model of the tower and compare ideal performance vs. the actual performance in the plant. As expected, significant discrepancy was found, i.e. the tower should have been able to do the job. After this determination, the team generated several hypotheses, including maldistribution and inerts loading. The plant did gamma scans and did not find anything suspicious, so maldistribution was not considered initially. However, the plant did simple chord gamma scans that are powerless to detect a radially symmetric (donut like) maldistribution problems. A CAT scan performed on the tower indicated a donut-shaped maldistribution problem. When this was accounted for in the model, there still was unexpected difference in performance. Therefore, the other potential problem, inerts loading was investigated. All instruments in this system were nitrogen purged. A significant leak of nitrogen can lead to inert loading that can significantly decrease the tower performance. A walk in the field revealed an instrument tap from which the instrument was long ago removed, but the nitrogen purge was still on and going full blast. That discovery and several other instrument purges running high accounted for the rest of the performance discrepancy. Closing the nitrogen purge valve on the old instrument tap improved the tower performance by 15%. In this case, Separations was the main discipline used to solve the problem. It is also important to note that the field walk is regularly needed when solving problems. The process owners may be acclimated to the present state and issues like the N_2 flow may not be identified, even when a remote FPS person specifically asks. The "fresh eyes" a new person brings can see issues that can be missed by people who are acclimatized to the present condition.

22.2.3 Continuous Crystallizer Is Having a Hard Time

In this case, a continuous crystallizer plant was having serious problems in the back end where the crystals are separated from the mother liquor and then dried and packaged. The back end was constantly plugging leading to costly shutdowns and continuous frustration. The initial investigation discovered that the back end was getting clogged by fused crystalline mass. This coincided with poor centrifuge performance. Further investigation discovered that the crystal size distribution during the times of problems has a long tail toward fine crystals. The fine crystals generally do not drain well in centrifuges, and this leads to higher than normal amount of mother liquor retained in the centrifuge cake. This cake then fuses in the back end, leading to complete plugging. The main question was why the crystallizer makes fines. There are various reasons why this could happen and one is impurities. Detailed analysis of the incoming streams initially did not discover anything abnormal. But the fundamentals indicated there had to be something wrong with the feeds, so the team kept on digging. When the analytical method used to analyze the incoming stream was investigated, it was found out that the lab standards were 10 years old and the method was using the wrong peaks in the chromatogram. The faulty composition analysis created a significant bias in the results. Correcting backward in time for the errors, it was discovered that indeed there was a significant amount of impurity in the feed. Digging even further, it was discovered that the plant that generates the feed to the crystallizer requires very tight ratio control at its own front end to keep that impurity in check. The flowmeters responsible to measure the flows of the reactants were redundant (two flowmeters per stream). On closer inspection, it was determined that one flowmeter drifted significantly off. On top of this, it was reading slightly higher than the healthy flowmeter. Since the control code was made to pick the higher reading out of the two due to safety reasons, it consistently picked the wrong flow rate. The data indicated the flow rate was fine, but the actual flow rate in the process was much lower. This let the plant believe that everything was fine, but it was in fact the perfect storm of wrong measurements – the flowmeter and the analytical being both wrong and canceling each other, so everything looks great on the outside. Various disciplines were utilized in this highly complex case, including Solids Processing, Process Analytical, Process Control, Reactive Chemicals (to check crystal properties via Thermo Gravitation Analysis), etc.

22.2.4 Plant Makes Smelly Solvent

In this example, a plant making an industrial solvent for years without any issues suddenly started to make smelly product due to trace contamination of a particular sulfur compound. Turning various knobs around the plant and the separation train did not lead to any improvement, and things became worse with time to the point no product could be sold due to the disagreeable odor. A sampling campaign and a mass balance of the smelly component revealed a large recycle loop in the plant that was accumulating it. The next steps involved the tracking of the sources and the drains of the smelly compound. It was then revealed that activated carbon beds installed at the plant feed lines were removed several years ago due to cost-saving activities. Similar cost-saving activities led to the recycle back to the process, a small stream that happened to be the purge of the smelly compound. So after "costs were reduced," the smelly compound precursor (sulfur) was allowed to enter the plant and at the same time the purge was recycled back to the plant. After releasing the purge back to waste, the problem slowly disappeared, as the smelly compound now had an exit point and could leave the plant without accumulating in the final product. In this case, one major discipline was involved to solve the problem, Separations, but on multiple fronts. It involved a mathematical model of the separation train to see where the smelly component will end up, a study of the performances of various activated carbons as a final product polishers, and another activated carbon study to evaluate the sulfur-eliminating efficiency of the feed carbon beds.

22.2.5 Batch Solvent Stripper Optimization

This example shows a case where the solution is quite simple, if one uses the fundamental disciplines appropriately. A batch solvent stripper was having a hard time making production rates due to long cycle times. An investigation revealed that the process control system asked the stripper temperature to ramp up to a certain value and then hold it. The ramp and hold job was assigned to a PI controller that was having an extremely hard time meeting the set point target smoothly. Typically, the controller would overshoot the ramp, then a control system interlock blocked further heating for 20 minutes to let the temperature go down below a threshold and then the story repeats itself. This led to substantial delays in the stripper operation. Instinctively, one may think this is a control-tuning problem.

Indeed, this is a control problem, but not a tuning one. In fact, the system is inherently stable due to it being a boiling system. (Boiling occurs at constant temperature for pure components. For mixtures, there is a boiling point elevation, but the temperature is bound by the boiling point elevation curve.) After quick check of the product temperature sensitivity and the capacity of the solvent condenser, it was determined that the system can handle all of the solvent vapor produced with the steam valve almost completely open (80%). The problem was ultimately solved by eliminating the fancy PI ramp controller and instructing the control system to simply turn the steam valve at full blast (80% open) right at the start of the stripping cycle. The boiling point elevation provides a nice temperature ramp all by itself and the temperature never reaches values that can lead to product damage (the final spec for the solvent in the product allows the stopping temperature to be well below product damage temperature). In this case, Heat Transfer and Process Control were the main disciplines involved, and when properly used, they solved in about an hour a problem that was plaguing the plant for months. No one could figure out why the stripper was configured with a ramp and a PI controller to maintain it in the first place.

22.3 Summary

In summary, in the authors experience, it has been clear that most of the hard problems encountered in the chemical industry are very complex and cannot be resolved by using a single engineering discipline. This requires a good operating knowledge of the engineering disciplines and good connections to the corresponding subject matter experts (SMEs). One person cannot cover all engineering disciplines in depth, but it is important to recognize which ones are needed and get them involved. Another important topic is the maintenance of sufficiently advanced expertise of these disciplines. Different organizations may have different ways of achieving this, but for efficient problem solving, it is important to maintain that capacity.

23

Six Sigma, LEAN, and Continuous Improvement

The introduction of Six Sigma and soon after LEAN into many companies has run through a progression of steps and mis-steps in many companies. Six Sigma and LEAN methodology, when sensibly implemented, is very valuable to the problem-solving process. Fundamental problem solving is complementary to Six Sigma and LEAN. As illustrated in the problem-solving continuum, many problems are simple enough to be resolved with one of these approaches.

Typically, a LEAN/Six Sigma (LSS) project is approached empirically without fully considering fundamentals. For single effect or simpler problems, this often leads to an improvement. Much of the criticism for these methods stems from a tendency of using the familiar approach rather than using the best suited approach. Misapplied LEAN can lead to projects that are analogous to eating unpeeled green bananas.

The methods are valuable and have a place in problem solving. The training and systematic approach is valuable to many employees as they develop. In an ideal implementation, there will be a sensible evaluation of which approach is suitable, be it LEAN or Six Sigma.

23.1 LSS Comparison with FPS

When introducing FPS to an audience, a frequently asked question of the FPS authors is how FPS is related to Six Sigma or LEAN or the scientific method? We will use the LEAN Six Sigma label "LSS" in the discussions that follow. When teaching FPS, the more pointed question we can hear, "isn't FPS the same as Six Sigma?" The authors, who include two Six Sigma Master Black Belts, agree that all these approaches (LEAN, DMAIC, and FPS) are at their core similar in many aspects; however, as typically practiced in industry, they are not the same. In

Fundamentals of Industrial Problem Solving: A Practitioner's Guide, First Edition.
Zdravko I. Stefanov, Eldad Herceg, Carla Schmidt, David M. Jacobson, Dana Livingston, J.P. Chauvel, Sunil Kumar Chaudhary, and Christopher Paul Christenson.

principle, the methods can be very similar, but in accepted practice, there are key differences. All the methods utilize a "plan – do – check – act" approach. All recommend the selection and use of different organizing, prioritizing, and communicating tools and approaches. Most of these tools are "best practice" and very helpful when used correctly. Generally, the tools can have significant overlap of actual function. At the same time, all the methodologies emphasize that not all tools will be appropriate to any problem.

> ***Wisdom Tidbit – If all you know how to use is a hammer, everything looks like a nail.***

LEAN Six Sigma trained and certified Green Belts and Black Belts are excellent candidates for FPS training, thanks to the emphasis on many valuable problem-solving behaviors which can be taught well in a Six Sigma training program.

Here, we reiterate the phases of the FPS process in Figure 23.1. Tables 23.1 and 23.2 below compare DMAIC key steps with FPS main steps. We feel in principle that DMAIC and FPS are indeed highly aligned. When teaching FPS, we generally don't use a detailed deliverables list as we do when teaching DMAIC. However, the DMAIC best practice is arguably a universal project management best practice. Thus, for most FPS projects, the Project Leader and core team will likely follow many of the same steps. What we feel are the key differences are noted in the Tables 23.1 and 23.2.

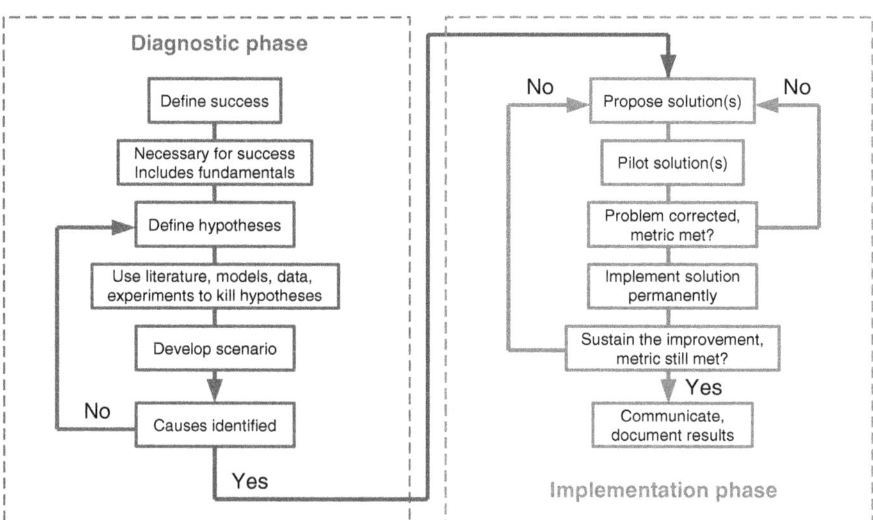

Figure 23.1 Phases of the FPS process.

Table 23.1 Comparison of DMAIC key deliverables with FPS.

DMAIC Phase	DMAIC deliverables	FPS Phase	FPS deliverables
Define	Team Charter	Diagnostic	Team Charter
Define	Action Plan	Diagnostic	Same
Define	Process Maps	Diagnostic	Same
Define	Quick Win Opportunities	Diagnostic	Move in the direction of good
Define	Critical Customer Requirements	Diagnostic	Describe Success
Define	Prepared Team	Diagnostic	Same
Define	Financial Impact	Diagnostic	Burn rate estimate
Measure	Input, Process and Output Indicators	Diagnostic	Learn the process
Measure	Operational Definitions	Diagnostic	Learn the process and fundamentals
Measure	Data Collection Formats and Plans	Diagnostic	DQO (Hypothesis kill driven; done after hypothesis are proposed)
Measure	Baseline Performance	Diagnostic	Same
Measure	Productive Team Atmosphere	Diagnostic	Same
Measure	Measurement System Evaluation	Diagnostic	Same
Measure	Current Sigma Level	Diagnostic	Describe Success

FPS uses two documents to formally write down planned work. The two documents have similarities in the questions that need to be asked and answered. One document is the overall project charter and a project will normally only have one charter document. Important note here is that the project charter is also a DQO, as it incorporates the three main FPS questions (What is success? What is necessary for success? and What is rate limiting?). The second kind of document will be a question or problem-specific DQO. The main points are:

- Both documents have many similar sections
- Different target audience and purposes
  - The charter is for core team members to have agreement with stakeholders.
  - The specific DQO is much less formal and concentrates on a specific problem or question in the hypothesis testing step and aims to get data about that. An excellent specific DQO example is given in Table 23.3.

Table 23.2 Analyze – Realization steps compared with FPS.

DMAIC phase	DMAIC deliverables	FPS comments
Analyze	Validated Root Causes	Validated Fundamental Root Causes
Analyze	Key Input/Process Variables	Same
	Validated Targets for Financial, Customer, and/or Business Impact	Same
Analyze	Finalized Project Goals	Describe Success
Improve	Validated Improvement and new Sigma level	Propose/Pilot Solution(s)
Improve	Change Management Strategy	Same
Improve	Validated Improved Performance for Financial, Customer, and/or Business Impact	Same
Improve	Implementation Plan	Same
Improve	Leverage Opportunities Identified	Same
Improve	Validated Key Input/Process Variable Limits	Same
Control	Control Plan Implemented	Sustain the Improvement
Control	Project Transfer to Process Owner	Same
Realization	Demonstrate Successful Performance	Same
Realization	Final Documentation – Portfolio/Storyboard/Report	Document and communicate
Realization	Project Audit/Key Learnings	Same

The more important distinctions between DMAIC and FPS are made in how the approach is commonly applied. In the end, a key difference can be distinguished as empirical versus fundamental (see Figure 23.2). Six Sigma does not prohibit the use of fundamentals, nor even directly recommend fundamentals be ignored. As commonly practiced, LSS looks to develop correlations between inputs and responses, and this approach frequently results in an empirical solution. Doing so is an acceptable and efficient way to improve or solve many problems. We support trying an empirical approach first in many cases as illustrated on the Problem-Solving Continuum (see Figure 1.1). Having an established LEAN Six Sigma program in an organization teaches valuable skills, develops important problem-solving experience, and solves many problems without the need to escalate the problem to a larger or more experienced team.

Table 23.3 Specific DQO example.

Background	Samples were pulled prior to washing the packaging equipment throughout the process. We are interested in understanding what material is accumulating and the appearance of those samples
Define success	Understand particle shapes through microscopy (optical)
	Understand accumulation based on MW. NOTE: Samples may not be homogeneous
	Review impurity data to see if different points in the process have different impurity levels
What is necessary for success?	Microscopy to see if we can see agglomerates (5–10 micrographs/sample plus extras for "interesting" features)
	GPC to look at MW to understand the source of buildup
	HPLC – Impurity levels
What is rate limiting?	Sample locating; Analysis
Samples	1. Main Chamber – South Manhole
	3. Cyclone 2 – rotary valve
	5. Baghouse – West Manhole
	6. E-Conveyor
	8. Bagger line – Vibrating bin
	All samples are expected to be heterogeneous and clumpy. Team would prefer to dissolve several (at least 3) parts of the sample (~10 g) for the GPC and LC analyses. Could we pre-dissolve in ACN or THF and then aliquot from that same solution for the LC analysis?
Resources	**In Building D6401:**
	Bonnie Jennings – Sample Location, Maximillian Mendez – Impurity analysis, Felix Dixon – GPC
	In Building D6403:
	Elizabeth Donaldson – Microscopy
Timeline	Microscopy by 11/20; Other analytical by 11/20?
Consequences of acting (What is the decision to be made?)	Potential to impact hypothesis around mechanical energy
Value	$$$$ – Business Critical
Experimental details	TBD by SMEs, described loosely above
Author(s)	Philip Walker, Eliza Le, Roy Brennan, Aleena Gibbons
Date	4 November 2020

Six Sigma MAIC	FPS
Empirical process (based more on experiment and observation rather than theory or scientific principles).	A problem-solving implementation of the scientific process, highly grounded by scientific principles such as chemistry and physics.
Applicable to solving manufacturing problems (yield, product quality, etc.) and for facilitating root cause investigations of serious unplanned events.	General-purpose problem-solving approach applicable to large variety of problems, such as manufacturing, supply chain, safety, and so on.

Figure 23.2 Compare typical Six Sigma work with FPS.

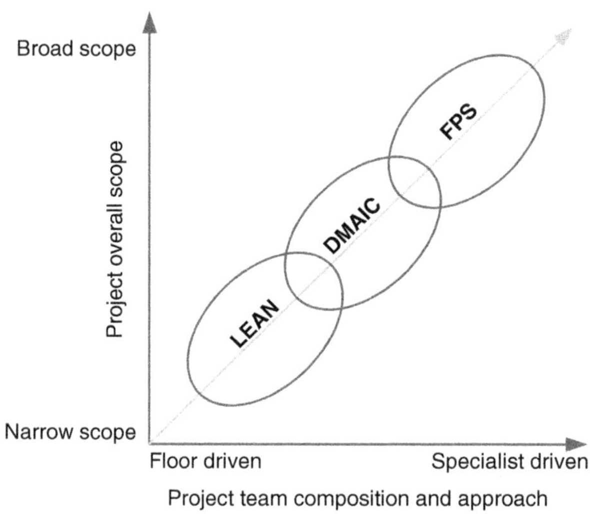

Figure 23.3 Comparison of problem-solving approaches.

Another way to compare LEAN, Six Sigma, and FPS is shown in Figure 23.3. It mimics the Problem-Solving Continuum from Figure 1.1.

23.2 Criticisms of LSS in Industry at Large, Beyond FPS

In the last decade, significant criticisms have emerged within our parent organizations and in industry at large.

- A widespread LSS training and certification program is too expensive to deliver
- LSS did not return enough benefits to pay for the costs
- A one-size-fits-all implementation approach is too rigid and inflexible to the needs of parts of the organization.

Consequently, many firms and organizations are reducing or eliminating their LSS programs. While the high resource commitment some firms While the high resource commitment some firms made that lead to the criticisms above was arguably valid, in most cases, there appears to be an overreaction leading to under sizing or even elimination of all LSS works.

Our criticisms are with the implementation, not the methodology.

We believe that a correctly sized and implemented LSS program is valuable and necessary to sustain an optimally capable organization over the long term. We are fans of the methodology within these boundaries.

- **Too Extensive.** Our criticisms are largely with the implementation plan followed in many cases. If too large, a fraction of employees is "run through the training" and required to get training or certification for promotion that will lead to many counterproductive behaviors. The people participating in many cases were not truly interested or engaged and just went through the motions to comply with management instructions. That results in high expenses for training delivery and participation without enough return on investment to the parent company.
- **Canceled Programs.** Unfortunately, many firms that began perhaps with too large a people commitment to an LSS program have followed with the reduction and in many cases elimination of their LSS programs entirely. The valuable skills and problem-solving methods taught within an LSS program are missing from the development of many newer employees.
- **Not Integrated.** In some companies, LEAN and Six Sigma were initially implemented separately and not managed in an aligned cooperative approach. After years, the programs remained separate and not integrated in management nor in subject matter experts. People within either the LEAN or the Six Sigma programs lacked adequate cross training and had severe bias to apply their familiar approach when a different approach was likely much more suitable to a specific case.
- **Poor Fit.** An additional counterproductive outcome occurred when all departments were required to implement Six Sigma. In that case, work functions like engineering or R&D tried to find defects in their work processes. This was a significant waste of time in most cases and resulted in "make work" projects that did not create value.

Below are given some industrial fables as examples of poor implementation of LSS.

- A plant where the rate-limiting step was a batch-curing process needed to increase production as demand exceeded demonstrated capacity. When a LEAN Consultant worked with the plant on increasing capacity, it was decided to end the batch earlier while ignoring the need to complete the curing.
- In the same plant, a DMAIC green belt trainee did not consider reducing a non-value-added step. In the batch-curing process, a sample and analyses step is in place from time immemorial. It requires a 10% delay of the whole batch step time before proceeding to the next step. A key question needed was "how often is the result of the analysis used to change how the material is managed?" In this case, the batch is always forwarded, regardless of the analytical result. The actual purpose of the analysis is not to determine the end of the batch has been reached, rather the analysis is used to document composition. Thus, the delay can be eliminated.
- Again, in the same plant where a LEAN SME chose to shorten a curing step that cannot be shortened arbitrarily and a DMAIC Green Belt does data analysis to look at existing data and finds nothing to change, an FPS SME learns the process, reads the literature, and applies the Theory of Constraints to (i) reduce the curing cycle time by adding a curing accelerator and (ii) eliminate the analysis wait time.

23.3 Recommendations for the Problem-Solving Community in a Company

Develop an integrated problem-solving approach including LSS and FPS, as summarized in Figure 23.4.

- Continue to train and develop LSS practitioners selectively. Right size the program avoiding the too broad error, avoid making it mandatory eliminating that trap, and be flexible as applied to different functions to meet their specific needs. Be selective and develop enough people as problem solvers to sustain the talent pool of problem solvers AND solve the flow of less complex problems.
- Solve many problems via sound problem solving and then LSS as illustrated in the Continuum Curve.
- Teach FPS, the purpose of this book.
- Teach an integrated selection method to select what problem-solving approach to apply to any problem following a selection process illustrated in Figure 23.4.

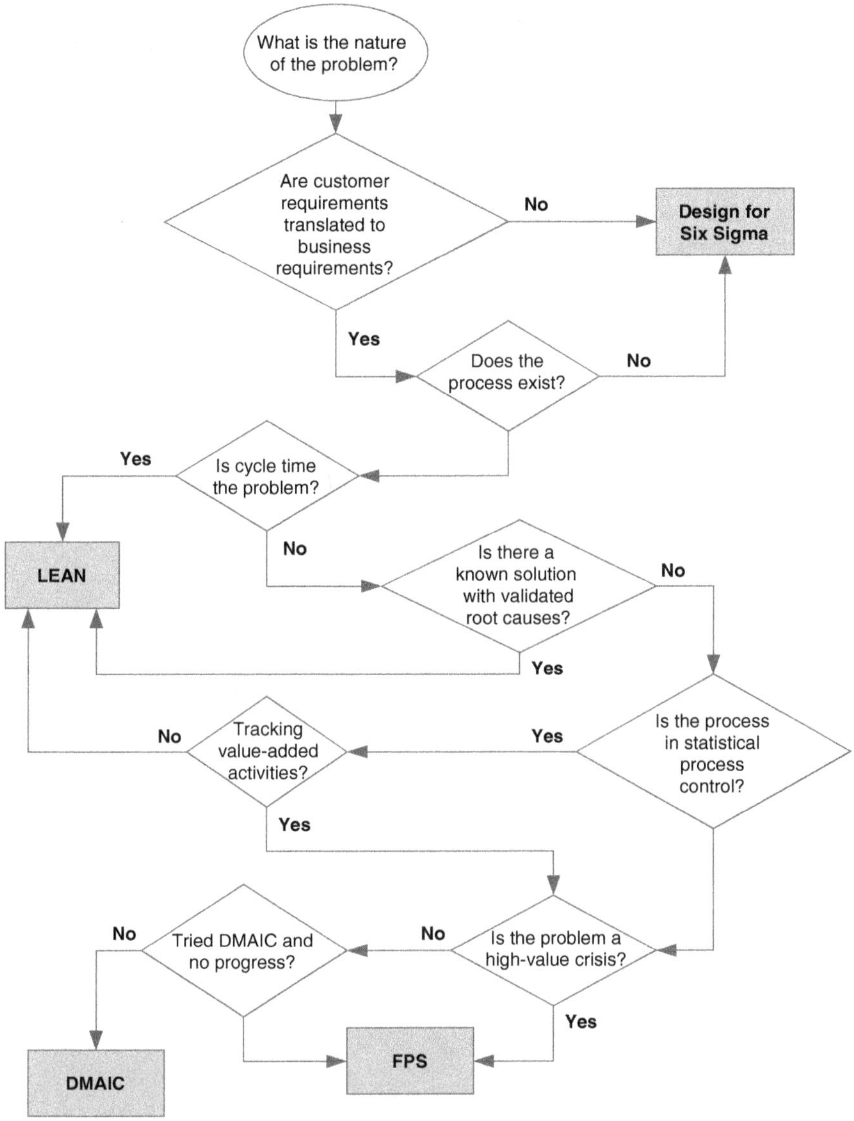

Figure 23.4 Problem-solving selection process.

We suggest that the ideal result is for an organization to have enough people with enough training and experience to start problem solving with a purposely selected initial likely approach. Then do an early validation of approach to formally consider if the initial approach is actually preferred and if indicated change to a different more valid approach.

In many cases, the most effective approach to solving a problem will be a combination of LEAN, DMAIC, and FPS where the fractions vary to suit the project. Most FPS projects will be heavy in FPS, but some projects will contain elements of LEAN, Six Sigma, or even Design for Six Sigma. As an example, an FPS project found that a contaminant caused by the washout procedures in a key raw material was causing unwanted side chemistry. The FPS team chose to set up a Kanban (LEAN tool) using dedicated tankers to eliminate the washout procedures. This solved the problem and demonstrates how methodologies and tools should be mixed where needed to maximize the impact of the project.

24

Theory of Constraints

Theory of Constraints (TOC) is an improvement philosophy that first identifies the critical path within a process, then identifies the limiting constraint along the critical path, and finally prioritizes efforts and resources to improve the rate-limiting step so that it is no longer the limiting factor to the system. A TOC overview is given in (Figure 24.1).

The TOC approach is usually attributed to being taught in the book, *The Goal*, by Eliyahu M. Goldratt. We won't attempt to teach the methodology of applying TOC to projects in general. TOC is taught well from multiple sources the reader is encouraged to reference and study. A bibliography is found at the end of this chapter. What is important within the context of this book is to understand TOC well enough that you apply the approach in the management of your problem-solving project. The fundamental problem-solving approach emphasizes integrating TOC in the project (Figure 24.2). Fundamental problem-solving approach is often being used on projects that have a high burn rate. They are costing a lot of money. Putting excessive work into non-bottle necks, non-rate-determining steps will be very costly indeed. Knowing how to avoid this pitfall and focus on the rate-determining step will make your project cheaper, better, and faster.

In a simple form, the approach taught by Goldratt is to identify the rate-limiting step and focus efforts and resources on that step to improve it. In Figure 24.3, a simple linear process is shown. The numbers displayed are the monthly throughput each step is presently producing products. Clearly, step (B) is the slowest step. If either steps A or C are improved, the whole process time will not be improved. The whole process throughput continues to be determined by step B.

> ***Wisdom Tidbit – Often if a system lacks accurate metrics or the metrics are out of date, the team could work on improving the wrong step.***

Fundamentals of Industrial Problem Solving: A Practitioner's Guide, First Edition.
Zdravko I. Stefanov, Eldad Herceg, Carla Schmidt, David M. Jacobson, Dana Livingston, J.P. Chauvel, Sunil Kumar Chaudhary, and Christopher Paul Christenson.

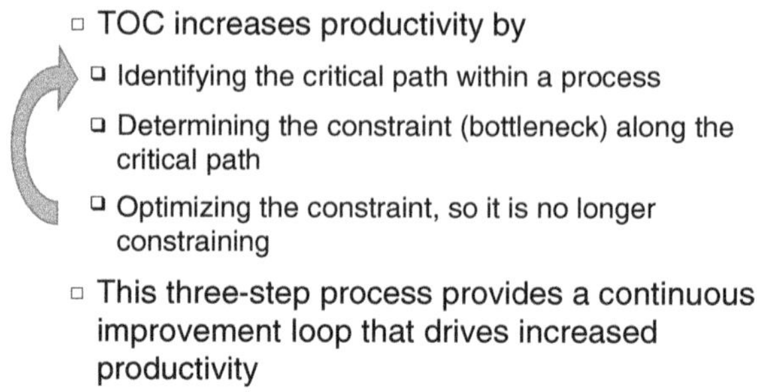

Figure 24.1 Theory of Constraints overview.

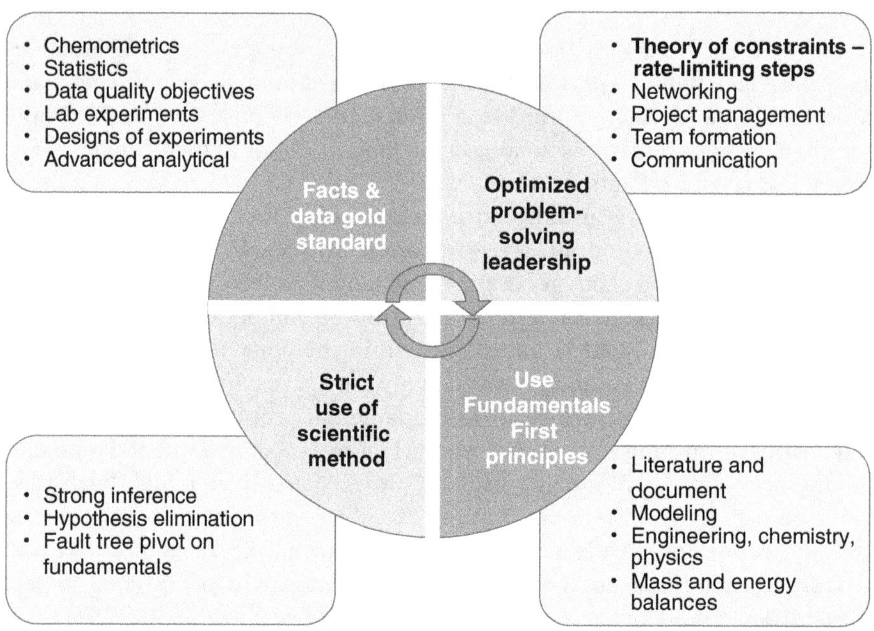

Figure 24.2 ToC place in the FPS framework.

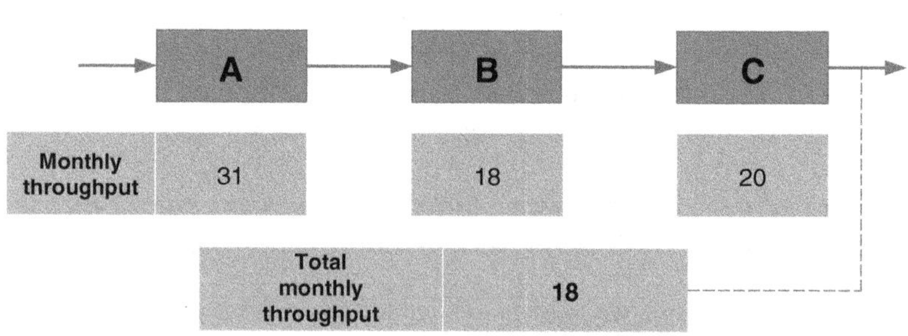

Figure 24.3 Simple linear process.

Fable: When my boss ran this process two years ago, the RATE-LIMITING STEP was A, so he insists that the team work to improve A now. Once we developed accurate metrics and shared these with him, we changed the priority to B, since today that is the RATE-LIMITING STEP.

A more complex example is to include the reliability of each step with the instantaneous capacity of each step, labeled as the unit sprint rate. In this example, there is only one path so that is the critical path. Then you need to identify the slowest step, the rate-limiting step. Teams frequently overlook the impact of reliability or undervalue reliability in prioritizing the improvement process. Consider the example in Figure 24.4, where the apparent RATE-LIMITING STEP is step B, if we ignore the low reliability of step A. However, if we consider the actual throughput, then the RATE-LIMITING STEP is A, more specifically, the RATE-LIMITING STEP is the low reliability of Step A. In this example, the team needs two complementary metrics, Sprint Rate (the rate the step can produce if it has 100% reliability) and Reliability (the percent of the time the system is operating when it is needed to operate).

Often identifying the preferred step to concentrate improvement efforts can be challenging even in apparently simple systems. The system constraint is the least productive step.

> ***Wisdom Tidbit – Increasing the sprint rate of a low reliability step is often not helpful. Instead, if less effort can increase the reliability, the capacity gains can be better. Often the team may want to work on improving the reliability of low reliability steps and later improve the sprint rate of slow steps as the true RATE-LIMITING STEP moves back and forth in the process due to successful improvements.***

An FPS team works to increase the production rate of the process in Figure 24.4 while applying the TOC methodology. The baseline condition is that Step A is the rate-limiting step with a capacity of 14. Apparently, the sprint rate of Step A is very high, while the reliability of Step A is poor. For Round 1, the team determined that the reliability of Step A needed to improve. Work was done, which improved Step A reliability to 90%, a doubling of the baseline reliability. This changed Step A from the slowest step to the fastest step. Now the rate-limiting step has moved, but to where? As illustrated in Figure 24.5, steps B and C appear equivalent in capacity after Round 1. Should the team focus on improving Step B rate, Step B reliability, Step C rate, or Step C reliability?

The team agreed to prioritize improving the reliability of Step C. This selection was done expecting the improvement of Step C reliability was easier and faster to complete than the other options. After Round 2, the Step C reliability was increased to 90% and the Step C capacity was now 18, as illustrated in Figure 24.6. After Round 2, the rate-limiting step was B, at 16. Now the team needed to

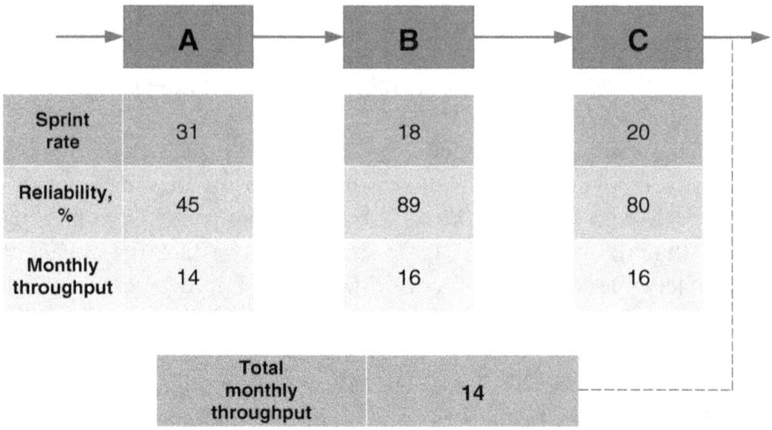

Figure 24.4 Process capacity baseline condition.

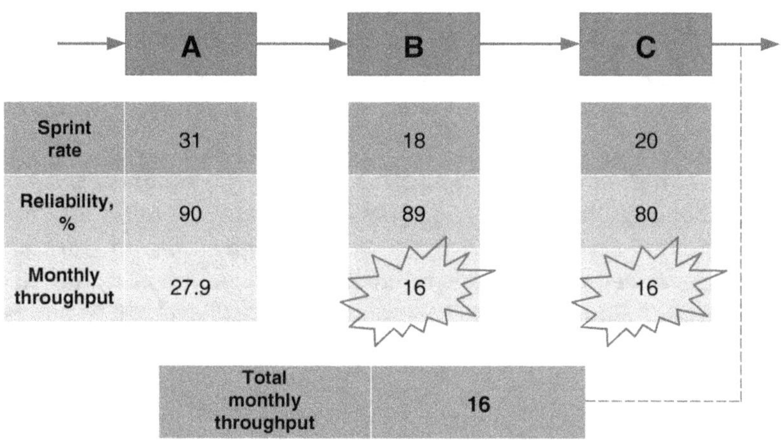

Figure 24.5 Results after Round 1 – Step A reliability improvement to 90%.

determine what to work on within Step B, the B sprint rate or the B reliability. The team agreed to improve the sprint rate, and as a lucky coincident, the Step B reliability increased slightly due to the same changes.

Figure 24.7 shows the increased system capacity after Step B sprint rate was improved along with a minor improvement in the Step B reliability. After Round 3, the system throughput was increased to 18 and the rate-limiting step was shifted to Step C.

The example above illustrates the need to iteratively measure the improvement metrics as improvements are made in the process. The rate-limiting step moved from Step A to Step C, then to Step B, and then to Step C again. By identifying the

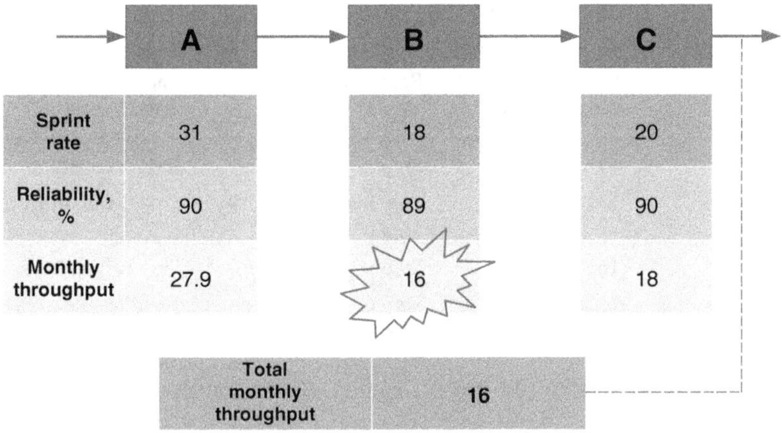

Figure 24.6 Results after Round 2 – Step C reliability improvement to 90%.

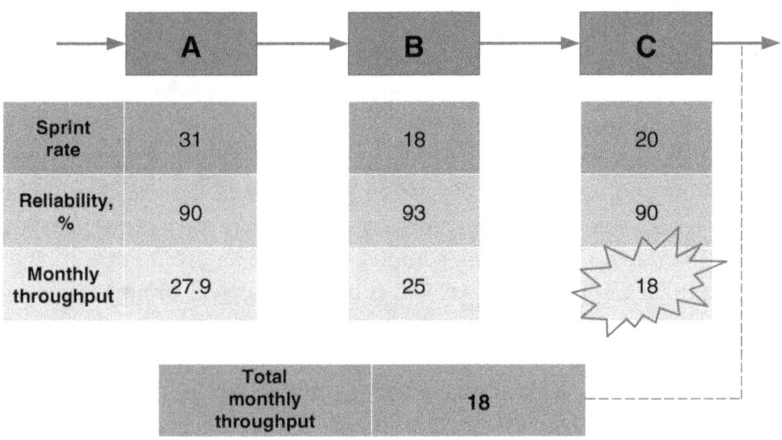

Figure 24.7 Results after Round 3 – Step B sprint increase.

present rate-limiting step and focusing improvement efforts on the rate-limiting step, the system capacity is optimally increased (Table 24.1).

> ***Wisdom Tidbit – The authors have experienced cases where changes in other parts of the system due to inter-relationships or due to random events caused both the critical path and the RATE-LIMITING STEP to move. Only regular re-validation of the performance metrics of the system can prevent prioritizing work to improve the non-RATE-LIMITING STEP.***

Table 24.1 Summary of Improvements during three rounds of improvement work.

Improvement round	System capacity	Rate-limiting step	Improvement priority selected
Baseline	14	Step A	Step A reliability
Round 1	16	Step B and Step C	Step C reliability
Round 2	16	Step B	Step B sprint rate
Round 3	18	Step C	Need to select

The three best questions that FPS project leaders will constantly ask throughout the life of the project are:

- What is success?
- What is necessary for success?
- What is rate limiting?

The third question is the one used to drive TOC thinking into the project teams. Other chapters discuss questions 1 and 2. Here is an example of how TOC thinking change the course and speed of a project.

A team was working very hard on a big costly problem. The entire waste treatment plant for a major chemical complex was on the verge of shutting down. The limits imposed by law on organics in the wastewater were very close to being exceeded. Shutting down the chemical complex would be an event large enough to warrant shareholder notification. An FPS project leader was brought into the project with the purpose of bringing a more systematic science-based project management approach.

During the first meeting with the new FPS team leader, the three questions were put to the team. To answer the first question, they decided that success was that the wastewater treatment plant would return back to historical organic levels and avoid exceeding the legal limits. When asked what was necessary for success, they decided that they needed to understand why the "bugs" (microbes) that are used to consume and control the organics level were failing to do their job. Then the FPS project leader asked the last of the three best questions. What is rate limiting?

The team after much back and forth discussion struggled with the question. The team was frustrated and not sure why the question was even being asked. The project leader reframed the question and said that what was necessary for success was more knowledge. What is slowing down our getting more knowledge? After much discussion, the engineer assigned to the plant said, "I know some things about these microbes because I have worked supporting the waste treatment plant for a few years, but I am no expert in bugs. For me, my personal rate limiting step would be time to study more about what causes bugs to work better or worse." The project leader smiled and asked everyone there how many of them already had studied microbes for years? The silence from the team answered the question. There were no microbiologists or specialist on the team. Immediately, the rate-limiting step was understood soon after the sponsor and project leader hired a consultant and also flew a corporate expert into fill that gap and get the right people there as quick as possible. Although the existing team had worked on trying a number of known levers to try and improve the situation, they had not been thinking about what was really holding them back. The FPS project leader helped them understand what was rate limiting and used TOC thinking to speed up the project.

When introducing the TOC approach, people often more easily understand the TOC concepts for tangible systems, especially within their normal areas of work. For example, engineers understand, when looking at a flowsheet, the idea to identify the bottleneck in a process. Construction project managers understand looking for the critical path and slowest step in managing their project schedule. Chemists understand the rate-limiting step in a multiple step chemical reaction. The opportunity for this chapter is to apply that approach and do the same analysis to problem solving. People often don't understand and consequently don't apply TOC to intangible concepts, which is to say knowledge, facts, or calculations you are missing.

> ***Wisdom Tidbit – In your team, you need to regularly ask "What is rate limiting our understanding? Our knowing?"***

For tangible processes, the TOC approach is already summarized in Figure 24.1. Success for this chapter is for the reader to transfer TOC thinking from the tangible aspects of their project to the intellectual aspects of their project.

> ***Wisdom Tidbit – In fundamental problem solving, generation of first principles fundamentals that govern the problem is of paramount importance – particularly to assure that the problem does not reappear later. When a team is stuck, always return to first principles – what are the governing laws of physics and engineering behind the problem? Then, develop these into hypotheses. This chapter discusses TOC applied both to knowledge generation and hypothesis killing.***

For knowledge and understanding process and problem solving, the TOC approach needs to change its focus to identifying and getting the knowledge that is missing, which when available will speed the teams progress.

- TOC increases problem-solving speed and efficiency by
  - Identifying the critical knowledge path within a problem
  - Determining the missing knowledge/fact constraint (bottleneck) along the critical path
  - Prioritizing resources and efforts to get the priority knowledge so that it is no longer constraining
- This three-step process provides a continuous improvement loop that minimizes the project schedule and resource costs

Figure 24.8 TOC applied to problem-solving speed and efficiency.

As summarized in Figure 24.8, TOC increases problem-solving speed and efficiency by:

First – Identifying the critical knowledge path within a problem (e.g. composition of a sample, consultation with an SME, completing a deep dive into applicable reference literature, measuring a flow rate, completing a mass balance, building a Hypothesis Onion, etc.)
Second – Determining the missing knowledge/fact constraint (bottleneck) along that critical path.
Third – Prioritizing resources and efforts to get the priority knowledge so that it is no longer constraining.

TOC plays a role in hypothesis testing as well. In a typical project, 10–15 independent hypotheses are usually generated. These major independent hypotheses might have several "dependent" hypotheses – close cousins, listed underneath a main hypothesis.

Typically, the project team does not have the resources or "bandwidth" to evaluate all the hypotheses simultaneously. The following approach is recommended. The hypotheses are placed in the following buckets:

- **M:** M means MUST EVALUATE NOW. These are the rate limiting, critical few hypotheses. The team should allocate itself only 3–5 Ms.
- **E:** E means it is really easy to kill this hypothesis; <1 day's work. Examples are read the literature, talk to an SME who can give you the answer straight away, look at the process data.
- **L:** L means do later. As Ms are killed and knowledge is generated, it is often necessary to elevate some Ls to Ms or even to gather the team and generate additional new hypotheses.
- **D:** D means Dead. Avoid tendency to keep discussing or working on hypothesis that have been killed.

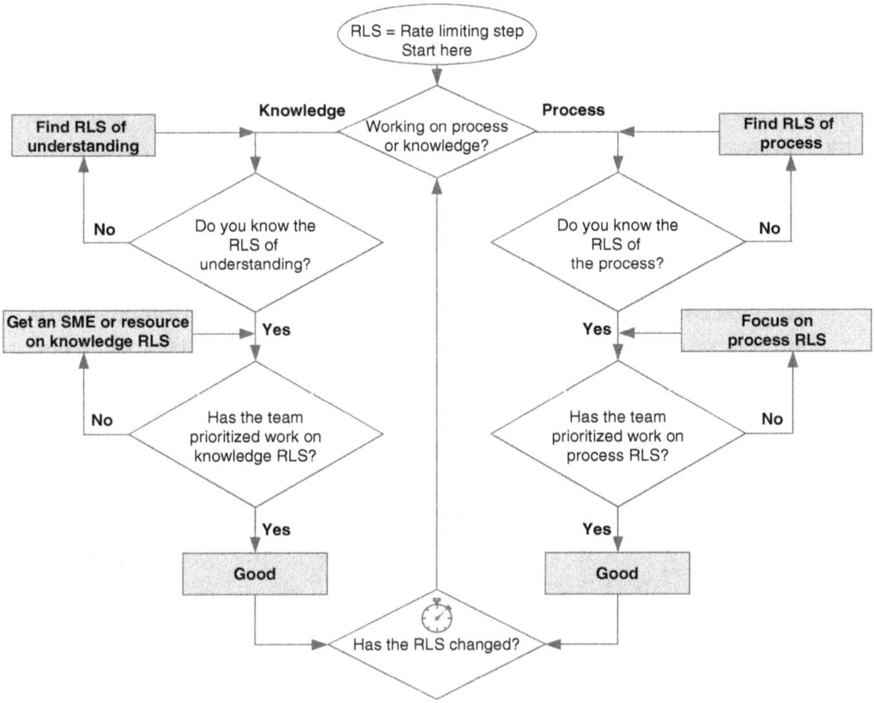

Figure 24.9 List of steps to follow in case of tangible and intangible rate-limiting step.

A very good side-by-side comparison of the list of steps to follow in case of tangible and intangible rate limiting step is provided in Figure 24.9.

Teams often don't prioritize working on the steps that are the most rate limiting to solving the problem. Why does a team work on the knowledge and information steps that are not the RLS?

- Some teams work on the steps that are easy first. This generates an appearance of progress. It is important to realize that activity is not necessarily the same as effectiveness.
- Some teams have favorite ideas and hypotheses that have been handed down over many years and tend to refocus on those rather than digging into the really hard work that is required to fundamentally understand their problem (system).
- The team does not have metrics for the knowledge work. One way this is identified by simply asking, "What is most limiting to the team, not the process or tangible system, but the team?"
- The team has metrics that are out of date. Progress has been made, what was previously rate limiting is gone, and the team fails to identify the next rate-limiting issue.

- Have up-to-date metrics but are not applying TOC thinking to the problem solving.
- Team thinks they can't get the data, fact, or calculation needed to improve the RLS, so they work on other steps. The team knows what they need, but feel it is too hard to get so they rationalize and accept working on other steps rather than the RLS.

Wisdom Tidbit – People usually don't know (realize) what they don't know or need to know. The fastest way through is determining what fundamental principles are involved or if there is a problem similar to the current one that has been solved before (use SMEs in problem solving, consultants, and literature).

Wisdom Tidbit – People often will not try to find out the fundamentals because they perceive they don't have the ability, funding, time, resources, etc. to investigate things fully, i.e. it will be too hard to find out. Some people will use the perceived difficulty as an excuse to give a reasonable hypothesis a lower priority.

Wisdom Tidbit – Of course, after working through several rate-limiting steps at some point the original success statement for the project will be achieved. For example, "Success is to increase throughput of the plant from 80 to 100 metric tons/hour final product without spending capital or compromising on safety or quality by September of this year." Once success is achieved, it is time to check with the sponsor. If there is an opportunity to push onto 115 metric tons/hour (i.e. market exists, price would support it, etc.), then a new Charter should be written for that extension of the original work.

In many problems, the KNOWLEDGE is insufficient to even know where the RLS is. For example, a team is working on a tubular reactor (10,000 tubes) that runs between 200 and 250 °C. The plant has three such reactors in a unique pseudo-parallel arrangement. When demand is slow, they "slow roll" one or two reactors – that is do not add reagents, so they run about 145 °C. They started doing this several years ago and to their surprise, a several decades old carbon steel reactor that had never experienced corrosion had 30% wall thickness removed (all gas phase). The "knowledge" in the business is that liquid water is bad but gas-phase water, which is present at about 0.4%, is not. They were not reading the literature about rust and carbon steel. Water certainly helps, but humid air with O_2 can corrode carbon steel without any liquid present. The point is once they understood the fundamentals, then the RLS became apparent – lack of any data for these conditions, so experiments are underway to confirm or kill this hypothesis.

The Wisdom Tidbit for them was: there are entire journals, societies, large meetings about gas phase corrosion of carbon steel. It turns out it makes a hydrate structure on the surface that is stable up to about 180 °C. When the reactor operating temperature dropped below this value, hydrates formed – then at 20 bar with 0.4% water and 10% O_2 lots of bad stuff can happen. The answer (contingent on experimental confirmation) is to run the heat transfer fluid hotter – above the hydration point – for these idled reactors (while the plant is running just 1 of 3).

In summary – knowledge first – how the plant really runs, what physics/engineering/science say, then move into problem.

24.1 TOC – Recommended Reading List

The authors recommend reading the following books, preferably in the order listed. It will give a very thorough illustration of the art and practice of the Theory of Constraints in business. There is necessarily some overlap in content.

Manufacturing and Production

The Goal by Eliyahu M. Goldratt and Jeff Cox
The Race by Eliyahu M. Goldratt and Robert E. Fox
Breaking the Constraints to World-Class Manufacture by H. William Dettmer
Production the ToC Way by the Goldratt Institute
(Manufacturer's Guide to) Implementing the Theory of Constraints by Mark J. Woeppel
Deming and Goldratt by Domenico Lepore and Oded Cohen

General Management Practices and Improvement

Thinking for a Change by Lisa J. Scheinkopf
It's Not Luck by Eliyahu M. Goldratt
Necessary But Not Sufficient by Eliyahu M. Goldratt
Introduction to the Theory of Constraints Management System by Thomas B. McMullen
The Measurement Nightmare by Debra Smith
Management Dilemmas by Eli Schragenheim
Securing the Future by Gerald I. Kendall
Goldratt's Theory of Constraints: A Systems Approach to Continuous Improvement by H. William Dettmer
Thinking Smart: Applying the Theory of Constraints in Development Thinking Skills by Kwah Choon Ean

25

Visual Presentation

Words, equations, paragraphs, and reports are essential to conveying ideas. But graphical presentations of complex systems are frequently much quicker to understand and more clearly show the complex relationships that abound in real-world problems. This is particularly true in today's environment where the team is dispersed geographically and must work remotely.

As the proverbial saying goes, "A picture is worth a thousand words." A picture can convey a lot more information than a verbal description. The same goes for numerical data, where a graph is much easier and faster to understand, compared to a spreadsheet full of numbers. The use of graphics in science and engineering is ubiquitous just for that reason. In the practice of solving of complex problems, the proper use of the graphical representation tools is critical. Such projects are usually under stringent timelines. Any tool that can facilitate the efficient representation, transfer, and understanding of information should be utilized at the greatest extent possible. This is valid for project team meetings (sharing of information about current work) and stakeholder meetings (sharing progress information). The efficient graphical representation at stakeholder meetings is even more important, as these meetings are usually time constrained and the audience may not have the same level of technical understanding of the problem as the project team. A quick and efficient transfer of information to the stakeholders may make the difference between funds and resources being allocated or not.

Graphical representation of information is a great tool; however, there can be good graphical representations and very bad graphical representations. Some of the best references on the topic are by Tufte (1991, 1997, 2001). In this chapter, we are not going to go to such depths, but will share some examples of particularly good graphical representations of data.

Fundamentals of Industrial Problem Solving: A Practitioner's Guide, First Edition.
Zdravko I. Stefanov, Eldad Herceg, Carla Schmidt, David M. Jacobson, Dana Livingston, J.P. Chauvel, Sunil Kumar Chaudhary, and Christopher Paul Christenson.

25.1 The Napoleon Chart

A beautiful example is the "Napoleon Chart" developed by Charles Joseph Minard to represent Napoleon's disastrous campaign in Russia in 1812–1813 (Figure 25.1). The graph describes the movement of the Napoleon's troops through Russia and combines various variables in one space. It shows the constant decrease in the army numbers, along with the distance traveled and the ambient temperatures on the way back. It is a brilliant way to expose the consequences of a disastrous decision.

In our practice, we have discovered that a similar approach to mass balancing a plant is also extremely helpful to understand a tough problem. A nominal flowsheet of the plant is given in Figure 25.2. The same flowsheet, in the form of a "Napoleon diagram," is given in Figure 25.3.

In the flowsheet in Figure 25.3, the black arrows are a mass balance of an impurity. The thickness of the arrow is proportional to the mass flow of the impurity. From the graph, several things become instantly clear. First, most of the impurity is made in the reactor. Second, the only exit for the impurity is with the product stream. Third, there is a large recycle loop, where this impurity gets infinite residence time in the plant. So if there is any chemistry that can lead to fouling or corrosion related to the impurity, it has infinite time to complete and the effects can accumulate with time even at very low reaction rates. Another example of a very useful graphical representation in the style of the Napoleon chart is where on the plant flowsheet, one adds various sample points, analytical information, etc. The approach of superimposing all relevant sources of information in a single graph provides fast access to the available information and helps substantially with the problem solving. This type of graph helps communicate this kind of message and its seriousness to stakeholders in an instant.

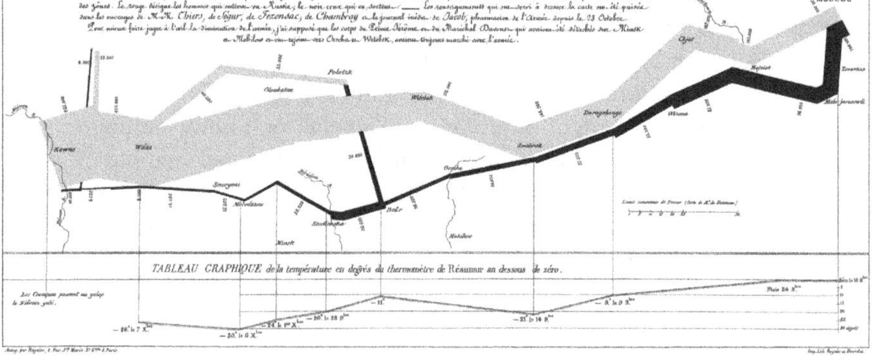

Figure 25.1 Map of Napoleon's Russian campaign.

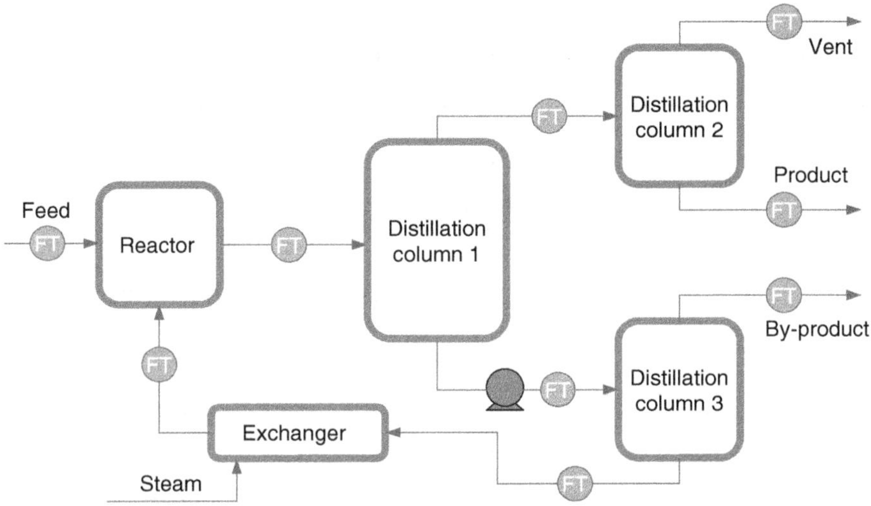

Figure 25.2 Nominal flowsheet. (*See insert for colour representation of the figure.*)

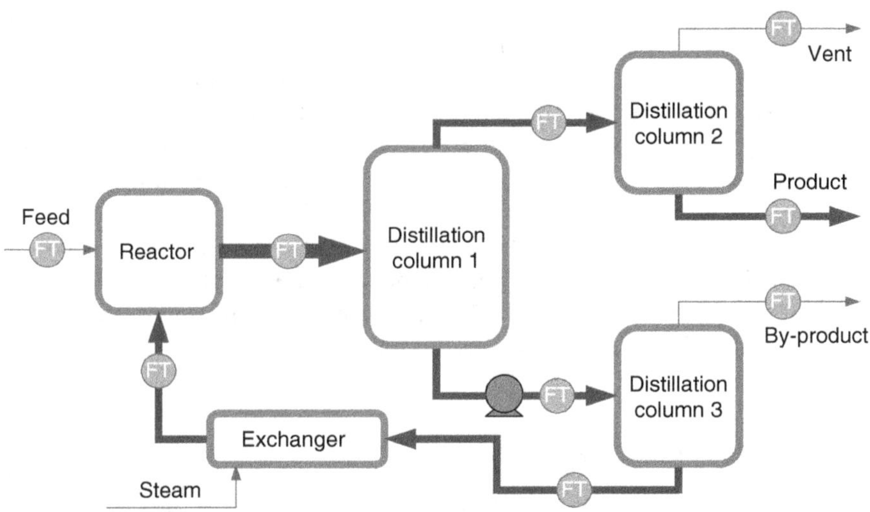

Figure 25.3 "Napoleon" chart flowsheet. (*See insert for colour representation of the figure.*)

25.2 The Hypothesis Matrix

The hypothesis matrix is a rectangular grid, where in one dimension one lists the hypotheses associated with a problem, and on the other dimension are the available facts (Figure 25.4). Where a hypothesis and a fact cross, the square is

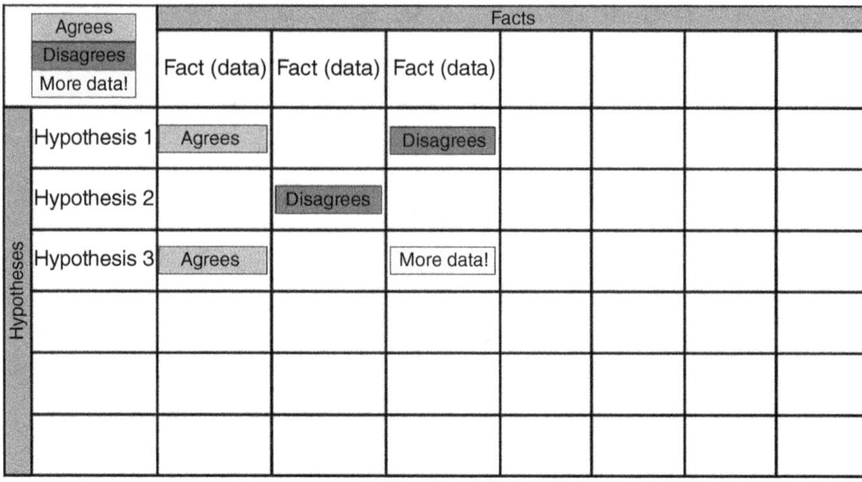

Figure 25.4 The hypothesis matrix is also a graphical tool. (*See insert for colour representation of the figure.*)

color-coded depending if the fact supports the hypothesis, disagrees with the hypothesis, it is neutral, or it is still to be obtained.

Although the hypothesis matrix can also be viewed as an approach to test hypotheses in a structured fashion, it is also a powerful graphical tool where it is immediately clear what hypotheses are supported by the available data, what hypotheses are not, where there is more work to be done (obtaining additional facts), and finally it encompasses the complexity of the problem, providing an insight to the stakeholders about the actual amount of work involved in solving a complicated problem. It can be implemented in various platforms, but most widely used in practice is Microsoft™ Excel, given its ubiquity and flexibility to incorporate additional features such as links, data etc.

25.3 The Issue Map

The issue map is another kind of tool that unifies multiple sources of information in a graphic environment. (A great reference on its use in problem solving is Chevalier (2016)). In essence, the map provides an environment to display the complex interconnections between various topics, facts, hypotheses, etc. There are various software solutions available that allow the addition of notes, images, files, links, etc. All this functionality makes such a map an extremely powerful problem solving tool. The issue map is particularly useful for very complex problems that include multitude of hypotheses, data, results, facts, observations, literature information, etc. It allows for efficient management of all of the project data. It is usually the next level after the hypothesis matrix, where the matrix may become extremely crowded and unreadable. The use of the hypothesis matrix does not exclude the use of the issue map, though. They can be used in concert to visualize different aspects of the project work. An example is shown in Figure 25.5.

25.4 Photos and Videos

In the current day and age, the advancements of the electronic technology has made the acquisition of images and videos extremely easy and convenient. High-quality images and videos can be acquired even by using one's phone. Therefore, it is not justified to not use this capability, given its great potential to assist in problem solving. As an example, internal fouling was encountered in one unit operation; however, the shutdown procedures required complete washing before opening. This washing thus erases the evidence about where the fouling was forming. During one shutdown, the run plant engineer took a short video of the insides of the equipment immediately after opening the manway. When the vessel was visually inspected, no fouling was observed. When the video was inspected later, it was discovered that a weld seam had a small piece of fouling retained on it. Just several minutes later, it had already slid off and had fallen to the bottom of the vessel to be dissolved in the pool of wash water. The evidence of the fouling retained on the weld was observed only in a few frames of the video. If the run plant engineer had not taken that video, the team would have not been able to determine that the weld seam was providing rough surface on which to initiate the growth of the foulant.

Photos and videos can be integrated in the other tools mentioned here quite successfully. For example, it can be very helpful to have images of fouling found in different locations around a plant on a "Napoleon diagram." This can serve as a very quick reference to the locations of the fouling and also what type of fouling is observed in which locations. Integration in an issue map is also very simple and useful, as one can immediately explore images related to different hypotheses.

In conclusion, the use of proper visual representation tools contributes to the successful execution of problem-solving projects. Therefore, it is a topic of diligence of the skilled problem solver to continuously search for better tools in that area and stay on top of what is the best.

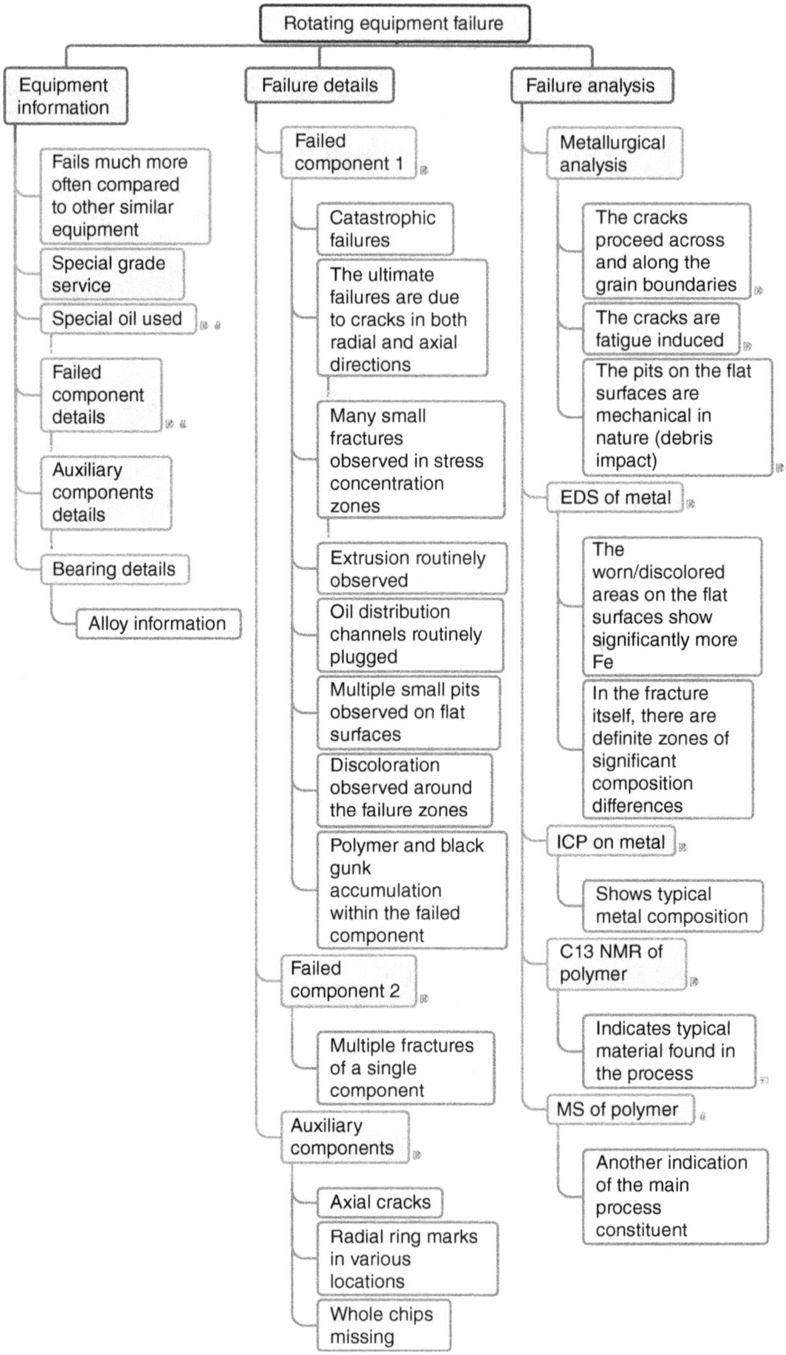

Figure 25.5 Issue map example.

References

Chevalier, A. (2016). *Strategic Thinking in Complex Problem Solving*. UK: Oxford University Press.

Tufte, E. (1991). *Envisioning Information*. Cheshire, CT: Graphics Press.

Tufte, E. (1997). *Visual Explanations: Images and Quantities, Evidence and Narrative*. Cheshire, CT: Graphics Press.

Tufte, E. (2001). *The Visual Display of Quantitative Information*. Cheshire, CT: Graphics Press.

Appendix: Scientific Method Practice Levels

The practice levels described here were developed with respect to laboratory work in R&D laboratories, but the levels of proficiency are applicable to the FPS teams as well. Stage I is the lowest proficiency stage, stage V is the highest.

- **Stage I**
  - **Define the problem:** The experimentalists usually define the problem alone. A clear specific definition of the problem is seldom documented.
  - **Hypothesize the governing laws:** Experimentalists seldom use the literature prior to experimentation and the application of theory to the solution is often after the fact.
  - **Design the experiment:** Experimental design is random in nature and empirical based.
  - **Do the experiment:** Experimental and analytical procedures seldom exist and are not defined. Experiments are executed randomly.
  - **Analyze the data:** The analysts are inadequately trained in experimental data analysis and statistical methodology. Normal data analysis is two-dimensional and empirical based.
  - **Document the results:** Documentation is by letter mostly and reports are not widely used as the primary source of documentation. Databooks are sometimes used, but usually after the experiment.
  - **Implement the solution:** Projects typically jump from marginally defined problems to experiments to implementation. Other functions are expected to follow through on the implementation.
  - **Training in the Scientific Method:** Experimentalists are not trained in experimental design, simple statistics, or brainstorming techniques. Resources for the personnel are limited or restricted, (i.e. library and literature, computer software capabilities, report writing, training, etc.)

Fundamentals of Industrial Problem Solving: A Practitioner's Guide, First Edition.
Zdravko I. Stefanov, Eldad Herceg, Carla Schmidt, David M. Jacobson, Dana Livingston, J.P. Chauvel, Sunil Kumar Chaudhary, and Christopher Paul Christenson.

- **Stage II**
  - **Define the problem:** A team of several experts within the department defines the problem at an early stage in the project.
  - **Hypothesize the governing laws:** Some library work is done early in the project. Some experimentalists seek advice from scientists and consultants at an early stage. The "Not invented here" syndrome begins to dissipate.
  - **Design the experiment:** Experimentalists are trained on only a few design techniques and use these exclusively on projects. Most experimental design is empirical and an error analysis of the anticipated results is seldom made prior to experimentation.
  - **Do the experiment:** Experimental and analytical procedures exist in the databooks and the experiments are planned and methodically executed.
  - **Analyze the data:** A few experimentalists in the lab are trained and use analysis techniques (i.e. regression analysis). Data are normally analyzed empirically and via linearization.
  - **Document the results:** Reports are not detailed and/or completion time is sometimes slow. Reports seldom provide an assessment of the literature, hypothesis of the governing theories, or the detailed design of the experiments.
  - **Implement the solution:** Laboratory personnel rely primarily on documentation to convey implementation instructions.
  - **Training in the Scientific Method:** Most persons are trained in several types of experimental design. Conducting a literature search, experimental error analysis, a working knowledge of the company technology centers are taught and made available.

- **Stage III**
  - **Define success:** Multifunctional teams begin to function with common goals. Customer input drives success definition. The definition and the goals are documented.
  - **Hypothesize the governing laws:** A good prior art review is evident in reports. Consultants comments and literature are defined in a bibliography in databooks and reports. The key governing physical science laws are considered.
  - **Design the experiment:** Experiments are almost always designed. Self-initiated training in statistical methods and data analysis is broadly evident.
  - **Do the experiment:** Lab practice is documented, shared, and reviewed between experimentalists. Good checklists and files are established. The laboratory utilizes the "good lab practices" guidelines.
  - **Analyze the data:** Lab experts in data analysis are often involved in determining the next experiment. Experimentalists utilize statistical data analysis techniques in evaluating data significance and accuracy.

- **Document the results:** Report writing is routinely completed on time and balanced between the Scientific Method steps. Databooks are up-to-date with reviews by the experts. Patent disclosures are encouraged and completed.
- **Implement the solution:** Personnel are involved in implementation throughout start-up of the technology. Training in other functional responsibilities and jobs is often required to effectively complete the implementation smoothly.
- **Training in the Scientific Method:** Experimentalists are well trained in the Scientific Method. The experts in the field are known and consulted with for guidance. Experimental design and statistical methods are utilized in projects widely.

- **Stage IV**
  - **Define success:** Project teams with multidisciplinary members work on most of the projects and success is defined through workshops including experts, customers, manufacturing, and marketing input. No boundaries exist between functions or geography.
  - **Hypothesize the governing laws:** Extensive prior art review and documented hypothesis of fundamental principles exists on each project. Experts or consultants are used frequently. The form of each set of governing equations is established.
  - **Design the experiment:** Experimental design is used on all projects, and theory-based designs are common. A thorough error analysis is completed prior to the first experiment and documented. The basis for math models are considered.
  - **Do the experiment:** Experimental procedures are detailed and improved continuously. Equipment files are maintained, and training on equipment operation is widespread. Documentation of the experiments, the results, and the unique observations are routinely made in the databook daily.
  - **Analyze the data:** Challenges of scale-up, time dependency, and variation are understood and addressed. Experimentalists are well trained in regression analysis and statistical methods. The utilization of computer software is widespread in evaluating the data. The next set of experiments is considered. The theory is proven or extended, and documented in reports.
  - **Document the results:** Project reviews and reports are on-going with project effort and publishing completed soon thereafter. Expert team reviews of the patent portfolio, the potential patents, and the competitive patent position are routine. Most projects have a number of reports to document progress throughout the project. The documentation includes a detailed recommendation for further research and extends technology to other related fields.

  - **Implement the solution:** Lab personnel work in implementation teams with customers and other functions to plan critical path technology needs at an early design stage. Most projects involve direct participation in the implication as a consultant and a member of the start-up team.
  - **Training in the Scientific Method:** Laboratory has developed several experts in advanced experimental design techniques and data analysis. Personnel rely on team member experts from multidisciplinary fields for direction. Modeling techniques are learned and utilized in most projects. All personnel are well trained in report writing and the execution of Patent Disclosures.

- **Stage V**
  - **Define success:** Teamwork is prevalent in each major project and rewarded by management. Team autonomy is high and team decisions are normally implemented. The first priority in defining success is the assembling of experts in varied fields. Success statements are well documented.
  - **Hypothesize the governing laws:** Lab members perform literature search extensively. Experts are consulted with for the development of the hypothesis of the theory. Results of the experiments often lead to a new broader understanding of the theory. The fundamental equations are documented prior to the first experiment. Lab members subscribe to technical awareness surveys in the field.
  - **Design the experiment:** Experimental design is model based and fundamentally grounded. The math model is often developed and run prior to the first experiments.
  - **Do the experiment:** Procedures for development, standards, and files on experimental procedures are used. Physical and chemical model systems are studied routinely with results related to more complex systems. Analytical power is utilized to microscopically inspect observations.
  - **Analyze the data:** Linear Programming, queuing theory, expert system expertise exists in lab and is used routinely. Some lab members actively participate in statistical and quality profession organizations. The experimental results successfully answer the "why" and confirm the theory. Math models are used to extend the data range past empirical extrapolation.
  - **Document the results:** Reports are arts of professional work. Work is critiqued by expert review. Reports are published throughout the project in a timely manner. Reports are well balanced in the Scientific Method steps. Databooks, reports, patent disclosures, and other documentation are clear, detailed, and well distributed to share technical results.

- **Implement the solution:** Technology is implemented with minimal changes in design, and start-ups proceed smoothly and on-time and budget. The experimentalist is recognized as the consultant and a key member of the implementation team.
- **Training in the Scientific Method:** Experts are developed to assist in major product and process development teams. Reports mirror the completeness of practicing The Scientific Methodology. Literature searching the databases is systematic and most experimentalists can access directly.

Index

a

adiabatic reactor scheme 301–302
aldol condensation reaction hypothesis 253–257
ammonia adsorption, in SAR32 mordenite 299–300
analysis of competing hypotheses (ACH) 350
analytical challenges 233–234
analytical data 303
  hypotheses, DQOs and 305–308
  mass balance data 313–315
  preparing an analytical or experimental DQO 308–313, 315
  in solving problems 313–315
analytical techniques 304
Aspen models 294
asymmetric workload balance 79–80

b

bad sensors 279
batch processes data analysis
  different durations of batches 285–286
  lack of steady state 285
  3D data set 285
  time 285
  unfolding methods 285, 286
batch solvent stripper optimization 322–323
benchmarking solutions 155–156
big data analysis 11–12, 21
Blackbelt Lean Six Sigma team 16, 17
Black Swan events 7–9
boxing/grouping 278
brainstorming, channeled 157
bridging solutions 151–153
business cycle failures 206
business reviews 128

c

calorimetry 228
catalysts 229–230
centralized research archives 266
change management 201–202
  plan 182–183
change threshold 276
channeled brainstorming 157
chronic problems 5–6, 32
  solution implementation 178–179
closing a project 217–219
cluster analysis 251
color problems 224, 249–257
  multivariate data analysis 286–270

Fundamentals of Industrial Problem Solving: A Practitioner's Guide, First Edition.
Zdravko I. Stefanov, Eldad Herceg, Carla Schmidt, David M. Jacobson, Dana Livingston, J.P. Chauvel, Sunil Kumar Chaudhary, and Christopher Paul Christenson.

common problems
  color 224
  odor 224–225
  out of specification product 223–224
  trace components 225–226
communication
  e-mail 100
  lack of 262–263
  stakeholders' 91–93
  streamlining 262
  tools 129
    elevator speech 134–135
    standard project e-mail updates 133–134
    storyboard 129–133
communication plans 139–140
  example of 141–142
  5Ws and an H technique 117–118
    HOW 138–139
    WHAT 121–129
    WHEN 137–138
    WHERE 135–137
    WHY 118–121
  patterns 140
  putting together 139–140
  solution-phase 183
complex solutions 158, 163
computational chemistry 226
  modeling methods 299–300
computational fluid dynamics (CFD) 320
continuous crystallizer, problems in 321–322
control charts 280
control plans 183, 185
  coffee process example SIPOC 209–210
  complex 198
  effective 193
  FPS 209, 211
  fundamental problem 208
  perception problem components of 190
  reality problem components of 207–208
  typical 209–211
conventional thinking 195–196
copy*.* solutions *see* standardized solutions
core team 35
corrosion problems 237
  corrosion catalyzed fouling 239–240
  drum 237–238
  reactor tubes 238
  rotating equipment seals 239
Cosorb 234–235
cost
  discussion 177
  management, in crisis projects 174–177
  of solution 164, 165, 170–171
cost–benefit analysis 111–115, 178
costly symptoms 18, 26–27
counterintuitive (nature of) problems 18–19, 28–29, 193–196
crisis
  control 57
  identification of 31
crisis projects
  cost management in 174–177
  team building 59–61
    leveraging situational interaction opportunities 69–70
    for life of the plant projects 77
    low tolerance for second guessing 70–72
    regular progress reminders 73–75
    respect 75–76
    taking action at every opportunity 72–73
    techniques for 64
    three best questions 67–68

understanding the consequences of crisis 64–67
WIIFM question 68–69
critical path 108, 337
customer/customer interface 82–83, 97
customer impact 114–115

d

data preprocessing
bad sensors 279
boxing/grouping 278
data with large difference in sampling frequency 278
historian compression considerations 276–277
importance of 275
lags 278
linearization and other transforms 278
process dynamics considerations 276
sampling 275–276
transitions, shutdowns, missing data 277–278
Data Quality Objective (DQO) 132
analytical 305–313, 315
EPA's 308–309
examples of 35–39
problem-solving aspect of 33–34
problem-specific 327, 329
project wrap-up 217
team dynamics aspect of 34
three main pillars of 67–68
dead easy solutions 153–154
decision making 177
decision timing 171
define, measure, analyze, improve, and control (DMAIC) 17, 326–328, 330, 332–334
descriptive modeling 292
difficult stakeholders, managing 140, 143–145
documentation 211
5Ws & H question 213–215
motivation 211–213
WIIFM game 211
DQO *see* Data Quality Objective (DQO)
drum, corrosion issues 237–238
dynamic modeling 299–301

e

elevator speech 134–135
E-mail
communication 100
project updates 133–134
empirical approach 12, 13
empirical-piloting 167–170
engineering disciplines 317
applications of
batch solvent stripper optimization 322–323
plant making smelly solvent 322
poor performance of isomer tower 320–321
problems in continuous crystallizer 321–322
problems in separation section 319–320
Fluid Mechanics and Mixing 318
heat transfer 319
process analytical 319
process control 319
Reaction Engineering 317
Reactive Chemicals 318
separations 317–318
solids processing 318
Environmental Protection Agency (EPA) 308–309
ethylene glycol (EG) process 249–250

f

face-to-face meetings 77–78, 135
facts *vs.* hypotheses 46–47
final stakeholder update 217, 218

financial communication 111–114
first principles 12
  focus on zone 4 characteristic problems
    costly symptoms 26–27
    counterintuitive nature 28–29
    life-of-the-plant problems 27–28
    managerial interferences 25–26
    multifactor problems 18–23
    multifunctional teams 26
    new science 29
    nonlinear relationships 23–25
    perceived personal loss 29–30
  modeling 297
  questions 25
5Ws and an H communication plan 117–118
  documentation 213–215
  HOW 138–139
  putting together 139–140
  WHAT 121–122
    communication tools to use 129–135
    formal stakeholder project milestone review 127–129
    in-progress project work 122
    meeting minutes 123–126
    standardized stakeholder project updates 126–127
    team working meetings 123
  WHEN 137–138
  WHERE 135–137
  WHY 118–121
Fluid Mechanics and Mixing discipline 318
formal stakeholder reviews 127–129
fouling 229, 234, 351
  corrosion catalyzed 239–240
  first principles models for 297
  of heat exchangers 319
  literature search 268
front-line problem-solving 14
full-scale piloting 168–169
functional analysis 158–159
fundamental models 291, 294–295
fundamental problem solving (FPS) 3–4
  DMAIC and 326–328
  integrated problem-solving approach 332–334
  LSS comparison with 325–330
  phases of 326
  projects
    parallel processing in 86–88
    phases of 6
    roles and responsibilities for 82–86
    stakeholders in 96–98
    strategies in team management 88–90
  ToC place in FPS framework 335, 336
  use of 5
fundamental understanding, need for 226–227, 235–236

g

GC-MS 225
generalized/typical team building approach 78–79
Gibbs energy estimation 299–300
good hypothesis 42–44
grade transitions 277
graphical representation 347
  hypothesis matrix 349–350
  issue map 350–351, 353
  Napoleon Chart 348–349
  photos and videos 351
Greenbelt Lean Six Sigma team 15
ground rules 122, 136

h

hard changes 198–199
heat exchangers
  mechanical failures 247–248
  fouling of 319
heat transfer discipline 319

high burn rates 59
high-energy solutions 197–198
high-risk problem types 190
  counter intuitive 193–196
  low frequency 191–193
  reward conflicting 196–197
high-risk solutions 191
  high-energy solutions 197–198
  lone wolves 199–200
  soft changes 198–199
high-skill training challenge 203–204
historian compression considerations 276–277
history of problem 32
homogeneous catalysts 229–230
Hotelling T^2 statistic 283
hypothesis 41, 51–52
  analysis of competing hypotheses 350
  analytical data and 305–308, 313
  facts *vs.* 46–47
  full clear statement of 132
  generation 109
  good 42–44
  implied 307–308
  MELD ranking 48
  onion 47–48, 52
  parallel processing 86–88
  qualified 44–46
  sacred cows 51
  testing 269, 272
    in problem-solving 41–50
    strong inference 42
    TOC in 342
  title of 131
  tracking 47–48

i
images 351
incomplete stakeholder's list
  company churn as people change jobs 102–103
  initial stakeholder analysis incomplete 99–100
    e-mail trails 100
    hallway/breakroom interactions 100–101
    meeting attendance adjustments 101
    phone calls or instant messages 101
  phase changes on project 102
  problem severity shifts 103–104
  scope change due to problem location understanding 101–102
information control 126
initial bias
  needs *vs.* 106–107
  for stakeholders 104–106
innovative solutions 156–161
instant messaging 101
instrumental analytical techniques 304
integrated problem-solving approach 332–334
internal never published/paper-only sources 267
isomer tower, poor performance of 320–321
issue map 350–351, 353

j
just in time planning 181

k
kinetics of products 227

l
lab-scale piloting 168
lags 278
large-scale reactor models 295
leadership
  in business 82, 96
  of team members 82, 96
leakage 5

LEAN Six Sigma (LSS)
  criticisms of 330–332
  vs. FPS 325–330
  integrated problem-solving approach 332–334
  poor implementation of 332
  team 15–17
lean time planning 181
life-of-the-plant problems 18, 27–28
life of the plant project type team building 76–77
  asymmetric workload balance 79–80
  crisis project techniques 77
  doing something new 81
  face-to-face kickoff meetings 77–78
  generalized/typical approach 78–79
  opportunity pairing of outsiders with insiders 80
  revisiting past 80–81
linearization 278
literature search
  benefits of 265–266
  errors 269–270
  lost knowledge 266
  patience 269
  pitfalls 269–270
  process 268–269
  sources
    age 270
    internal never published/paper-only 267
    patents 267–268
    peer-reviewed external 267
    peer-reviewed internal 266–267
    plant libraries 267
    unpublished works 268
lone wolf solutions 199–200
low frequency of problems 191–193
low tolerance for second guessing 70–72
lubrication degradation hypothesis 245

m

macro-scale modeling 296
managerial interferences 18, 25–26
mass balance data 313–315
materials failure 231
materials of construction 246–247
mechanical issues 241
  heat exchangers 247–248
  materials of construction 246–247
  rotating equipment 242–246
meetings 123
  face-to-face 77–78, 135
  generation of stakeholder list 94
  minutes 123–126
  mixed 135–136
  remote web/telecom-based systems 136–137
  team 123
MELD hypothesis ranking 48
meso-scale modeling 296
milestone review 109, 110, 127–129
mini storyboards 131–133
minutes of meeting 123–126
missing data 277–278
missing of key measurements 233–234
mixed meetings 135–136
model(ing)
  Aspen 294
  decisions 293
  descriptive 292
  dynamic 299–301
  fundamental 294–296
  industrial applications 292
  large-scale reactor 295
  macro-scale 296
  meso-scale 296
  molecular 295–296
  plant-scale 294, 296
  role of 291
  selection criteria 292–293
  statistical 294

supply chain 294
theoretical modeling techniques 297–302
molecular modeling 295–296
money communication 111–114
money control problems 58
motivation 74
documentation 211–213
positive feedback 212
motivational issue 68
multifactor interaction problems 18–23
multifunctional teams 18, 26
multiple cause/effect problems 234
multivariate data analysis
advantage of 281
PCA 281–284
of plant data 286–290
PLS 284

n

natural planning 181
new product/process development 226–232
need for fundamental understanding 226–227, 235–236
new science problem 19, 29
nonlinear relationships 18, 23–25

o

obvious solutions 153–154
odor problems 224–225
opportunity identification and initial team formation
charter/DQO
problem-solving aspect of 33–34
team dynamics aspect of 34
crisis and chronic issue 31–32
history of the problem 32
team structure 35
optimum long-term staffing model concept 9
organizational complexity 15, 16
overplanning 180–181

p

parallel processing 86–88
Pareto 80/20 principle 111
partial least squares (PLS) 251, 284
passive sampling 232
patents 267–268
PCA *see* principal component analysis (PCA)
peer-reviewed external sources 267
peer-reviewed internal sources 266–267
perceived personal loss 29–30
perception problem 186–190
perfection 213
piloting a solution
full-scale 168–169
lab-scale 168
measurement plan value in 169–170
PDCA cycle 170
success criteria 167–168, 170
pilot plants 235–236
Plan-Do-Check-Act (PDCA) cycle 170, 184
plan(ning)
change management 182–183
communication 183
control 183
detailed 181
lean/just in time 181
natural 181
overplanning 180–181
solution implementation 179–183
stress testing 182
theory of constraints 181
underplanning 181
plant data analysis 286–290
plant libraries 267
plant making smelly solvent 322
plant problem solving 13–14

plant-scale modeling 294, 296
plant shutdowns 277
PLS *see* partial least squares (PLS)
plugging problem 297
polymers 228–229
principal component analysis (PCA) 281–282
  advantages of 283–284
  decomposition problem 284
  extraction of PCs 283
  Hotelling T^2 statistic 283
  principle of 282
  Q^2 283
problem severity shifts 103–104
problem-solving continuum 5, 328
problem-solving selection process 333
problem-specific DQO 327, 329
problem understanding 11, 265
procedure changes 199
process analytical discipline 319
process chemometrics 251
process control discipline 319
process data analysis 275
  batch processes considerations
    different durations of batches 285–286
    lack of steady state 285
    3D data set 285
    time 285
    unfolding methods 285, 286
  data preprocessing
    bad sensors 279
    boxing/grouping 278
    data with large difference in sampling frequency 278
    historian compression considerations 276–277
    importance of 275
    lags 278
    linearization and other transforms 278
    process dynamics considerations 276
    sampling 275–276
    transitions, shutdowns, missing data 277–278
  example of 286–290
  multivariate methods
    advantage of 281
    PCA 281–284
    PLS 284
  univariate methods 279
    advanced statistics aspects of 281
    control charts 280
    statistical distribution analysis 280
    trending of process variables 279–280
process owners 82, 97
progress reminders 73–75
project charter/DQO 132, 327
  examples of 35–39
  problem-solving aspect of 33–34
  team dynamics aspect of 34
  three main pillars of 67–68
project leaders 59, 82, 96, 145
project management 55
project planning 109
project sponsors *see* sponsors
project updates
  e-mail 133–134
  one-page 134
  stakeholder 126–127
project wrap-up 217–219

q

Q^2 (distance-to-the-model) 283
Qualified Hypotheses 44–46
questions 25
  identifying stake holders 93–94
  three best questions 67–68
  WIIFM 68–69, 211

r

rare problems 234–235
rate-determining steps 108
rate-limiting step (RLS) 343–345
  illustration of 335, 337–340
  requests 35
  tangible and intangible 343
Reaction Engineering discipline 317
reactive chemicals 230–231
Reactive Chemicals discipline 318
reactor tubes, corrosion issues 238
reality sustainability problems 190–191
  high-risk problem types
    counter intuitive 193–196
    low frequency 191–193
    reward conflicting 196–197
  high-risk solutions
    high-energy solutions 197–198
    lone wolves 199–200
    soft changes 198–199
  outside forces conspiring against sustainability 204–205
    business cycle failures 206
    reorganization disease 206–207
    rotten apples 205
  poor solution implementation 200
    high-skill training breakdown 203–204
    poor change management and buy-in 201–202
    roles and responsibility breakdown 202–203
recognition 219
regular progress reminders 73–75
reliability 337
reorganization disease 206–207
Research and Development (R&D) 105
resilience engineering (RE) 33
respect 75–76
revisiting past 80–81
reward conflicting 196–197
risk mitigation 159–161
roles and responsibility breakdown 202–203
rotating equipment
  mechanical issues 242–246
  seals, corrosion of 239
rotten apples 205
route lists 95

s

sampling 275–276
sampling frequency, data with large difference in 278
scenario development 51–52
science of problem 160–161
scientific method
  goal of 271
  practice levels 355–359
  steps of 271–273
second guessing 70–72
sensor failures 279
separations discipline 317–318
separation train 317–318
  problems in 319–320
serial processing 86, 87
Shannon's sampling theorem 276
shared systems approach 95–96
shelf life 227–228
Shewhart charts 280
shutdowns 277
simple multivariate analysis 280
soft changes 198–199
solids processing discipline 318
solution implementation 173–174, 200–201
  in chronic problem project type 178–179
  in crisis project type 174–177
  high-skill training challenge 203–204
  planning 179–180
    change management 182–183
    communication 183

solution implementation (*cont'd*)
control plan generation and validation 183
detailed 181
lean/just in time 181
natural 181
overplanning 180–181
stress testing 182
success criteria definition 179–183
theory of constraints 181
underplanning 181
poor change management and buy-in 201–202
roles and responsibility breakdown 202–203
solutions 147–148, 184
adjustment 184
benchmarking 155–156
bridging 151–153
closure of options 172
complex 158, 163
concept of science 160–161
dead easy/obvious 153–154
decision timing 171
failure of 172
generation of 150–161
high-energy 197–198
high-risk 191, 197–200
implementation (*see* solution implementation)
innovative 156–161
joyful moments 148–150
lone wolf 199–200
ownership
individual 159–160
team 160
selection of 164–165
cost 164, 165, 170–171
final selection 171–172
from generated multiple solutions 161–164
quality 165–170
reversibility 163
timing 162–165, 170–171
soft change 198–199
standardized 154–155
successful 165
testing
empirical-piloting (*see* piloting a solution)
theoretical-paper exercises 165–167
validation 184
sponsors 35, 56, 82, 96–97, 145
sprint rate 337
staffing, optimum long-term staffing model concept 9
stakeholders 35, 56
bad apple example story 140, 143–145
changing nature of 99
communication 91–93
customer impact 114–115
impact on employees 115–116
money 111–114
of rate-determining steps 108
of timing 108–111
conflict 144
decision making 177
discussion, tips for 106
final review with 217, 218
5Ws and an H technique 118
HOW 138–139
WHAT 121–135
WHEN 137–138
WHERE 135–137
WHY 118–121
identification 104
chart 95
in FPS project 96–98
questions for 93–94
initial bias for 104–107
list

incomplete (*see* incomplete stakeholder's list)
information on 94
route lists 95
shared systems approach 95–96
storage 94, 95
technique for generating 94
listening to 107
management 140, 143–145
needs 107–116
concept of 107
vs. initial bias 106–107
review 127–129
steering team 98–99
Win-Win solutions 144
standardized e-mail project updates 133–134
standardized solutions 154–155
standardized stakeholder project updates 126–127
statistically designed experiments 227
statistical models 294
steering team 35, 88–89, 98
storyboards 129–133
stress testing planning 182
strong inference hypothesis testing 42
subject matter experts (SMEs) 10, 83, 97, 100
success criteria
definition 180
piloting a solution 167–168, 170
suppliers, inputs, process, outputs, and customers (SIPOC) maps 169
supply chain modeling 294
sustaining the gains
and control plans 208–211
documentation in 211–215
perception problem 186–190
reality problem 190–191
high-risk problem types 191–197
high-risk solutions 197–200
outside forces conspiring against sustainability 204–207
poor solution implementation 200–204

t

taking actions 72–73
tar formation, chemical mechanism for 297
team building
in crisis projects 59–61
leveraging situational interaction opportunities 69–70
low tolerance for second guessing 70–72
regular progress reminders 73–75
respect 75–76
taking action at every opportunity 72–73
techniques for 64
three best questions 67–68
understanding the consequences of crisis 64–67
WIIFM question 68–69
life of the plant projects 76–77
asymmetric workload balance 79–80
crisis project techniques 77
face-to-face kickoff meetings 77–78
generalized/typical approach 78–79
opportunity pairing of outsiders with insiders 80
revisiting past 80–81
something new 81
team member's leadership 82, 96
team(s)
management 88–90
outsider issue in 59
recognition 219
solution ownership 160
steering 35, 88–89, 98

team(s) (*cont'd*)
  sub team structures 89
  technical 35
theoretical approach 12–13
theoretical modeling techniques 297–302
theoretical-paper exercises 165–167
Theory of Constraints (TOC) 335
  in FPS framework 335, 336
  in hypothesis testing 342
  illustration of 335, 337–340
  overview 336
  planning 181
  problem-solving speed and efficiency 342
  recommend reading 345
  for tangible processes 341
three-factor interaction 20
time-control problems 57
timelines 108–111
timing 174
  batch processes 285
  decision 171
  planning 181
  of solution 164, 165, 170–171
  in solution selection process 162–163
total installed cost (TIC) factor 171
trace components 225–226
transitions 277
trial and error 292
two-factor interaction 19–20

u

underplanning 181
understanding the problem 11, 265
univariate data analysis 279
  advanced statistics aspects of 281
  control charts 280
  statistical distribution analysis 280
  trending of process variables 279–280
unpublished works 268
updates 126–127
UV spectra 256–257

v

value engineering techniques 158
value of project 55–58
variable importance for prediction (VIP) metrics 251, 252, 288
variation problem 187–189
verb-noun thinking 158–159
videos 351
virtual meetings 136–137
visual inspection 303
visual presentation 347
  hypothesis matrix 349–350
  issue map 350–351, 353
  Napoleon Chart 348–349
  photos and videos 351

w

wastewater treatment plant (WWTP) 231–232, 234
web meetings 136–137
What's In It For Me (WIIFM) question 68–69, 211
workload balance 79–80
wrap-up 217–219
writing meeting minutes 123–126

z

Zone 1 problems 13–14
Zone 2 problems 15–16
Zone 3 problems 16–17
Zone 4 problems 17–18
  characteristics of 18–19
  first principle focus on
    costly symptoms 26–27
    counterintuitive nature 28–29
    life-of-the-plant problems 27–28
    managerial interferences 25–26
    multifactor problems 18–23
    multifunctional teams 26
    new science 29
    nonlinear relationships 23–25
    perceived personal loss 29–30

www.ingramcontent.com/pod-product-compliance
Lightning Source LLC
Chambersburg PA
CBHW072019120726
47787CB00003B/77
9781119543183